247
Advances in Polymer Science

Advances in Polymer Science

Recently Published and Forthcoming Volumes

Polymers in Nanomedicine
Volume Editors: Kunugi, S., Yamaoka, T.
Vol. 247, 2012

Biomedical Applications of Polymeric Nanofibers
Volume Editors: Jayakumar, R., Nair, S.V.
Vol. 246, 2012

Synthetic Biodegradable Polymers
Volume Editors: Rieger, B., Künkel, A., Coates, G.W., Reichardt, R., Dinjus, E., Zevaco, T.A.
Vol. 245, 2012

Chitosan for Biomaterials II
Volume Editors: Jayakumar, R., Prabaharan, M., Muzzarelli, R.A.A.
Vol. 244, 2011

Chitosan for Biomaterials I
Volume Editors: Jayakumar, R., Prabaharan, M., Muzzarelli, R.A.A.
Vol. 243, 2011

Self Organized Nanostructures of Amphiphilic Block Copolymers II
Volume Editors: Müller, A.H.E., Borisov, O.
Vol. 242, 2011

Self Organized Nanostructures of Amphiphilic Block Copolymers I
Volume Editors: Müller, A.H.E., Borisov, O.
Vol. 241, 2011

Bioactive Surfaces
Volume Editors: Börner, H.G., Lutz, J.-F.
Vol. 240, 2011

Advanced Rubber Composites
Volume Editor: Heinrich, G.
Vol. 239, 2011

Polymer Thermodynamics
Volume Editors: Enders, S., Wolf, B.A.
Vol. 238, 2011

Enzymatic Polymerisation
Volume Editors: Palmans, A.R.A., Heise, A.
Vol. 237, 2010

High Solid Dispersion
Volume Editor: Cloitre, M.
Vol. 236, 2010

Silicon Polymers
Volume Editor: Muzafarov, A.
Vol. 235, 2011

Chemical Design of Responsive Microgels
Volume Editors: Pich, A., Richtering, W.
Vol. 234, 2010

Hybrid Latex Particles – Preparation with Emulsion
Volume Editors: van Herk, A.M., Landfester, K.
Vol. 233, 2010

Biopolymers
Volume Editors: Abe, A., Dušek, K., Kobayashi, S.
Vol. 232, 2010

Polymer Materials
Volume Editors: Lee, K.-S., Kobayashi, S.
Vol. 231, 2010

Polymer Characterization
Volume Editors: Dušek, K., Joanny, J.-F.
Vol. 230, 2010

Modern Techniques for Nano- and Microreactors/-reactions
Volume Editor: Caruso, F.
Vol. 229, 2010

Complex Macromolecular Systems II
Volume Editors: Müller, A.H.E., Schmidt, H.-W.
Vol. 228, 2010

Complex Macromolecular Systems I
Volume Editors: Müller, A.H.E., Schmidt, H.-W.
Vol. 227, 2010

Shape-Memory Polymers
Volume Editor: Lendlein, A.
Vol. 226, 2010

Polymer Libraries
Volume Editors: Meier, M.A.R., Webster, D.C.
Vol. 225, 2010

Polymer Membranes/Biomembranes
Volume Editors: Meier, W.P., Knoll, W.
Vol. 224, 2010

Polymers in Nanomedicine

Volume Editors: Shigeru Kunugi
Tetsuji Yamaoka

With contributions by

T. Akagi · M. Akashi · Y. Arima · M. Baba · R.K. Dutta · Y. Ikeda · K. Ishihara · H. Iwata · K. Kato · H. Kobayashi · T. Konno · J.J. Mercuri · K. Nagahama · Y. Nagasaki · Y. Ohya · A.C. Pandey · P.K. Sharma · D.T. Simionescu · A. Takahashi · Y. Teramura · E. Wagner · Y. Xu

Editors
Prof. Shigeru Kunugi
Kyoto Institute of Technology
Department of Biomolecular Engineering
1 Hashigami-cho, Matsugasaki, Sakyo-ku
Kyoto, 606-8585
Japan
kunugi@kit.ac.jp

Dr. Tetsuji Yamaoka
National Cerebral and Cardiovascular Center
Research Institute
Department of Biomedical Engineering
5-7-1 Fujishiro-dai, Suita
Osaka, 565-8565
Japan
yamtet@ri.ncvc.go.jp

ISSN 0065-3195 e-ISSN 1436-5030
ISBN 978-3-642-43253-6 ISBN 978-3-642-27856-3 (eBook)
DOI 10.1007/978-3-642-27856-3
Springer Heidelberg Dordrecht London New York

Printed on acid-free paper

Springer is part of Springer Science+Business Media (www.springer.com)

Volume Editors

Prof. Shigeru Kunugi
Kyoto Institute of Technology
Department of Biomolecular Engineering
1 Hashigami-cho, Matsugasaki, Sakyo-ku
Kyoto, 606-8585
Japan
kunugi@kit.ac.jp

Dr. Tetsuji Yamaoka
National Cerebral and Cardiovascular Center
Research Institute
Department of Biomedical Engineering
5-7-1 Fujishiro-dai, Suita
Osaka, 565-8565
Japan
yamtet@ri.ncvc.go.jp

Editorial Board

Advances in Polymer Sciences Also Available Electronically

Advances in Polymer Sciences is included in Springer's eBook package *Chemistry and Materials Science*. If a library does not opt for the whole package the book series may be bought on a subscription basis. Also, all back volumes are available electronically.

For all customers who have a standing order to the print version of *Advances in Polymer Sciences*, we offer free access to the electronic volumes of the Series published in the current year via SpringerLink.

If you do not have access, you can still view the table of contents of each volume and the abstract of each article by going to the SpringerLink homepage, clicking on "Browse by Online Libraries", then "Chemical Sciences", and finally choose *Advances in Polymer Science*.

You will find information about the

- Editorial Board
- Aims and Scope
- Instructions for Authors
- Sample Contribution

at springer.com using the search function by typing in *Advances in Polymer Sciences*.

Color figures are published in full color in the electronic version on SpringerLink.

Aims and Scope

The series presents critical reviews of the present and future trends in polymer and biopolymer science including chemistry, physical chemistry, physics and material science. It is addressed to all scientists at universities and in industry who wish to keep abreast of advances in the topics covered.

Review articles for the topical volumes are invited by the volume editors. As a rule, single contributions are also specially commissioned. The editors and publishers will, however, always be pleased to receive suggestions and supplementary information. Papers are accepted for *Advances in Polymer Science* in English.

In references *Advances in Polymer Sciences* is abbreviated as *Adv Polym Sci* and is cited as a journal.

Special volumes are edited by well known guest editors who invite reputed authors for the review articles in their volumes.

Impact Factor in 2010: 6.723; Section "Polymer Science": Rank 3 of 79

Preface

Nanotechnology involves the manipulation of matter on atomic and molecular scales. This technology combines nanosized materials in order to create entirely new products ranging from computers to micromachines and includes even the quantum level operation of materials. The structural control of materials on the nanometer scale can lead to the realization of new material characteristics that are totally different from those realized by conventional methods, and it is expected to result in technological innovations in a variety of materials including metals, semiconductors, ceramics, and organic materials.

The application of nanotechnology in advanced and high-tech medical care is known as nanomedicine. It covers a wide range of scientific and technological fields ranging from fundamental aspects related to the creation of new materials to actual applications in clinical medicine. The term nanomedicine was first mentioned in literature in 1990 by Drexler. The National Nanotechnology Initiative in the USA (2000) listed improved healthcare as one of the implications of nanotechnology, and the European Technology Platform "Nanomedicine" started in 2005. In this decade, more than 300 review articles have been published on nanomedicine. At present, nanomedicine is advocated to have the following promising applications: (1) the development of medication using nanosized devices including nanospheres, nanomicelles, nanocapsules, and nanofibers, (2) nano-order regulation of the interface between materials and cells/tissues, and (3) nanoimaging, which includes single-molecule imaging, using optical systems in particular, as well as whole body imaging at a very high resolution.

Nanosized objects perform various functions in the biomedical field. In the human body, nanosized particulate substances behave very differently from larger particles. In 1986, Maeda et al. found that the stained albumin, having a size of several nanometers, naturally accumulates in the region of cancerous tissues, which is now well known as the enhanced permeability and retention (EPR) effect. Many studies in the field of nanoparticles are based on this finding. Another application of nanoparticles is the delivery system using various polyplexes that are composed of carrier molecules and plasmid DNA or nucleic acid drugs such as antisenses and siRNA. In addition, nanofibers are mainly used for biodegradable scaffolds in tissue engineering because of their good cell adhesion properties. The second area is the nanoscale regulation of the interface between materials and cells/tissues. Recently, regulation of the microenvironment for stem cells, referred to as stem cell niches, has attracted much attention. The elasticity of matrices, nanopatterning of cellular adhesion machinery, and mobility of the interfaces are known to be very important "cues" for cell functions. The third area is nanoimaging. Quantum dots, which are strong tools for in vivo imaging, are promising and useful inorganic fluorophores, and they are one of the most important nanomaterials. Other modalities in molecular imaging, such as magnetic resonance imaging using super-paramagnetic iron oxide and microcomputed tomography, have also improved significantly.

Polymers and polymeric materials are important organic materials and have played a very important role in the research and development of nanomedicines. The polymeric materials that are useful in this field can be classified by their chemical nature as follows: (1) water-soluble (compatible) synthetic polymers, (2) polyelectrolytes and polyion complexes, (3) natural polymers, and (4) biodegradable synthetic polymers. On the other hand, higher-order structures and morphology of the used materials such as nanowires, particles, vesicles, capsules, shells, gels, cages, skeletons, and film membranes can also be considered as key issues. For instance, tissue engineering requires two- or three-dimensional structures that are made from biodegradable polymers, and the synthesis of nanoparticles requires the self-assembling nature of polymers or pre-polymers.

Several review articles focusing on the contribution of polymeric materials to nanomedicine have been published. This volume of "Polymers in Nanomedicine" in the series *Advances in Polymer Science* will be one of the pioneering review books dedicated to the study of polymer science for medical nanotechnology.

In this book, we consider the importance of the chemical nature of polymeric materials and target the three major themes leading to their actual application in nanomedicines. With regard to polymeric delivery systems, specific nucleic acids and vaccine–antigen delivery systems are reviewed. Furthermore, general drug delivery systems using biodegradable synthetic polymers and poly(ethylene glycol)-modified drugs are discussed. The understanding and control of biological responses against artificial materials are important for the development of medical devices and tissue engineering therapies. The control of cell adhesion, modification of cell surfaces, and the development of biopolymer scaffolds are reviewed in the context of tissue engineering. The last and the most important area addressed in this book concerns imaging and therapeutic modalities.

The editors believe that incorporating contributions that cover topics ranging from molecular design to tissue architecture into one book will prove to be helpful and promote research in this rapidly developing area. We would like to thank all the contributing authors and colleagues assigned to work with us during the editing process.

Kyoto, Autumn 2011

Shigeru Kunugi
Tetsuji Yamaoka

Contents

Adv Polym Sci (2012) 247: 1–30
DOI: 10.1007/12_2011_148

Published online: 14 September 2011

Functional Polymer Conjugates for Medicinal Nucleic Acid Delivery

Ernst Wagner

Abstract Medicinal nucleic acids like antisense oligonucleotides, antagomirs, gene vectors, mRNA, or siRNA are exciting novel drugs for manipulating gene expression in a controlled, therapeutically useful way. Because of their physicochemical nature, medicinal nucleic acids cannot freely diffuse to the intracellular target sites to exert their therapeutic function. Polymers have been developed as carriers, which package nucleic acids and protect them against degradation. These carriers specifically attach their nucleic acid cargo to cells via targeting ligands and trigger intracellular uptake. They participate in intracellular delivery steps including endosomal escape. Depending on the intracellular site of action, they may play an important role in cytosolic migration, nuclear import, and subsequent presentation of the nucleic acid in active form. Ideally, polymers act in a bioresponsive way to overcome the different delivery steps. Therapeutic developments with medicinal nucleic acid polyplexes, including recent clinical trials, are discussed.

Keywords Gene transfer · Plasmid DNA · Polyplex · siRNA · Targeting

Contents

E. Wagner
Pharmaceutical Biotechnology, Center for System-based Drug Research, and Center for Nanoscience, LMU University of Munich, Butenandtstrasse 5-13, 81377 Munich, Germany
e-mail: ernst.wagner@cup.uni-muenchen.de

1 Introduction

Polymers can be designed in non-immunogenic and biocompatible form, incorporating different physicochemical properties and attachment sites for (covalent or noncovalent) modification. In theory, they are an excellent drug carrier platform for targeted, repeated drug applications resulting in long-lasting therapeutic effects [1]. Medicinal nucleic acids present a novel class of drugs with exciting possibilities. From the functional perspective, at least three subcategories can be defined. The first category induces gene expression in a sequence-derived way; plasmid DNA (pDNA) expression vectors [2–8], natural or chemically modified messenger RNA (mRNA) [9, 10], exon-skipping oligonucleotides [11], and microRNA antagonists (antagomirs) [12, 13] belong to this category. The second category triggers shutdown of gene expression and includes antisense oligonucleotides [14, 15], synthetic short interfering RNA (siRNA) [16–23] and microRNA [24], ribozymes [25], and DNA decoys [26]. In the third category, medicinal nucleic acids may act by direct interaction with proteins in form of RNA or DNA aptamers [27, 28], or immune-stimulating and apoptosis-inducing RNA [29–31]. Targeted delivery is the major bottleneck in the further development of such medicinal nucleic acids.

In the field of polymer-enhanced nucleic acid therapy, the road towards widespread clinical use is still filled with many obstacles and challenges, despite significant improvements in polymer-based carriers over the last three decades [32–35]. The major challenges include the following:

1. Precise chemical synthesis and analysis of polymers, which are macromolecular structures more complex than small chemical molecules
2. Controlled supramolecular assembly of polymeric carriers with the nucleic acid into uniform nanoparticles
3. Better understanding of the mechanisms and alternative pathways required for further optimization of extracellular and intracellular targeted delivery of medicinal nucleic acids.

This chapter reviews the main delivery barriers for medicinal nucleic acids and the strategies to overcome these barriers by using functionalized polymer conjugates (Sect. 2). Chemical strategies will be presented to illustrate how polymers can be designed to be bioresponsive (Sect. 3). The advantages of such dynamic polymer systems responding to the biological microenvironment of the various delivery steps will be discussed. Current therapeutic strategies using medicinal nucleic acids in preclinical and clinical setting will be presented (Sect. 4).

2 Functions Required Within Polyplexes for Overcoming Delivery Barriers

Medicinal nucleic acids are too polar and too large to passively enter the intracellular target tissue compartments where they should exert their therapeutic function. In addition, they are rapidly recognized by the host defense system and easily enzymatically degraded by nucleases. Cationic polymers can package nucleic acids into polyplexes [36] and protect them against degradation. Cationic polymers primarily bind nucleic acids via electrostatic interactions with the negatively charged phosphate backbone. Nucleic acid binding depends on charge density, size, flexibility, and topology of polymers. For example, linear polymers such as poly(L-lysine) (PLL) [37, 38], linear polyethylenimine (LPEI) [39], and many others linear structures [40, 41]; branched polymers such as branched PEI (brPEI) [42]; and dendrimers including polyamidoamines (PAMAMs) [43–46] have been used. In addition to electrostatic interaction, hydrogen bonding [47] and hydrophobic polymer interactions [48, 49] can increase the stability of polyplexes. Also, caging [50–52] or covalent coupling to the nucleic acid [53–57] has been pursued. Polymers can directly or indirectly (via targeting ligands) attach the cargo to cells and trigger intracellular uptake. They can participate in intracellular delivery steps including endosomal escape and cytosolic transport, nuclear uptake, and cytosolic or nuclear presentation of the nucleic acid in bioactive form.

2.1 Extracellular Transport

Locoregional administration of gene vectors and other therapeutic nucleic acids, such as aerosol delivery into the lung, various injections into muscle [2, 5, 58], brain [59], the eye [60], or into isolated tumors [4, 8, 61] can be quite useful for the treatment of some diseases. For many therapeutic applications, intravenous systemic treatment would be preferred. However, vascular barriers and numerous unintended interactions with biological surfaces including blood proteins, extracellular matrix, and immune cells and other non-target cells mean that only a tiny fraction (in the low percentage range or even less) reaches the actual target tissue. Cationic polymers used for polyplex formation activate the alternative pathway of the complement system, which is part of the innate immune system [62], and systemically administered positively charged polyplexes are rapidly cleared by the reticuloendothelial system [63, 64]. Also, dissociation of polyplexes by serum proteins or extracellular matrix presents a significant problem [65–67].

Several of these problems can be solved by polyplex modification with polyethylene glycol (PEG). PEGylation has been broadly explored for surface shielding ("stealthing") of many liposomal and nanoparticulate carriers. In the case of cationic polymers, Plank et al. [62] demonstrated that complement activation can be reduced when the polymers are PEGylated. Such a modification can be introduced by direct, covalent incorporation of PEG into the polymeric carrier [68–72], direct PEGylation of the nucleic acid [55, 73, 74], PEGylation after polyplex formation [75–77], or by attachment in a noncovalent manner [78, 79]. PEGylation of polyplexes improves solubility, reduces the interaction with blood cells and serum proteins, provides a better biocompatibility, and prolongs blood circulation times [64, 80, 81]. In addition to PEG, other hydrophilic molecules like poly(hydroxypropyl methacrylate) (pHPMA) [82, 83], hyaluronan/hyaluronic acid [84, 85], various polyanions such as γ-polyglutamic acid [86, 87], or the receptor-targeting ligand serum transferrin [88] have been applied to reduce the positive surface charge of polyplexes.

Shielded polyplexes with improved blood circulating properties are interesting tools for systemic cancer therapy (see Sect. 4.2). Nanoparticles can take advantage of the "enhanced permeability and retention" (EPR effect) [89] for passive tumor targeting. The EPR effect is based on the leakiness of tumor vasculature, due to neovascularization in growing tumors, combined with an inadequate lymphatic drainage. Nanoparticles with an elongated plasma circulation time can extravasate and passively accumulate at the tumor site.

Polyplex surface shielding solves several crucial problems, but may also create new problems. Shielding can strongly reduce the efficiency of subsequent cellular steps of the delivery process [68, 69], and also can negatively alter other polyplex characteristics. For pDNA/PEI polyplexes with optimum medium size of PEI, PEG was found to reduce the polyplex stability in vivo [64, 65, 81]. For a discussion of these aspects see Sect. 3.1.

2.2 *Cell Targeting and Intracellular Uptake*

By targeting cell surface receptors, defined target cells in various tissues can be addressed. Targeting may be essential for efficiency, mediating cell-binding of shielded polyplexes, and triggering enhanced polyplex uptake by receptor-mediated endocytosis or related uptake pathways. Different ligands have been found to be suitable for targeted nucleic acid delivery [90–94]. These can be vitamins, carbohydrates, peptides, proteins and glycoproteins, antibodies in various modifications, or nucleic acid aptamers. It has to be kept in mind that the mere presence of targeting ligands does not guarantee targeting; there is no chemotaxis involved. Polyplexes have to first reach the receptor by other means (e.g., by passive targeting) to be available for biochemical receptor docking. It has been observed by several groups that passive accumulation at the target site of targeted and nontargeted polyplexes can be very similar. Active retention and endocytosis at the target cells has been seen as a major functional difference and advantage of receptor-targeted nanocarriers.

The 80 kDa glycoprotein transferrin (Tf) is responsible for intracellular iron transport via the transferrin receptor (TfR), using a clathrin-dependent endocytosis process. Tumor tissues frequently overexpress the TfR. While natural Tf recycles to the cell surface, Tf conjugates commonly are efficiently delivered to various intracellular locations. For these reasons, Tf (or TfR binding antibodies and peptides) has been used as targeting ligand for drug, protein, and gene delivery [95], including the delivery of pDNA [68, 88, 96–105] and siRNA [106–108].

In most cases Tf has been incorporated into polyplexes by direct conjugation to a DNA binding molecule. Ethidium homodimer [99] and many polycations, including PLL and protamine [96–98], PEI [68, 100], polypropylenimine (PPI) dendrimer [104], or PAMAM dendrimer [103], have been used for this purpose. Coupling has been performed by modification of Tf lysine amines with bifunctional linkers [97] or via periodate-oxidized carbohydrates [99, 102] of Tf. In some cases, a bridging PEG linker has been applied [68, 103]. Tf has been attached to polyplexes also in a more indirect way. Activated Tf has been added to polyplexes after polyplex formation [109], Tf has been conjugated to polyplex surface-shielding copolymers [82] or applied as a PEG–adamantane conjugate for noncovalent attachment to cationic β-cyclodextrin-based polymers [101, 110].

Epidermal growth factor receptor (EGFR) is another receptor that is upregulated in several cancers, such as glioblastoma, lung or colorectal cancer. Various EGFR binding molecules have been explored as targeting ligands in polyplexes, including recombinant EGF protein [111–114], EGFR binding antibodies [115], and peptides [116, 117]. For EGF/PEG/LPEI polyplexes, the cellular uptake process was evaluated in detail by real-time single particle fluorescence microscopy. HUH7 hepatoma cells expressing high EGFR levels were used in recombinant version, with either enhanced green fluorescent protein (eGFP)-tagged actin (for visualization of the actin cytoskeleton network) or eGFP-tagged tubulin (for visualization of the microtubules). In the initial phase, both targeted and untargeted polyplexes dock to the cell surface and slowly diffuse along the cell surface simultaneously with the actin cytoskeleton, presumably via transmembrane contacts (mediated by transmembrane adaptor molecules). The presence of the receptor ligand EGF strongly accelerated the intracellular uptake of polyplexes. After 5 min, 50% of the EGF-targeted polyplexes were internalized, whereas less than 10% of untargeted polyplexes were internalized at this time point [114]. Within the cells, polyplexes were found within endosomes, which migrate along the microtubule cytoskeleton.

Numerous other ligands have been evaluated for polymer-based tumor cell targeting [118–124]. For example, bombesin, a peptide that binds bombesin receptors expressed in tumors such as small cell lung carcinoma and gastric cancer, was used as ligand for siRNA polyplex delivery [122]. The folate receptor is upregulated in various cancers. Thus, the natural low molecular weight targeting molecule folic acid was applied for delivery of polymer-based pDNA [123] and siRNA, with siRNA covalently bound to folate via a short linker [124].

The neoangiogenic tumor vasculature overexpresses certain integrins and other surface markers, which can also be used for targeting of polyplexes. The RGD peptide motif has been successfully applied for integrin-targeted pDNA [125–128] and siRNA [129, 130] delivery. In many cases, the PEG motif-containing peptide was attached to the polycation via a PEG spacer. For RGD-PEG-PEI/pDNA polyplexes, an optimum grafting with RGD-PEG was required because transfection efficiency decreased when the degree of PEG-RGD grafting onto PEI was further increased [127]. RGD-PEG-PEI conjugates have been successfully used for systemic anti-angiogenic siRNA therapy [129]. Repeated intravenous administration into neuroblastoma-bearing mice resulted in sequence-specific inhibition of tumor growth. Analogously, RGD-PEG grafting to a defined, oligomerizable lipopolymer (named EHCO) was applied for systemic targeted delivery of an anti-HIF-1α siRNA. The treatment was effective in specific silencing of HIF-1α gene expression and blocking tumor growth in mice [130].

The NGR peptide, which shows high affinity for aminopeptidase N (CD13), has been investigated as a peptide for targeting angiogenic tumor blood vessels. CD13-targeted PEG-PEI/pDNA polyplexes displayed enhanced transgene expression at the tumor site. When applied to p53 gene transfer, the polyplexes showed encouraging antitumoral efficacy, such as significant regressions of lung carcinoma and improved survival of treated mice [131, 132].

Cell targeting has also been applied in several tissues besides tumors. Hepatocyte targeting has been pioneered by Wu and Wu [133–135], using asialoorosomucoid as a liver-specific targeting ligand conjugated with PLL. As the asialoglycoprotein receptor recognizes trimeric galactoside with high affinity, many synthetic carbohydrate targeting ligands have been subsequently tested, with encouraging effects for pDNA and siRNA delivery in vitro and in vivo [56, 136–140].

Both Tf and lactoferrin (Lf) have been evaluated for brain-targeted delivery of PAMAM-PEG polyplexes [103, 141]. Anti-TfR antibodies had been previously extensively investigated for liposomal brain targeting. In a direct comparison of PAMAM-PEG/pDNA polyplex delivery, Lf-targeting was found to be more effective than Tf-targeting, resulting in more than twofold higher brain accumulation and gene expression [141]. Repeated systemic injections of Lf-modified pDNA polyplexes delivering the glial cell line-derived neurotropic factor (GDNF) gene provided effective neuroprotection in a rat brain lesion Parkinson model [142].

Various targeting ligands have been tested for pDNA delivery to the lung. The receptor-mediated endocytosis pathway of airway epithelial polymeric immunoglobulin receptor was utilized for in vivo intravenous targeted delivery of PLL-Fab conjugate/pDNA polyplexes [143]. Peptide-PLL conjugates targeting the serpin–enzyme complex receptor at the apical side of the airway epithelium was successfully used for topical delivery of CFTR-expressing pDNA [144, 145]. Lf receptors are present at high levels on bronchial epithelial cells but not alveolar epithelial cells. Consistently, only in this cell type do Lf-PEI conjugates mediate enhanced gene expression levels as compared to PEI polyplexes [146]. Conversely, insulin receptors are expressed on alveolar but not bronchial epithelial cells. pDNA/PEI polyplexes coated with insulin increased gene transfer to alveolar epithelial cells up to 16-fold [147].

Also, small chemical drugs such as clenbuterol, an agonist specifically binding the β2-adrenoceptor (β2-AR) [148], or prostaglandin I2 analogs iloprost and treprostinil, targeting the prostacyclin receptor IP1 [149], have been coupled to PEI and successfully applied for improved receptor-mediated gene transfer of pDNA/PEI polyplexes to the lung. Gene expression in the lungs of mice after aerosol delivery of pDNA with iloprost-grafted PEI was 14-fold higher than for plain PEI polyplexes.

The selection of the receptor ligand can have an impact both on cell binding and intracellular uptake. "Dual targeting" options have been developed with the option to combine two targeting ligands in an effective way; for example, combining ligands with unique cell binding characteristics but moderate internalization with co-ligands mediating efficient endocytosis. In transfections of prostate cancer cells, synergistic dual targeting characteristics were observed with DNA polyplexes containing two different peptide ligands. PEGylated PEI/DNA polyplexes were decorated with RGD peptide ligands for integrin targeting and with peptide B6 for TfR targeting. In a series of flow cytometer experiments, either cell association or cell internalization with and without ligand competition were evaluated. RGD was found to play a major role in cell surface binding, whereas B6 had a major role in intracellular uptake [150].

2.3 *Intracellular Transport*

Several enveloped viruses, and some physical gene transfer techniques such as electroporation, deliver the nucleic acid into the cell by direct crossing of the cell membrane. Lipid-based, enveloped systems can do this by a physiological, self-sealing membrane fusion process, avoiding physical damage of the cell membrane. For cationic lipid-mediated delivery of siRNA, most material is taken up by endocytotic processes. Recently, direct transfer into the cytosol has been demonstrated to be the bioactive delivery principle for certain siRNA lipid formulations [151].

Other systems like electroporation have no lipids that might help in membrane sealing or fusion; for direct transfer of the nucleic acid across membranes they have to generate transient pores, a process where efficiency is usually directly correlated with membrane destruction and cytotoxicity. Alternatively, like for the majority of polymer-based polyplexes, cellular uptake proceeds by clathrin- or caveolin-dependent and related endocytic pathways [152–156]. The polyplexes end up inside endosomes, and the membrane disruption happens in intracellular vesicles. It is noteworthy that several observed uptake processes may not be functional in delivery of bioactive material. Subsequent intracellular obstacles may render a specific pathway into a dead end [151, 154, 156]. With time, endosomal vesicles become slightly acidic (pH 5–6) and finally fuse with and mature into lysosomes. Therefore, polyplexes have to escape into the cytosol to avoid the nucleic acid-degrading lysosomal environment, and to deliver the therapeutic nucleic acid to the active site. Either the carrier polymer or a conjugated endosomolytic domain has to mediate this process [157], which involves local lipid membrane perturbation. Such a lipid membrane interaction could be a toxic event if occurring at the cell surface or mitochondrial membrane. Thus, polymers that show an endosome-specific membrane activity are favorable.

PEI or PAMAM dendrimers are particularly effective in nucleic acid transfer because of their intrinsic dynamic behavior in response to endosomal acidification. These polymers are "proton sponges," displaying only moderate (about 20%) protonation of nitrogens at neutral pH, which increases with endosomal acidification. The increased density of positive charges leads to an influx of chloride and water into the endosome [158]. As a consequence of the "proton sponge effect," the endosome bursts due to the elevated osmotic pressure and the membrane destabilizing effect of positively charged polymer domains. Efficient endosomal escape, however, still presents a bottleneck. Especially for PEI, the window between effective dose for endosome disruption and cytotoxic dose is very narrow. Less cytotoxic, biodegradable proton sponge polymers containing ethylenimine units have been developed, for example diaminoethyl residue-containing polyapartamide [41]. Similarly, the cytotoxicity of PEI can be reduced by modifications. Introduction of propionic acid or succinic acid residues into branched PEI 25 kDa diminishes positive charges at neutral pH by conversion into negatively charged carboxylate groups. The modified PEI polymers are still proton sponges, containing both protonable carboxylates and nitrogens. Because of a far lower cytotoxicity, it was possible to apply higher doses of polymer for improved endosomal escape. Efficient siRNA delivery was demonstrated with acid-modified PEI derivatives under conditions where standard PEI was inactive [159, 160].

Other polymers like PLL cannot efficiently destabilize endosomes at nontoxic concentrations. Incorporation of virus-derived inactivated particles [161, 162], fusion proteins derived from bacteria such as listeriolysin (LLO) [163], synthetic membrane-active peptides derived from viral fusion sequences (influenza virus, rhinovirus) and analogs thereof [138, 164–168], either by covalent conjugation or ionic association, has strongly enhanced the gene transfer efficiency of polyplexes. Also, various synthetic versions derived from the highly lytic bee venom component melittin (Mel) have been used in polyplexes [169–171]. The fusion agents can be classified into two categories: those with a very pH-specific endosomal activity (e.g., peptides containing acidic residues) that are not lytic at normal physiological pH, and those displaying a pH-independent lytic activity associated with considerable toxicity, such as free Mel or LLO. Bioresponsive modifications and the mode of incorporation have to be performed to reduce such unwanted side effects (see Sect. 3.2).

Instead of peptide- and protein-based fusion elements, other membrane-active agents can also be applied. Examples include a series of pH-sensitive copolymers designed for the delivery of drugs or biomacromolecules [54, 172, 173]. Amphiphilic copolymers consisting of methacrylic acid and butyl acrylate display pH-specific lytic properties [54]. Alternatively, copolymers containing hydrophobic, membrane-disruptive methacrylate blocks were reversibly masked with PEG chains through pH-sensitive benzaldehyde acetal linkers [172]. Various pH-sensitive polyglycidol derivatives with dicarboxylic acids of varying hydrophobicities (glutaric acid, 3-methyl glutaric acid, cyclohexane-1,2-dicarboxylic acid) were investigated for their ability to destabilize liposomes in an endosomal pH-sensitive manner. Some of these pH-regulated polymers, although mainly designed for incorporation into liposomal drug delivery systems [174, 175], might be also useful in the context of polyplexes.

Intracellular pathways after escape from the endolysosomal system into the cytosol are less clear. Obvious bottlenecks include, in the case of gene transfer (pDNA delivery), cytosolic transport to the perinuclear area, nuclear uptake, and nuclear presentation of the pDNA to the transcriptional machinery in bioactive form. In the case of siRNA (or mRNA and some other nucleic acids such as oligonucleotides), cytosolic accessibility for the required function is essential. Besides cytosolic transport [176, 177] and the nuclear import of large nucleic acid molecules [178–180], incorporation of functional nuclear import peptide domains has been evaluated [181–186]. Another bottleneck, nucleic acid unpackaging [187], i.e., partial or complete dissociation from the polymeric carrier, which is required for biological accessibility of the delivered nucleic acid, will be discussed in Sect. 3.3.

3 Polymer Design for Bioresponsive Activity

Natural viruses provide us with perfect demonstrations of how effective nucleic acid transfer into mammalian cells can proceed. The "secret" of their efficiency is their dynamic, bioresponsive behavior during delivery, which distinguishes them from classic synthetic nanoparticles. Thus, it has been tempting for us and many research colleagues [69, 92, 164, 188–194] to design nucleic acid nanoparticles with virus-like characteristics ("synthetic viruses").

Polymers can be designed to be bioresponsive and to change their properties in various biological compartments, for example their conformation and charge. They may contain chemical bonds that are cleaved under temporal or spatial control, or they may associate/dissociate in a controlled fashion. Bioresponsiveness of the carrier is required in several steps of the transport (Fig. 1).

A highly stable and shielded polyplex should circulate in the blood stream without undesired interactions until it reaches the target cell. At that location, specific interactions with the cell surface should trigger intracellular uptake. While lipid membrane interaction is undesired at the cell surface, it should happen subsequently within the endosomal vesicle and mediate polyplex delivery into the cytosol. During or after intracellular transport to the site of action, the polyplex stability should be weakened to an extent that the nucleic acid is accessible to exert its function.

The following sections discuss how polymers and polyplexes can be chemically designed to be bioresponsive in three key delivery functions: (1) polyplex surface shielding, (2) interaction with lipid bilayers, and (3) polyplex stability.

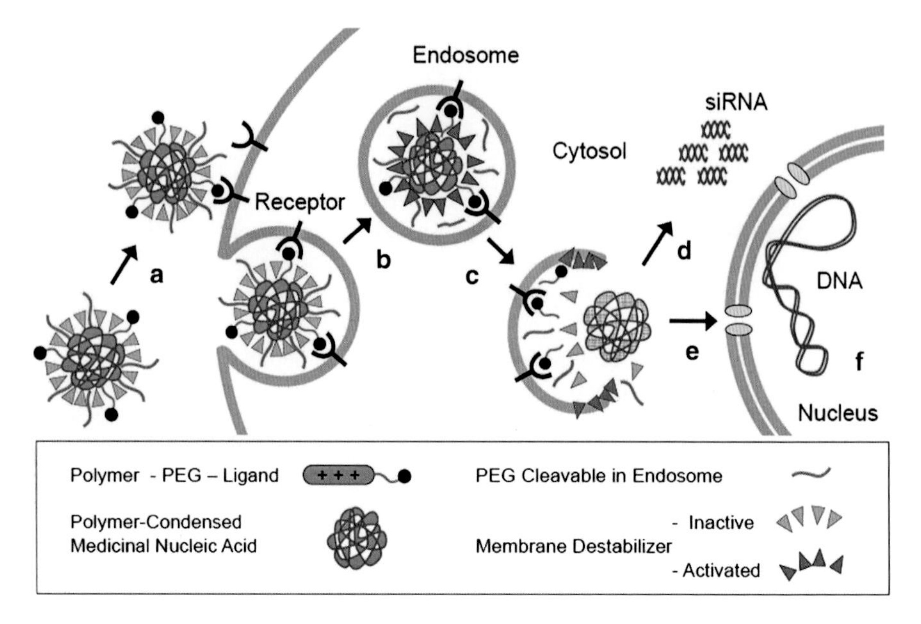

Fig. 1 Bioresponsive polyplexes. (*a*) Systemic circulation of shielded polyplexes in blood stream and attachment to cell surface receptor; (*b*) endocytosis into endosomes, deshielding by cleavage of PEG linkers and activation of membrane-destabilizing component by acidic pH or other means; (*c*) endosomal escape into cytosol; (*d*) siRNA transfer to form a cytosolic RNA-induced silencing complex complex; (*e*) cytosolic migration and intranuclear import of pDNA; (*f*) presentation of pDNA in accessible form to the transcription machinery

3.1 Bioresponsive Polyplex Shielding

In the blood circulation, polyplexes need to be shielded against numerous possible biological interactions, but they should actively interact with the target cell surface by electrostatic or ligand receptor interactions. In this regard, shielding with PEG or other hydrophilic polymers provides many benefits (see Sect. 2.1). Irreversible, stable surface shielding, however, may also be disadvantageous for the transfection efficiency in at least two aspects: it may block interactions of the polyplex with the target cell surface, and it may also prevent lipid membrane-destabilizing interaction within endosomes. Targeting ligands can restore the cell surface–polyplex interaction in a receptor-specific manner [68, 75, 82], but it cannot restore cationic polymer interactions with the cellular lipid membranes involved in endosomal escape. As a solution for this problem, bioresponsive deshielding strategies (i.e., removal of the shielding molecules) have been developed (see Fig. 2). Location-specific changes such as in pH [55, 56, 69, 76, 195–197], enzymatic activity [198], or disulfide reducing potential [83] have been applied for timely removal of the hydrophilic shield after nucleic acid nanoparticles have reached their target tissue.

pH-triggered deshielding strategies [199] include pH-labile acetal linkages [172, 196, 200], and dialkylmaleic acid [56] or pyridylhydrazone [69, 76] bonds

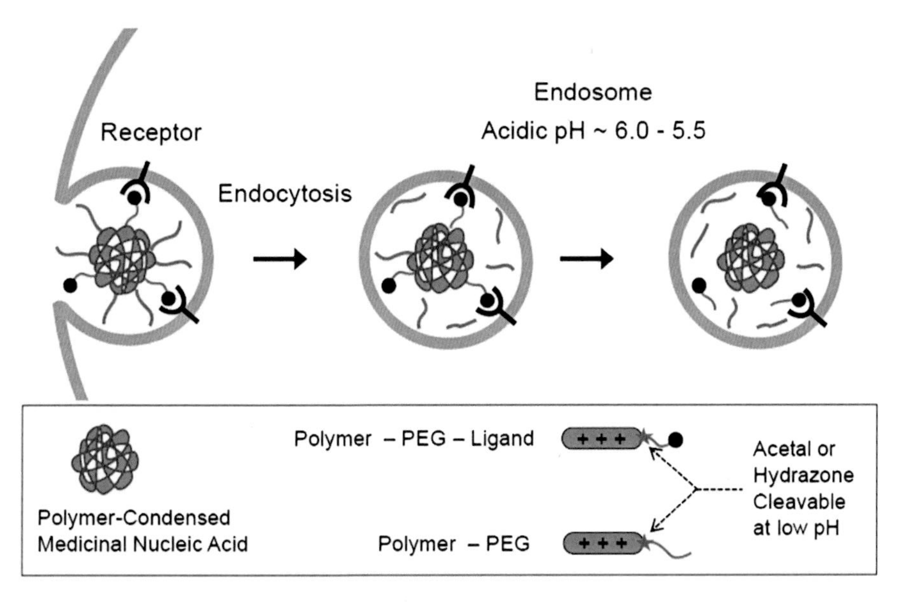

Fig. 2 Deshielding of polyplexes. After endocytosis of polyplexes into endosomes, deshielding by cleavage of PEG hydrazone or acetal linkers

incorporated into PEG–polymer conjugates. Deshielding at endosomal pH strongly (up to 100-fold) enhanced gene transfer of targeted PEG-PEI/pDNA polyplexes in vitro and in vivo [69, 76]. The most plausible explanation for this positive effect is cleavage of the PEG inside the endosomes, exposing cationic PEI domains with endosomolytic properties. In analogous fashion, DNA/PEI lipopolyplexes containing PEG linked with the lipid layer via pyridyl hydrazone linkages were far more effective than their pH-stable analogs [201]. Dynamic siRNA polyconjugates [56] contain an endosomal-sensitive dialkylmaleic acid linkage between a cationic amphipathic (butyl-amino-modified) polyvinyl ether and PEG.

A different pH-triggered deshielding concept with hydrophilic polymers is based on reversing noncovalent electrostatic bonds [78, 195, 197]. For example, a pH-responsive sulfonamide/PEI system was developed for tumor-specific pDNA delivery [195]. At pH 7.4, the pH-sensitive diblock copolymer, poly(methacryloyl sulfadimethoxine) (PSD)-*block*-PEG (PSD-*b*-PEG), binds to DNA/PEI polyplexes and shields against cell interaction. At pH 6.6 (such as in a hypoxic extracellular tumor environment or in endosomes), PSD-*b*-PEG becomes uncharged due to sulfonamide protonation and detaches from the nanoparticles, permitting PEI to interact with cells. In this fashion PSD-*b*-PEG is able to discern the small difference in pH between normal and tumor tissues.

Tumor tissues overexpress matrix metalloproteinases (MMPs). A liposomal pDNA carrier (MEND) was developed containing PEG conjugated to lipid via a peptide linker that is a target sequence for MMPs. In this strategy, PEG is removed from the carrier via MMP-triggered cleavage [198]. Intravenous administration in mice resulted in enhanced pDNA expression in tumor tissue as compared with a conventional, stable PEG-modified MEND.

3.2 Bioresponsive Lipid Bilayer Interaction

As outlined in previous sections, escape of polyplexes from endosomes to the cytosol can be a major bottleneck in delivery. Membrane-active polymer domains or other conjugated molecules can help to overcome this barrier (see Sect. 2.3), but they may trigger cytotoxicity when acting extracellularly or at the cell surface. Therefore membrane-crossing agents either have to be inherently specific for endosomal compartments (for example by pH-specificity), or they have to be modified to be activated in endosomes. For example, the reducing stimulus of intracellular vesicles has been used to activate formulations containing less active disulfide precursors of LLO [163] or Mel [170].

Also, acid-labile masking of membrane-disruptive agents has been pursued. Copolymers consist of hydrophobic, membrane-disruptive methacrylate polymers were reversibly masked with PEG chains through pH-sensitive acetal linkers [172]. Dialkylmaleic acid derivatives were used for pH-reversible acylation of amino groups of a membrane-destabilizing polyvinylether polymer [56] or the lytic, but not pH-specific peptide Mel [57, 202–205]. Modification of Mel lysines with dimethyl maleic anhydride (DMMAn) blocks the lytic activity. At endosomal pH, DMMAn groups are released, unmasking highly lytic Mel for endosome disruption (Fig. 3). PLL when covalently modified with DMMAn had a 1,800-fold improved gene transfer activity over unmodified PLL [203] and, in a PEG-modified version, could also mediate efficient siRNA transfer [57, 204].

3.3 Bioresponsive Polyplex Stability

The polymeric carrier has to stably bind the medicinal nucleic acid outside the cell to compact and protect it from degradation. Inside the cell, the polyplex has to disassemble to such an extent that the nucleic acid can be biologically active. Different chemical characteristics outside the cell, in the endosome, the cytosol, and the nucleus can be utilized to manage the controlled disassembly process. Medicinal nucleic acids can be either noncovalently complexed or covalently conjugated to the carrier polymer. Release of the nucleic acid at the target location from the polymer may proceed, for example, by exchange processes against polyions such as intracellular RNA [206], by polymer degradation [207], or by cleavage of the nucleic acid from the polymer attachment sites [54].

High nucleic acid/polymer affinity does not necessarily directly correlate with high efficiency. Apparently, an optimum has to be reached. Also, the big difference in size of different medical nucleic acids (pDNA has several thousand negative

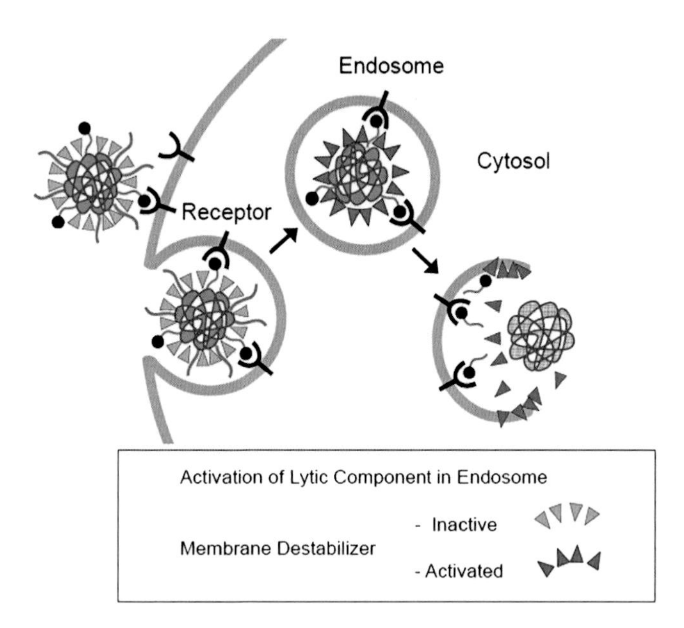

Fig. 3 Activation of membrane-destabilizing component by acidic pH, resulting in endosomal escape into the cytosol

charges in contrast to siRNA with 42 negative charges) has to be considered. For example, the intracellular fates of medium-sized LPEI (22 kDa) and brPEI (25 kDa) were compared using confocal fluorescence microscopy and FRET technology [208]. The LPEI polyplexes were found to be more effective but less stable; when reaching the cytosol, they release pDNA much faster than the more stable brPEI polyplexes. The same research group [209] discovered that insufficient decondensation of PEI/pDNA polyplexes in the nucleus was a major limiting factor for gene expression. In this respect, polymers with lower pDNA affinity seem to be preferable in transfections in vitro.

On the other hand, pDNA/PEI polyplexes were found to be not stable enough in the extracellular in vivo environment. Unpackaging of PEI and PEG-PEI polyplexes was observed [64, 65, 81], for example by serum proteins, soluble glycosaminoglycans, or extracellular matrix components. The situation is even worse in the case of siRNA polyplexes, where PEI polyplexes are dissociated in full human serum, as monitored by fluorescence fluctuation spectroscopy [66, 67].

Bioresponsive polymers are one logical solution for this "polyplex dilemma," i.e., the insufficient stability of polyplexes outside the cell, but too-high a stability inside the cell. Strategies include the design of biodegradable high molecular weight polymers that intracellularly degrade into low molecular weight nontoxic fragments. Cleavable bonds include acetal bonds that degrade in the acidic environment of endosomes [200, 210], disulfide bonds that are reduced in the cytosol [207, 211, 212], or hydrolyzable esters [213–215]. Polyplexes can also be reversibly stabilized by caging [50–52, 109, 216, 217], i.e., chemical crosslinking of the pDNA-bound polymers via cleavable linkers. For example, pseudodendrimers containing low molecular weight oligoethylenimine and biodegradable diacrylate esters [218, 219] can deliver pDNA with high efficiency and low cytotoxicity in vitro and in vivo upon systemic delivery in a murine tumor model. Gene expression (luciferase or the therapeutically more relevant sodium iodide symporter NIS) was primarily detected at the subcutaneous tumor site [218–220]. Stability of these polyplexes however was limited. Lateral stabilization by crosslinking surface amines via a bifunctional crosslinker resulted in improved stability. Best gene transfer results were obtained when a bioreducible disulfide-containing crosslinker was used for caging [109].

Electrostatic binding of siRNA or oligonucleotides (ONs) to cationic polymers is weaker than pDNA binding, due to the lower number of negative charges in the short phosphate backbones. To overcome this hurdle, ONs and siRNAs can be covalently bound to their carriers [53]. This can also be performed in a bioresponsive reversible way. For example, disulfide bonds, which can be easily cleaved in the cytosol due to the reducing environment inside the cell, were applied for ON conjugation with a pH-specific membrane-disruptive carrier [54] for the synthesis of dynamic siRNA polyconjugates [56], and for siRNA conjugated to a PLL that was modified with an endosomolytic peptide [57]. The covalent polyplexes are not cleavable by heparin in concentrations at which analogous electrostatic complexes are dissociated. Only combinations of heparin with a reducing reagent were able to release siRNA from the novel conjugates [57].

4 Therapeutic Strategies

Since the design of the first targeted polyplexes more than 20 years ago [97, 134], numerous efforts have been made to develop polyplexes for use in medical products, both in pharmacological animal studies and in human studies. Therapeutic modalities include ex vivo treatment of isolated human patient cells, localized in vivo treatments, and – currently the most challenging delivery scenario – in vivo targeted intravenous delivery.

4.1 Ex Vivo and Localized Therapies

First clinical human gene therapy trials with polyplexes were performed using cancer vaccines based on autologous patient tumor cells. These were modified ex vivo with interleukin-2 pDNA. To obtain high level transfection rates of patient's primary tumor cells, Tf-PLL/pDNA polyplexes linked with inactivated endosomolytic adenovirus particles were applied [221]. Polymer-based in vivo human gene transfer studies were performed with PEGylated PLL polyplexes, delivering CFTR pDNA to the airway epithelium of cystic fibrosis patients [222], and with PEGylated cholesterol-modified PEI polyplexes for intraperitoneal administration of interleukin-12 pDNA in ovarian cancer patients [7, 223]. In the latter case, a phase I trial demonstrated the safety of administration, and stable disease and reduction in serum CA-125 levels in some patients.

Intravesical infusion of linear PEI/pDNA polyplexes was evaluated in patients with superficial bladder cancer where intravesical therapy with bacillus Calmette-Guerin had failed [6, 224]. Patients had low grade superficial bladder cancer, which expressed H19. The therapeutic pDNA contains H19 gene regulatory sequences that drive the expression of an intracellular toxin. Escalating doses of 2–20 mg plasmid per intravesical treatment were applied, with responders continuing to receive polyplexes once a month every month for 1 year. The treatment resulted in complete ablation of the marker tumor, without any new tumors in four of the 18 patients (22% overall complete response rate). Eight of the 18 patients (44%) had complete marker tumor ablation or a 50% reduction of the marker lesion.

A localized treatment of glioblastoma was explored in the preclinical setting. Synthetic antiproliferative poly(I:C) RNA, a strong activator of apoptosis and interferon response, was targeted to U87MG-EGFR glioblastoma by PEI-PEG-EGF polyplexes [4]. Intratumoral delivery of EGFR-targeted poly(I:C) induced rapid apoptosis of the target cells and complete regression of intracranial tumors in nude mice, without opposing toxicity on normal brain tissue. Similarly, delivery of poly(I:C) completely eliminated EGFR-overexpressing breast cancer and adenocarcinoma xenografts.

4.2 *Systemic Therapies*

Systemic targeting of pDNA and siRNA polyplexes has been demonstrated in several animal models. In continuation of the work with localized antiproliferative and immunostimulatory poly(I:C) RNA, intravenous systemic delivery of EGFR-targeted PEG-modified polyplexes were successfully used for human carcinoma treatment in mice [225]. The therapeutic effect was most pronounced when intravenous delivery of poly(I:C) polyplexes was followed by intraperitoneal injection of peripheral blood mononuclear cells [226]. This induced the complete cure of SCID mice with pre-established disseminated EGFR-overexpressing tumors, without adverse toxic effects. Due to the chemokines produced by the internalized poly (I:C) in the tumor cells, the immune cells home to the tumors of the treated animal and contribute to the tumor destruction.

EGFR targeting was also used for systemic delivery of pDNA expressing the sodium iodide symporter (NIS) gene to liver cancer cells, followed by administration of radioactive isotope iodine-131, which accumulates in the tumor by NIS-mediated uptake in radiotherapeutic doses [227].

Various researchers have applied the receptor-targeted strategy in pharmacological models for tumor-targeted delivery of pDNA expressing tumor necrosis factor alpha (TNFα). For example, Tf- or Tf-PEG-shielded PEI polyplexes have been used for the delivery and expression of this immunostimulatory and cytotoxic gene product, which may negatively affect both the tumor endothelium and tumor cells. Repeated systemic application of such Tf-coated TNFα polyplexes induced tumor necrosis and inhibition of tumor growth in four tumor-bearing mouse models (MethA fibrosarcoma, Neuro2A neuroblastoma, M3 and B16F10 melanomas) [68, 228]. Due to the fact that TNFα gene expression was largely localized within the tumor, no significant systemic TNFα-related toxicities were observed. Therapeutic effects were enhanced by the combination of tumor-targeted TNFα gene therapy with the liposomal doxorubicin formulation DOXIL that passively accumulates within tumors [229].

Polyplex formulations basing on polypropylenimine (PPI) dendrimers were also effective in intravenous TNFα gene therapy, as shown in three other established tumor models (A431 epidermoid carcinoma, C33a cervix carcinoma, and LS174T colorectal adenocarcinoma) [230]. In those studies, a tumor-specific promoter was applied for selective expression in the tumor. The cationic PPI carrier was found to exhibit additional intrinsic antitumor activity, enhancing the therapeutic efficacy. Very recently, the investigators demonstrated that the conjugation of the PPI dendrimer to Tf triggers a selective gene delivery to A431 tumors after intravenous administration, leading to an increased therapeutic efficacy and sustained tumor regression and long-term survival of 100% of mice [104].

Tf-containing PEG-shielded polyplexes have also been applied for systemic tumor-targeted delivery of siRNA [106–108]. Systemic treatment of Neuro 2A tumor-bearing mice using Tf-PEG-shielded crosslinked oligoethylenimines for delivery of siRNA against Ras-related nuclear protein (Ran) led to >80% reduced Ran protein expression, associated with tumor apoptosis and reduced tumor growth [108].

Advanced studies were performed by Davis and colleagues [106, 107] who developed cationic β-cyclodextrin carriers containing Tf-PEG bound via a noncovalent adamantine/cyclodextrin guest/host complexation. The researchers demonstrated in several studies that Tf was an essential part of the particle formulation for siRNA-mediated gene silencing in mouse tumors [231]. Biodistribution data using multimodal in vivo imaging in mice demonstrated similar tumor localization (passive targeting) for both PEGylated nontargeted and Tf-PEG modified targeted siRNA particles [232]. Tf-containing siRNA nanoparticles, however, showed higher functional activity in tumors, presumably due to receptor-mediated cellular uptake into the tumor cells. Sequence-specific knockdown of EWS-FLI1 inhibited tumor growth in a murine model of metastatic Ewing's sarcoma [106]. The presence of Tf within the PEGylated polyplexes was a critical requirement for the therapeutic efficacy. Treatment of tumor-bearing mice with therapeutic siRNA targeting the M2 subunit of ribonucleotide reductase also resulted in a decrease in tumor growth [107].

Based on these findings and toxicology studies [233], human clinical phase I studies have been initiated [234]. In tumor biopsies from melanoma patients obtained after treatment, a reduction was found in both the specific mRNA (M2 subunit of ribonucleotide reductase, RRM2) and in the protein (RRM2) levels when compared to pre-dosed tissue. The presence of an mRNA fragment that demonstrates that siRNA-mediated mRNA cleavage occurs specifically at the predicted site was detectable in a high-dose treated patient [235]. Together, these data demonstrate that systemically administered siRNA can produce a specific gene inhibition (reduction in mRNA and protein) by an RNA interference mechanism in humans. These data present an important and encouraging step in the slow process of transferring ideas into medicines. It has to be mentioned that it took 20 years from the design of Tf-targeting polyplexes [97] to the demonstration of direct molecular effects in humans [235].

5 Conclusions

Over more than two decades of development, functional polymer conjugates have become useful carriers for drug and nucleic acid delivery. Dynamic, bioresponsive polymers have been designed that can better cope with the multiple sequential delivery steps. Improved chemistry is able to provide monodisperse polymer structures with defined size, topology, and transport domains [236, 237]. This will also be essential for practical development reasons, including production of clinical grade materials according to good manufacturing practice guidelines. The first polymeric carriers have already entered clinical testing, with encouraging results. It will be very interesting to see how successfully they will fill clinical development pipelines in the upcoming future.

Acknowledgments Many thanks to Olga Brück (LMU) for skillful assistance in preparing the review. Our own work in the reviewed research area was supported by the German DFG excellence cluster "Nanosystems Initiative Munich (NIM)" and the BMBF Munich Biotech cluster m4 project T12.

References

1. Duncan R (2003) The dawning era of polymer therapeutics. Nat Rev Drug Discov 2:347–360
2. Wolff JA, Malone RW, Williams P, Chong W, Acsadi G, Jani A, Felgner PL (1990) Direct gene transfer into mouse muscle in vivo 726. Science 247:1465–1468
3. Caplen NJ, Alton EW, Middleton PG, Dorin JR, Stevenson BJ, Gao X, Durham SR, Jeffery PK, Hodson ME, Coutelle C (1995) Liposome-mediated CFTR gene transfer to the nasal epithelium of patients with cystic fibrosis. Nat Med 1:39–46
4. Heller LC, Heller R (2006) In vivo electroporation for gene therapy. Hum Gene Ther 17:890–897
5. Tjelle TE, Rabussay D, Ottensmeier C, Mathiesen I, Kjeken R (2008) Taking electroporation-based delivery of DNA vaccination into humans: a generic clinical protocol. Methods Mol Biol 423:497–507
6. Sidi AA, Ohana P, Benjamin S, Shalev M, Ransom JH, Lamm D, Hochberg A, Leibovitch I (2008) Phase I/II marker lesion study of intravesical BC-819 DNA plasmid in H19 over expressing superficial bladder cancer refractory to bacillus Calmette-Guerin. J Urol 180:2379–2383

7. Anwer K, Barnes MN, Fewell J, Lewis DH, Alvarez RD (2010) Phase-I clinical trial of IL-12 plasmid/lipopolymer complexes for the treatment of recurrent ovarian cancer. Gene Ther 17:360–369
8. Bedikian AY, Richards J, Kharkevitch D, Atkins MB, Whitman E, Gonzalez R (2010) A phase 2 study of high-dose allovectin-7 in patients with advanced metastatic melanoma. Melanoma Res 20:218–226
9. Yamamoto A, Kormann M, Rosenecker J, Rudolph C (2009) Current prospects for mRNA gene delivery. Eur J Pharm Biopharm 71:484–489
10. Kormann MS, Hasenpusch G, Aneja MK, Nica G, Flemmer AW, Herber-Jonat S, Huppmann M, Mays LE, Illenyi M, Schams A, Griese M, Bittmann I, Handgretinger R, Hartl D, Rosenecker J, Rudolph C (2011) Expression of therapeutic proteins after delivery of chemically modified mRNA in mice. Nat Biotechnol 29:154–157
11. Yin H, Lu Q, Wood M (2008) Effective exon skipping and restoration of dystrophin expression by peptide nucleic acid antisense oligonucleotides in mdx mice. Mol Ther 16:38–45
12. Krützfeldt J, Rajewsky N, Braich R, Rajeev K, Tuschl T, Manoharan M, Stoffel M (2005) Silencing of microRNAs in vivo with 'antagomirs'. Nature 438:685–689
13. Fabani MM, Gait MJ (2008) miR-122 targeting with LNA/2'-O-methyl oligonucleotide mixmers, peptide nucleic acids (PNA), and PNA-peptide conjugates. RNA 14:336–346
14. Crooke ST (2004) Progress in antisense technology. Annu Rev Med 55:61–95
15. Veedu RN, Wengel J (2009) Locked nucleic acid as a novel class of therapeutic agents. RNA Biol 6:321–323
16. Elbashir SM, Harborth J, Lendeckel W, Yalcin A, Weber K, Tuschl T (2001) Duplexes of 21-nucleotide RNAs mediate RNA interference in cultured mammalian cells. Nature 411:494–498
17. Aagaard L, Rossi JJ (2007) RNAi therapeutics: principles, prospects and challenges. Adv Drug Deliv Rev 59:75–86
18. Meyer M, Wagner E (2006) Recent developments in the application of plasmid DNA-based vectors and small interfering RNA therapeutics for cancer. Hum Gene Ther 17:1062–1076
19. Aigner A (2006) Gene silencing through RNA interference (RNAi) in vivo: strategies based on the direct application of siRNAs. J Biotechnol 124:12–25
20. Gao K, Huang L (2009) Nonviral methods for siRNA delivery. Mol Pharmaceutics 6:651–658
21. Behlke MA (2008) Chemical modification of siRNAs for in vivo use. Oligonucleotides 18:305–319
22. Li L, Shen Y (2009) Overcoming obstacles to develop effective and safe siRNA therapeutics. Expert Opin Biol Ther 9:609–619
23. Frohlich T, Wagner E (2010) Peptide- and polymer-based delivery of therapeutic RNA. Soft Matter 6:226–234
24. Kota J, Chivukula RR, O'Donnell KA, Wentzel EA, Montgomery CL, Hwang HW, Chang TC, Vivekanandan P, Torbenson M, Clark KR, Mendell JR, Mendell JT (2009) Therapeutic microRNA delivery suppresses tumorigenesis in a murine liver cancer model. Cell 137:1005–1017
25. Aigner A, Fischer D, Merdan T, Brus C, Kissel T, Czubayko F (2002) Delivery of unmodified bioactive ribozymes by an RNA-stabilizing polyethylenimine (LMW-PEI) efficiently downregulates gene expression. Gene Ther 9:1700–1707
26. Tomita N, Azuma H, Kaneda Y, Ogihara T, Morishita R (2004) Application of decoy oligodeoxynucleotides-based approach to renal diseases. Curr Drug Targets 5:717–733
27. Rusconi CP, Scardino E, Layzer J, Pitoc GA, Ortel TL, Monroe D, Sullenger BA (2002) RNA aptamers as reversible antagonists of coagulation factor IXa. Nature 419:90–94
28. Ulrich H, Trujillo CA, Nery AA, Alves JM, Majumder P, Resende RR, Martins AH (2006) DNA and RNA aptamers: from tools for basic research towards therapeutic applications. Comb Chem High Throughput Screen 9:619–632

29. Shir A, Ogris M, Wagner E, Levitzki A (2006) EGF receptor-targeted synthetic double-stranded RNA eliminates glioblastoma, breast cancer, and adenocarcinoma tumors in mice. PLoS Med 3:e6
30. Poeck H, Besch R, Maihoefer C, Renn M, Tormo D, Morskaya S, Kirschnek S, Gaffal E, Landsberg J, Hellmuth J, Schmidt A, Anz D, Bscheider M, Schwerd T, Berking C, Bourquin C, Kalinke U, Kremmer E, Kato H, Akira S, Meyers R, Häcker G, Neuenhahn M, Busch D, Ruland J, Rothenfusser S, Prinz M, Hornung V, Endres S, Tüting T, Hartmann G (2008) 5′-Triphosphate-siRNA: turning gene silencing and Rig-I activation against melanoma. Nat Med 14:1256–1263
31. Besch R, Poeck H, Hohenauer T, Senft D, Hacker G, Berking C, Hornung V, Endres S, Ruzicka T, Rothenfusser S, Hartmann G (2009) Proapoptotic signaling induced by RIG-I and MDA-5 results in type I interferon-independent apoptosis in human melanoma cells. J Clin Invest 119:2399–2411
32. Pack DW, Hoffman AS, Pun S, Stayton PS (2005) Design and development of polymers for gene delivery. Nat Rev Drug Discov 4:581–593
33. Wagner E, Kloeckner J (2006) Gene delivery using polymer therapeutics. Adv Polym Sci 192:135–173
34. Schaffert D, Wagner E (2008) Gene therapy progress and prospects: synthetic polymer-based systems. Gene Ther 15:1131–1138
35. Wagner E (2008) The silent (R)evolution of polymeric nucleic acid therapeutics. Pharm Res 25:2920–2923
36. Felgner PL, Barenholz Y, Behr JP, Cheng SH, Cullis P, Huang L, Jessee JA, Seymour L, Szoka F, Thierry AR, Wagner E, Wu G (1997) Nomenclature for synthetic gene delivery systems. Hum Gene Ther 8:511–512
37. Wagner E, Ogris M, Zauner W (1998) Polylysine-based transfection systems utilizing receptor-mediated delivery. Adv Drug Deliv Rev 30:97–113
38. Zauner W, Kichler A, Schmidt W, Sinski A, Wagner E (1996) Glycerol enhancement of ligand-polylysine/DNA transfection. Biotechniques 20:905–913
39. Zou SM, Erbacher P, Remy JS, Behr JP (2000) Systemic linear polyethylenimine (L-PEI)-mediated gene delivery in the mouse. J Gene Med 2:128–134
40. van de Wetering P, Cherng JY, Talsma H, Crommelin DJ, Hennink WE (1998) 2-(Dimethylamino)ethyl methacrylate based (co)polymers as gene transfer agents 1031. J Control Release 53:145–153
41. Miyata K, Oba M, Kano MR, Fukushima S, Vachutinsky Y, Han M, Koyama H, Miyazono K, Nishiyama N, Kataoka K (2008) Polyplex micelles from triblock copolymers composed of tandemly aligned segments with biocompatible, endosomal escaping, and DNA-condensing functions for systemic gene delivery to pancreatic tumor tissue. Pharm Res 25:2924–2936
42. Boussif O, Lezoualc'h F, Zanta MA, Mergny MD, Scherman D, Demeneix B, Behr JP (1995) A versatile vector for gene and oligonucleotide transfer into cells in culture and in vivo: polyethylenimine. Proc Natl Acad Sci USA 92:7297–7301
43. Tang MX, Szoka FC (1997) The influence of polymer structure on the interactions of cationic polymers with DNA and morphology of the resulting complexes. Gene Ther 4:823–832
44. Fant K, Esbjorner EK, Lincoln P, Norden B (2008) DNA condensation by PAMAM dendrimers: self-assembly characteristics and effect on transcription. Biochemistry 47: 1732–1740
45. Kukowska-Latallo JF, Bielinska AU, Johnson J, Spindler R, Tomalia DA, Baker JR Jr (1996) Efficient transfer of genetic material into mammalian cells using Starburst polyamidoamine dendrimers. Proc Natl Acad Sci USA 93:4897–4902
46. Zhou J, Wu J, Hafdi N, Behr JP, Erbacher P, Peng L (2006) PAMAM dendrimers for efficient siRNA delivery and potent gene silencing. Chem Commun (Camb) 2006:2362–2364
47. Prevette LE, Kodger TE, Reineke TM, Lynch ML (2007) Deciphering the role of hydrogen bonding in enhancing pDNA-polycation interactions. Langmuir 23:9773–9784

48. Philipp A, Zhao X, Tarcha P, Wagner E, Zintchenko A (2009) Hydrophobically modified oligoethylenimines as highly efficient transfection agents for siRNA delivery. Bioconjug Chem 20:2055–2061
49. Creusat G, Zuber G (2008) Self-assembling polyethylenimine derivatives mediate efficient siRNA delivery in mammalian cells. Chembiochem 9:2787–2789
50. Trubetskoy VS, Loomis A, Slattum PM, Hagstrom JE, Budker VG, Wolff JA (1999) Caged DNA does not aggregate in high ionic strength solutions. Bioconjug Chem 10:624–628
51. Tamura A, Oishi M, Nagasaki Y (2009) Enhanced cytoplasmic delivery of siRNA using a stabilized polyion complex based on PEGylated nanogels with a cross-linked polyamine structure. Biomacromolecules 10:1818–1827
52. Oupicky D, Parker AL, Seymour LW (2002) Laterally stabilized complexes of DNA with linear reducible polycations: strategy for triggered intracellular activation of DNA delivery vectors. J Am Chem Soc 124:8–9
53. Leonetti JP, Degols G, Lebleu B (1990) Biological activity of oligonucleotide-poly(L-lysine) conjugates: mechanism of cell uptake 381. Bioconjug Chem 1:149–153
54. Bulmus V, Woodward M, Lin L, Murthy N, Stayton P, Hoffman A (2003) A new pH-responsive and glutathione-reactive, endosomal membrane-disruptive polymeric carrier for intracellular delivery of biomolecular drugs. J Control Release 93:105–120
55. Oishi M, Nagasaki Y, Itaka K, Nishiyama N, Kataoka K (2005) Lactosylated poly(ethylene glycol)-siRNA conjugate through acid-labile beta-thiopropionate linkage to construct pH-sensitive polyion complex micelles achieving enhanced gene silencing in hepatoma cells. J Am Chem Soc 127:1624–1625
56. Rozema DB, Lewis DL, Wakefield DH, Wong SC, Klein JJ, Roesch PL, Bertin SL, Reppen TW, Chu Q, Blokhin AV, Hagstrom JE, Wolff JA (2007) Dynamic PolyConjugates for targeted in vivo delivery of siRNA to hepatocytes. Proc Natl Acad Sci USA 104: 12982–12987
57. Meyer M, Dohmen C, Philipp A, Kiener D, Maiwald G, Scheu C, Ogris M, Wagner E (2009) Synthesis and biological evaluation of a bioresponsive and endosomolytic siRNA-polymer conjugate. Mol Pharmaceutics 6:752–762
58. Hagstrom JE, Hegge J, Zhang G, Noble M, Budker V, Lewis DL, Herweijer H, Wolff JA (2004) A facile nonviral method for delivering genes and siRNAs to skeletal muscle of mammalian limbs. Mol Ther 10:386–398
59. Lemkine GF, Goula D, Becker N, Paleari L, Levi G, Demeneix BA (1999) Optimisation of polyethylenimine-based gene delivery to mouse brain. J Drug Target 7:305–312
60. Lysik MA, Wu-Pong S (2003) Innovations in oligonucleotide drug delivery. J Pharm Sci 92:1559–1573
61. Bergen M, Chen R, Gonzalez R (2003) Efficacy and safety of HLA-B7/beta-2 microglobulin plasmid DNA/lipid complex (Allovectin-7) in patients with metastatic melanoma. Expert Opin Biol Ther 3:377–384
62. Plank C, Mechtler K, Szoka FC Jr, Wagner E (1996) Activation of the complement system by synthetic DNA complexes: a potential barrier for intravenous gene delivery. Hum Gene Ther 7:1437–1446
63. Ogris M, Wagner E (2002) Tumor-targeted gene transfer with DNA polyplexes. Somat Cell Mol Genet 27:85–95
64. Merdan T, Kunath K, Petersen H, Bakowsky U, Voigt KH, Kopecek J, Kissel T (2005) PEGylation of poly(ethylene imine) affects stability of complexes with plasmid DNA under in vivo conditions in a dose-dependent manner after intravenous injection into mice. Bioconjug Chem 16:785–792
65. Burke RS, Pun SH (2008) Extracellular barriers to in vivo PEI and PEGylated PEI polyplex-mediated gene delivery to the liver. Bioconjug Chem 19:693–704
66. Buyens K, Lucas B, Raemdonck K, Braeckmans K, Vercammen J, Hendrix J, Engelborghs Y, De Smedt SC, Sanders NN (2008) A fast and sensitive method for measuring the integrity of siRNA-carrier complexes in full human serum. J Control Release 126:67–76

67. Buyens K, Meyer M, Wagner E, Demeester J, De Smedt SC, Sanders NN (2010) Monitoring the disassembly of siRNA polyplexes in serum is crucial for predicting their biological efficacy. J Control Release 141:38–41
68. Kursa M, Walker GF, Roessler V, Ogris M, Roedl W, Kircheis R, Wagner E (2003) Novel shielded transferrin-polyethylene glycol-polyethylenimine/DNA complexes for systemic tumor-targeted gene transfer. Bioconjug Chem 14:222–231
69. Walker GF, Fella C, Pelisek J, Fahrmeir J, Boeckle S, Ogris M, Wagner E (2005) Toward synthetic viruses: endosomal pH-triggered deshielding of targeted polyplexes greatly enhances gene transfer in vitro and in vivo. Mol Ther 11:418–425
70. Malek A, Czubayko F, Aigner A (2008) PEG grafting of polyethylenimine (PEI) exerts different effects on DNA transfection and siRNA-induced gene targeting efficacy. J Drug Target 16:124–139
71. Zhang X, Pan SR, Hu HM, Wu GF, Feng M, Zhang W, Luo X (2008) Poly(ethylene glycol)-block-polyethylenimine copolymers as carriers for gene delivery: effects of PEG molecular weight and PEGylation degree. J Biomed Mater Res A 84:795–804
72. Brus C, Petersen H, Aigner A, Czubayko F, Kissel T (2004) Physicochemical and biological characterization of polyethylenimine-graft-poly(ethylene glycol) block copolymers as a delivery system for oligonucleotides and ribozymes. Bioconjug Chem 15:677–684
73. Kim SH, Jeong JH, Lee SH, Kim SW, Park TG (2006) PEG conjugated VEGF siRNA for anti-angiogenic gene therapy. J Control Release 116:123–129
74. Beh CW, Seow WY, Wang Y, Zhang Y, Ong ZY, Ee PL, Yang YY (2009) Efficient delivery of Bcl-2-targeted siRNA using cationic polymer nanoparticles: downregulating mRNA expression level and sensitizing cancer cells to anticancer drug. Biomacromolecules 10:41–48
75. Kircheis R, Schuller S, Brunner S, Ogris M, Heider KH, Zauner W, Wagner E (1999) Polycation-based DNA complexes for tumor-targeted gene delivery in vivo. J Gene Med 1:111–120
76. Fella C, Walker GF, Ogris M, Wagner E (2008) Amine-reactive pyridylhydrazone-based PEG reagents for pH-reversible PEI polyplex shielding. Eur J Pharm Sci 34:309–320
77. Taratula O, Garbuzenko OB, Kirkpatrick P, Pandya I, Savla R, Pozharov VP, He H, Minko T (2009) Surface-engineered targeted PPI dendrimer for efficient intracellular and intratumoral siRNA delivery. J Control Release 140:284–293
78. Finsinger D, Remy JS, Erbacher P, Koch C, Plank C (2000) Protective copolymers for nonviral gene vectors: synthesis, vector characterization and application in gene delivery. Gene Ther 7:1183–1192
79. Rudolph C, Schillinger U, Plank C, Gessner A, Nicklaus P, Muller R, Rosenecker J (2002) Nonviral gene delivery to the lung with copolymer-protected and transferrin-modified polyethylenimine 1. Biochim Biophys Acta 1573:75
80. Ogris M, Brunner S, Schuller S, Kircheis R, Wagner E (1999) PEGylated DNA/transferrin-PEI complexes: reduced interaction with blood components, extended circulation in blood and potential for systemic gene delivery. Gene Ther 6:595–605
81. Merkel OM, Librizzi D, Pfestroff A, Schurrat T, Buyens K, Sanders NN, De Smedt SC, Behe M, Kissel T (2009) Stability of siRNA polyplexes from poly(ethylenimine) and poly (ethylenimine)-g-poly(ethylene glycol) under in vivo conditions: effects on pharmacokinetics and biodistribution measured by Fluorescence Fluctuation Spectroscopy and Single Photon Emission Computed Tomography (SPECT) imaging. J Control Release 138:148–159
82. Fisher KD, Ulbrich K, Subr V, Ward CM, Mautner V, Blakey D, Seymour LW (2000) A versatile system for receptor-mediated gene delivery permits increased entry of DNA into target cells, enhanced delivery to the nucleus and elevated rates of transgene expression. Gene Ther 7:1337–1343
83. Carlisle RC, Etrych T, Briggs SS, Preece JA, Ulbrich K, Seymour LW (2004) Polymer-coated polyethylenimine/DNA complexes designed for triggered activation by intracellular reduction. J Gene Med 6:337–344

84. Hornof M, de la FM, Hallikainen M, Tammi RH, Urtti A (2008) Low molecular weight hyaluronan shielding of DNA/PEI polyplexes facilitates CD44 receptor mediated uptake in human corneal epithelial cells. J Gene Med 10:70–80
85. Ito T, Yoshihara C, Hamada K, Koyama Y (2010) DNA/polyethyleneimine/hyaluronic acid small complex particles and tumor suppression in mice. Biomaterials 31:2912–2918
86. Kurosaki T, Kitahara T, Fumoto S, Nishida K, Nakamura J, Niidome T, Kodama Y, Nakagawa H, To H, Sasaki H (2009) Ternary complexes of pDNA, polyethylenimine, and gamma-polyglutamic acid for gene delivery systems. Biomaterials 30:2846–2853
87. Kurosaki T, Kitahara T, Kawakami S, Higuchi Y, Yamaguchi A, Nakagawa H, Kodama Y, Hamamoto T, Hashida M, Sasaki H (2010) Gamma-polyglutamic acid-coated vectors for effective and safe gene therapy. J Control Release 142:404–410
88. Kircheis R, Wightman L, Schreiber A, Robitza B, Rossler V, Kursa M, Wagner E (2001) Polyethylenimine/DNA complexes shielded by transferrin target gene expression to tumors after systemic application. Gene Ther 8:28–40
89. Maeda H (2001) The enhanced permeability and retention (EPR) effect in tumor vasculature: the key role of tumor-selective macromolecular drug targeting. Adv Enzyme Regul 41:189–207
90. Wagner E, Culmsee C, Boeckle S (2005) Targeting of polyplexes: toward synthetic virus vector systems. Adv Genet 53:333–354
91. Hughes JA, Rao GA (2005) Targeted polymers for gene delivery. Expert Opin Drug Deliv 2:145–157
92. Boeckle S, Wagner E (2006) Optimizing targeted gene delivery: chemical modification of viral vectors and synthesis of artificial virus vector systems. AAPS J 8:E731–E742
93. Russ V, Wagner E (2007) Cell and tissue targeting of nucleic acids for cancer gene therapy. Pharm Res 24:1047–1057
94. Philipp A, Meyer M, Wagner E (2008) Extracellular targeting of synthetic therapeutic nucleic acid formulations. Curr Gene Ther 8:324–334
95. Wagner E, Curiel D, Cotten M (1994) Delivery of drugs, proteins and genes into cells using transferrin as a ligand for receptor-mediated endocytosis. Adv Drug Del Rev 14:113–136
96. Cotten M, Langle-Rouault F, Kirlappos H, Wagner E, Mechtler K, Zenke M, Beug H, Birnstiel ML (1990) Transferrin-polycation-mediated introduction of DNA into human leukemic cells: stimulation by agents that affect the survival of transfected DNA or modulate transferrin receptor levels. Proc Natl Acad Sci USA 87:4033–4037
97. Wagner E, Zenke M, Cotten M, Beug H, Birnstiel ML (1990) Transferrin-polycation conjugates as carriers for DNA uptake into cells. Proc Natl Acad Sci USA 87:3410–3414
98. Zenke M, Steinlein P, Wagner E, Cotten M, Beug H, Birnstiel ML (1990) Receptor-mediated endocytosis of transferrin-polycation conjugates: an efficient way to introduce DNA into hematopoietic cells. Proc Natl Acad Sci USA 87:3655–3659
99. Wagner E, Cotten M, Mechtler K, Kirlappos H, Birnstiel ML (1991) DNA-binding transferrin conjugates as functional gene-delivery agents: synthesis by linkage of polylysine or ethidium homodimer to the transferrin carbohydrate moiety. Bioconjug Chem 2:226–231
100. Kircheis R, Kichler A, Wallner G, Kursa M, Ogris M, Felzmann T, Buchberger M, Wagner E (1997) Coupling of cell-binding ligands to polyethylenimine for targeted gene delivery. Gene Ther 4:409–418
101. Bellocq NC, Pun SH, Jensen GS, Davis ME (2003) Transferrin-containing, cyclodextrin polymer-based particles for tumor-targeted gene delivery. Bioconjug Chem 14:1122–1132
102. Pons B, Mouhoubi L, Adib A, Godzina P, Behr JP, Zuber G (2006) omega-Hydrazino linear polyethylenimine: a monoconjugation building block for nucleic acid delivery. Chembiochem 7:303–309
103. Huang RQ, Qu YH, Ke WL, Zhu JH, Pei YY, Jiang C (2007) Efficient gene delivery targeted to the brain using a transferrin-conjugated polyethyleneglycol-modified polyamidoamine dendrimer. FASEB J 21:1117–1125

104. Koppu S, Oh YJ, Edrada-Ebel R, Blatchford DR, Tetley L, Tate RJ, Dufes C (2010) Tumor regression after systemic administration of a novel tumor-targeted gene delivery system carrying a therapeutic plasmid DNA. J Control Release 143:215–221
105. Curiel DT, Agarwal S, Romer N, Wagner E, Cotten M, Birnstiel ML, Boucher RC (1992) Gene transfer to respiratory epithelial cells via the receptor-mediated endocytosis pathway. Am J Resp Cell Mol Biol 6:247–252
106. Hu-Lieskovan S, Heidel JD, Bartlett DW, Davis ME, Triche TJ (2005) Sequence-specific knockdown of EWS-FLI1 by targeted, nonviral delivery of small interfering RNA inhibits tumor growth in a murine model of metastatic Ewing's sarcoma. Cancer Res 65:8984–8992
107. Heidel JD, Liu JY, Yen Y, Zhou B, Heale BS, Rossi JJ, Bartlett DW, Davis ME (2007) Potent siRNA inhibitors of ribonucleotide reductase subunit RRM2 reduce cell proliferation in vitro and in vivo. Clin Cancer Res 13:2207–2215
108. Tietze N, Pelisek J, Philipp A, Roedl W, Merdan T, Tarcha P, Ogris M, Wagner E (2008) Induction of apoptosis in murine neuroblastoma by systemic delivery of transferrin-shielded siRNA polyplexes for downregulation of Ran. Oligonucleotides 18:161–174
109. Russ V, Frohlich T, Li Y, Halama A, Ogris M, Wagner E (2010) Improved in vivo gene transfer into tumor tissue by stabilization of pseudodendritic oligoethylenimine-based polyplexes. J Gene Med 12:180–193
110. Pun SH, Tack F, Bellocq NC, Cheng J, Grubbs BH, Jensen GS, Davis ME, Brewster M, Janicot M, Janssens B, Floren W, Bakker A (2004) Targeted delivery of RNA-cleaving DNA enzyme (DNAzyme) to tumor tissue by transferrin-modified, cyclodextrin-based particles. Cancer Biol Ther 3:641–650
111. Xu B, Wiehle S, Roth JA, Cristiano RJ (1998) The contribution of poly-L-lysine, epidermal growth factor and streptavidin to EGF/PLL/DNA polyplex formation. Gene Ther 5:1235–1243
112. Blessing T, Kursa M, Holzhauser R, Kircheis R, Wagner E (2001) Different strategies for formation of pegylated EGF-conjugated PEI/DNA complexes for targeted gene delivery. Bioconjug Chem 12:529–537
113. Wolschek MF, Thallinger C, Kursa M, Rossler V, Allen M, Lichtenberger C, Kircheis R, Lucas T, Willheim M, Reinisch W, Gangl A, Wagner E, Jansen B (2002) Specific systemic nonviral gene delivery to human hepatocellular carcinoma xenografts in SCID mice. Hepatology 36:1106–1114
114. de Bruin K, Ruthardt N, von Gersdorff K, Bausinger R, Wagner E, Ogris M, Brauchle C (2007) Cellular dynamics of EGF receptor-targeted synthetic viruses. Mol Ther 15:1297–1305
115. Chen J, Gamou S, Takayanagi A, Shimizu N (1994) A novel gene delivery system using EGF receptor-mediated endocytosis 97. FEBS Lett 338:167–169
116. Liu X, Tian P, Yu Y, Yao M, Cao X, Gu J (2002) Enhanced antitumor effect of EGF R-targeted p21WAF-1 and GM-CSF gene transfer in the established murine hepatoma by peritumoral injection. Cancer Gene Ther 9:100–108
117. Li Z, Zhao R, Wu X, Sun Y, Yao M, Li J, Xu Y, Gu J (2005) Identification and characterization of a novel peptide ligand of epidermal growth factor receptor for targeted delivery of therapeutics. Faseb J 19:1978–1985
118. Chiu SJ, Ueno NT, Lee RJ (2004) Tumor-targeted gene delivery via anti-HER2 antibody (trastuzumab, Herceptin) conjugated polyethylenimine. J Control Release 97:357–369
119. Germershaus O, Merdan T, Bakowsky U, Behe M, Kissel T (2006) Trastuzumab-polyethylenimine-polyethylene glycol conjugates for targeting Her2-expressing tumors. Bioconjug Chem 17:1190–1199
120. Moffatt S, Papasakelariou C, Wiehle S, Cristiano R (2006) Successful in vivo tumor targeting of prostate-specific membrane antigen with a highly efficient J591/PEI/DNA molecular conjugate. Gene Ther 13:761–772
121. Wood KC, Azarin SM, Arap W, Pasqualini R, Langer R, Hammond PT (2008) Tumor-targeted gene delivery using molecularly engineered hybrid polymers functionalized with a tumor-homing peptide. Bioconjug Chem 19:403–405

122. Wang XL, Xu R, Lu ZR (2009) A peptide-targeted delivery system with pH-sensitive amphiphilic cell membrane disruption for efficient receptor-mediated siRNA delivery. J Control Release 134:207–213
123. Cheng H, Zhu JL, Zeng X, Jing Y, Zhang XZ, Zhuo RX (2009) Targeted gene delivery mediated by folate-polyethylenimine-block-poly(ethylene glycol) with receptor selectivity. Bioconjug Chem 20:481–487
124. Thomas M, Kularatne SA, Qi L, Kleindl P, Leamon CP, Hansen MJ, Low PS (2009) Ligand-targeted delivery of small interfering RNAs to malignant cells and tissues. Ann N Y Acad Sci 1175:32–39
125. Harbottle RP, Cooper RG, Hart SL, Ladhoff A, McKay T, Knight AM, Wagner E, Miller AD, Coutelle C (1998) An RGD-oligolysine peptide: a prototype construct for integrin-mediated gene delivery. Hum Gene Ther 9:1037–1047
126. Erbacher P, Remy JS, Behr JP (1999) Gene transfer with synthetic virus-like particles via the integrin-mediated endocytosis pathway. Gene Ther 6:138–145
127. Suh W, Han SO, Yu L, Kim SW (2002) An angiogenic, endothelial-cell-targeted polymeric gene carrier 1. Mol Ther 6:664–672
128. Kunath K, Merdan T, Hegener O, Haberlein H, Kissel T (2003) Integrin targeting using RGD-PEI conjugates for in vitro gene transfer. J Gene Med 5:588–599
129. Schiffelers RM, Ansari A, Xu J, Zhou Q, Tang Q, Storm G, Molema G, Lu PY, Scaria PV, Woodle MC (2004) Cancer siRNA therapy by tumor selective delivery with ligand-targeted sterically stabilized nanoparticle. Nucleic Acids Res 32:e149
130. Wang XL, Xu R, Wu X, Gillespie D, Jensen R, Lu ZR (2009) Targeted systemic delivery of a therapeutic siRNA with a multifunctional carrier controls tumor proliferation in mice. Mol Pharmaceutics 6:738–746
131. Moffatt S, Wiehle S, Cristiano RJ (2005) Tumor-specific gene delivery mediated by a novel peptide-polyethylenimine-DNA polyplex targeting aminopeptidase N/CD13. Hum Gene Ther 16:57–67
132. Moffatt S, Wiehle S, Cristiano RJ (2006) A multifunctional PEI-based cationic polyplex for enhanced systemic p53-mediated gene therapy. Gene Ther 13:1512–1523
133. Wu GY, Wu CH (1987) Receptor-mediated in vitro gene transformation by a soluble DNA carrier system. J Biol Chem 262:4429–4432
134. Wu GY, Wu CH (1988) Receptor-mediated gene delivery and expression in vivo 738. J Biol Chem 262:14621–14624
135. Frese J, Wu CH, Wu G (1994) Targeting of genes to the liver with glyoprotein carriers 216. Adv Drug Deliv Rev 14:137–152
136. Plank C, Zatloukal K, Cotten M, Mechtler K, Wagner E (1992) Gene transfer into hepatocytes using asialoglycoprotein receptor mediated endocytosis of DNA complexed with an artificial tetra-antennary galactose ligand. Bioconjug Chem 3:533–539
137. Merwin JR, Noell GS, Thomas WL, Chiou HC, DeRome ME, McKee TD, Spitalny GL, Findeis MA (1994) Targeted delivery of DNA using YEE(GalNAcAH)3, a synthetic glycopeptide ligand for the asialoglycoprotein receptor 440. Bioconjug Chem 5:612–620
138. Nishikawa M, Yamauchi M, Morimoto K, Ishida E, Takakura Y, Hashida M (2000) Hepatocyte-targeted in vivo gene expression by intravenous injection of plasmid DNA complexed with synthetic multi- functional gene delivery system. Gene Ther 7:548–555
139. Kim EM, Jeong HJ, Park IK, Cho CS, Moon HB, Yu DY, Bom HS, Sohn MH, Oh IJ (2005) Asialoglycoprotein receptor targeted gene delivery using galactosylated polyethylenimine-graft-poly(ethylene glycol): in vitro and in vivo studies. J Control Release 108:557–567
140. Jiang HL, Kwon JT, Kim YK, Kim EM, Arote R, Jeong HJ, Nah JW, Choi YJ, Akaike T, Cho MH, Cho CS (2007) Galactosylated chitosan-graft-polyethylenimine as a gene carrier for hepatocyte targeting. Gene Ther 14:1389–1398
141. Huang R, Ke W, Liu Y, Jiang C, Pei Y (2008) The use of lactoferrin as a ligand for targeting the polyamidoamine-based gene delivery system to the brain. Biomaterials 29:238–246

142. Huang R, Han L, Li J, Ren F, Ke W, Jiang C, Pei Y (2009) Neuroprotection in a 6-hydroxydopamine-lesioned Parkinson model using lactoferrin-modified nanoparticles. J Gene Med 11:754–763
143. Ferkol T, Perales JC, Eckman E, Kaetzel CS, Hanson RW, Davis PB (1995) Gene transfer into the airway epithelium of animals by targeting the polymeric immunoglobulin receptor 200. J Clin Invest 95:493–502
144. Ziady AG, Ferkol T, Dawson DV, Perlmutter DH, Davis PB (1999) Chain length of the polylysine in receptor-targeted gene transfer complexes affects duration of reporter gene expression both in vitro and in vivo. J Biol Chem 274:4908–4916
145. Ziady AG, Kelley TJ, Milliken E, Ferkol T, Davis PB (2002) Functional evidence of CFTR gene transfer in nasal epithelium of cystic fibrosis mice in vivo following luminal application of DNA complexes targeted to the serpin-enzyme complex receptor. Mol Ther 5:413–419
146. Elfinger M, Maucksch C, Rudolph C (2007) Characterization of lactoferrin as a targeting ligand for nonviral gene delivery to airway epithelial cells. Biomaterials 28:3448–3455
147. Elfinger M, Pfeifer C, Uezguen S, Golas MM, Sander B, Maucksch C, Stark H, Aneja MK, Rudolph C (2009) Self-assembly of ternary insulin-polyethylenimine (PEI)-DNA nanoparticles for enhanced gene delivery and expression in alveolar epithelial cells. Biomacromolecules 10:2912–2920
148. Elfinger M, Geiger J, Hasenpusch G, Uzgun S, Sieverling N, Aneja MK, Maucksch C, Rudolph C (2009) Targeting of the beta(2)-adrenoceptor increases nonviral gene delivery to pulmonary epithelial cells in vitro and lungs in vivo. J Control Release 135:234–241
149. Geiger J, Aneja MK, Hasenpusch G, Yuksekdag G, Kummerlowe G, Luy B, Romer T, Rothbauer U, Rudolph C (2010) Targeting of the prostacyclin specific IP1 receptor in lungs with molecular conjugates comprising prostaglandin I2 analogues. Biomaterials 31: 2903–2911
150. Nie Y, Schaffert D, Roedl W, Ogris M, Wagner E, Guenther M (2011) Dual-targeted polyplexes: one step towards a synthetic virus for cancer gene therapy. J Control Release 152:127–134. doi:10.1016/j.jconrel.2011.02.028
151. Lu JJ, Langer R, Chen J (2009) A novel mechanism is involved in cationic lipid-mediated functional siRNA delivery. Mol Pharmaceutics 6:763–771
152. Rejman J, Oberle V, Zuhorn IS, Hoekstra D (2004) Size-dependent internalization of particles via the pathways of clathrin- and caveolae-mediated endocytosis. Biochem J 377:159–169
153. Rejman J, Bragonzi A, Conese M (2005) Role of clathrin- and caveolae-mediated endocytosis in gene transfer mediated by lipo- and polyplexes. Mol Ther 12:468–474
154. von Gersdorff K, Sanders NN, Vandenbroucke R, De Smedt SC, Wagner E, Ogris M (2006) The internalization route resulting in successful gene expression depends on both cell line and polyethylenimine polyplex type. Mol Ther 14:745–753
155. Hufnagel H, Hakim P, Lima A, Hollfelder F (2009) Fluid phase endocytosis contributes to transfection of DNA by PEI-25. Mol Ther 17:1411–1417
156. Gabrielson NP, Pack DW (2009) Efficient polyethylenimine-mediated gene delivery proceeds via a caveolar pathway in HeLa cells. J Control Release 136:54–61
157. Wagner E (1998) Effects of membrane-active agents in gene delivery. J Control Release 53:155–158
158. Sonawane ND, Szoka FC Jr, Verkman AS (2003) Chloride accumulation and swelling in endosomes enhances DNA transfer by polyamine-DNA polyplexes. J Biol Chem 278:44826–44831
159. Zintchenko A, Philipp A, Dehshahri A, Wagner E (2008) Simple modifications of branched PEI lead to highly efficient siRNA carriers with low toxicity. Bioconjug Chem 19:1448–1455
160. Oskuee RK, Philipp A, Dehshahri A, Wagner E, Ramezani M (2010) The impact of carboxyalkylation of branched polyethylenimine on effectiveness in small interfering RNA delivery. J Gene Med 12:729–738
161. Wagner E, Zatloukal K, Cotten M, Kirlappos H, Mechtler K, Curiel DT, Birnstiel ML (1992) Coupling of adenovirus to transferrin-polylysine/DNA complexes greatly enhances receptor-

mediated gene delivery and expression of transfected genes. Proc Natl Acad Sci USA 89:6099–6103
162. Michael SI, Huang CH, Romer MU, Wagner E, Hu PC, Curiel DT (1993) Binding-incompetent adenovirus facilitates molecular conjugate-mediated gene transfer by the receptor-mediated endocytosis pathway. J Biol Chem 268:6866–6869
163. Saito G, Amidon GL, Lee KD (2003) Enhanced cytosolic delivery of plasmid DNA by a sulfhydryl-activatable listeriolysin O/protamine conjugate utilizing cellular reducing potential. Gene Ther 10:72–83
164. Wagner E, Plank C, Zatloukal K, Cotten M, Birnstiel ML (1992) Influenza virus hemagglutinin HA-2 N-terminal fusogenic peptides augment gene transfer by transferrin-polylysine-DNA complexes: toward a synthetic virus-like gene-transfer vehicle. Proc Natl Acad Sci USA 89:7934–7938
165. Haensler J, Szoka FC (1993) Polyamidoamine cascade polymers mediate efficient transfection of cells in culture. Bioconjug Chem 4:372–379
166. Plank C, Oberhauser B, Mechtler K, Koch C, Wagner E (1994) The influence of endosome-disruptive peptides on gene transfer using synthetic virus-like gene transfer systems. J Biol Chem 269:12918–12924
167. Zauner W, Blaas D, Kuechler E, Wagner E (1995) Rhinovirus-mediated endosomal release of transfection complexes. J Virol 69:1085–1092
168. Wyman TB, Nicol F, Zelphati O, Scaria PV, Plank C, Szoka FC (1997) Design, synthesis, and characterization of a cationic peptide that binds to nucleic acids and permeabilizes bilayers. Biochemistry 36:3008–3017
169. Boeckle S, Wagner E, Ogris M (2005) C- versus N-terminally linked melittin-polyethylenimine conjugates: the site of linkage strongly influences activity of DNA polyplexes. J Gene Med 7:1335–1347
170. Chen CP, Kim JS, Steenblock E, Liu D, Rice KG (2006) Gene transfer with poly-melittin peptides. Bioconjug Chem 17:1057–1062
171. Boeckle S, Fahrmeir J, Roedl W, Ogris M, Wagner E (2006) Melittin analogs with high lytic activity at endosomal pH enhance transfection with purified targeted PEI polyplexes. J Control Release 112:240–248
172. Murthy N, Campbell J, Fausto N, Hoffman AS, Stayton PS (2003) Design and synthesis of pH-responsive polymeric carriers that target uptake and enhance the intracellular delivery of oligonucleotides. J Control Release 89:365–374
173. Sakaguchi N, Kojima C, Harada A, Kono K (2008) Preparation of pH-sensitive poly (glycidol) derivatives with varying hydrophobicities: their ability to sensitize stable liposomes to pH. Bioconjug Chem 19:1040–1048
174. Sakaguchi N, Kojima C, Harada A, Koiwai K, Kono K (2008) The correlation between fusion capability and transfection activity in hybrid complexes of lipoplexes and pH-sensitive liposomes. Biomaterials 29:4029–4036
175. Yuba E, Kojima C, Sakaguchi N, Harada A, Koiwai K, Kono K (2008) Gene delivery to dendritic cells mediated by complexes of lipoplexes and pH-sensitive fusogenic polymer-modified liposomes. J Control Release 130:77–83
176. Lukacs GL, Haggie P, Seksek O, Lechardeur D, Freedman N, Verkman AS (2000) Size-dependent DNA mobility in cytoplasm and nucleus. J Biol Chem 275:1625–1629
177. Lechardeur D, Verkman AS, Lukacs GL (2005) Intracellular routing of plasmid DNA during non-viral gene transfer. Adv Drug Deliv Rev 57:755–767
178. Lechardeur D, Lukacs GL (2006) Nucleocytoplasmic transport of plasmid DNA: a perilous journey from the cytoplasm to the nucleus. Hum Gene Ther 17:882–889
179. Brunner S, Sauer T, Carotta S, Cotten M, Saltik M, Wagner E (2000) Cell cycle dependence of gene transfer by lipoplex, polyplex and recombinant adenovirus. Gene Ther 7:401–407
180. Brunner S, Furtbauer E, Sauer T, Kursa M, Wagner E (2002) Overcoming the nuclear barrier: cell cycle independent nonviral gene transfer with linear polyethylenimine or electroporation. Mol Ther 5:80–86
181. Zanta MA, Belguise VP, Behr JP (1999) Gene delivery: a single nuclear localization signal peptide is sufficient to carry DNA to the cell nucleus. Proc Natl Acad Sci USA 96:91–96

182. Brandén LJ, Mohamed AJ, Smith CI (1999) A peptide nucleic acid-nuclear localization signal fusion that mediates nuclear transport of DNA. Nat Biotechnol 17:784–787
183. Carlisle RC, Bettinger T, Ogris M, Hale S, Mautner V, Seymour LW (2001) Adenovirus hexon protein enhances nuclear delivery and increases transgene expression of polyethylenimine/plasmid DNA vectors. Mol Ther 4:473–483
184. Nagasaki T, Myohoji T, Tachibana T, Futaki S, Tamagaki S (2003) Can nuclear localization signals enhance nuclear localization of plasmid DNA? Bioconjug Chem 14:282–286
185. van der Aa MA, Koning GA, d'Oliveira C, Oosting RS, Wilschut KJ, Hennink WE, Crommelin DJ (2005) An NLS peptide covalently linked to linear DNA does not enhance transfection efficiency of cationic polymer based gene delivery systems. J Gene Med 7:208–217
186. Wagstaff KM, Jans DA (2007) Nucleocytoplasmic transport of DNA: enhancing non-viral gene transfer. Biochem J 406:185–202
187. Schaffer DV, Fidelman NA, Dan N, Lauffenburger DA (2000) Vector unpacking as a potential barrier for receptor-mediated polyplex gene delivery. Biotechnol Bioeng 67:598–606
188. Remy JS, Kichler A, Mordvinov V, Schuber F, Behr JP (1995) Targeted gene transfer into hepatoma cells with lipopolyamine-condensed DNA particles presenting galactose ligands: a stage toward artificial viruses. Proc Natl Acad Sci USA 92:1744–1748
189. Budker V, Gurevich V, Hagstrom JE, Bortzov F, Wolff JA (1996) pH-sensitive, cationic liposomes: a new synthetic virus-like vector. Nat Biotechnol 14:760–764
190. Behr JP (1997) The proton sponge: A trick to enter cells the viruses did not exploit. Chimia 51:34–36
191. Zuber G, Dauty E, Nothisen M, Belguise P, Behr JP (2001) Towards synthetic viruses. Adv Drug Deliv Rev 52:245–253
192. Wagner E (2004) Strategies to improve DNA polyplexes for in vivo gene transfer: will "artificial viruses" be the answer? Pharm Res 21:8–14
193. Martin B, Sainlos M, Aissaoui A, Oudrhiri N, Hauchecorne M, Vigneron JP, Lehn JM, Lehn P (2005) The design of cationic lipids for gene delivery. Curr Pharm Des 11:375–394
194. Wagner E (2008) Converging paths of viral and non-viral vector engineering. Mol Ther 16:1–2
195. Sethuraman VA, Na K, Bae YH (2006) pH-responsive sulfonamide/PEI system for tumor specific gene delivery: an in vitro study. Biomacromol 7:64–70
196. Knorr V, Allmendinger L, Walker GF, Paintner FF, Wagner E (2007) An acetal-based PEGylation reagent for pH-sensitive shielding of DNA polyplexes. Bioconjug Chem 18:1218–1225
197. Sethuraman VA, Lee MC, Bae YH (2008) A biodegradable pH-sensitive micelle system for targeting acidic solid tumors. Pharm Res 25:657–666
198. Hatakeyama H, Akita H, Kogure K, Oishi M, Nagasaki Y, Kihira Y, Ueno M, Kobayashi H, Kikuchi H, Harashima H (2007) Development of a novel systemic gene delivery system for cancer therapy with a tumor-specific cleavable PEG-lipid. Gene Ther 14:68–77
199. Meyer M, Wagner E (2006) pH-responsive shielding of non-viral gene vectors. Expert Opin Drug Deliv 3:563–571
200. Knorr V, Ogris M, Wagner E (2008) An acid sensitive ketal-based polyethylene glycol-oligoethylenimine copolymer mediates improved transfection efficiency at reduced toxicity. Pharm Res 25:2937–2945
201. Nie Y, Gunther M, Gu Z, Wagner E (2011) Pyridylhydrazone-based PEGylation for pH-reversible lipopolyplex shielding. Biomaterials 32:858–869
202. Rozema DB, Ekena K, Lewis DL, Loomis AG, Wolff JA (2003) Endosomolysis by masking of a membrane-active agent (EMMA) for cytoplasmic release of macromolecules. Bioconjug Chem 14:51–57
203. Meyer M, Zintchenko A, Ogris M, Wagner E (2007) A dimethylmaleic acid-melittin-polylysine conjugate with reduced toxicity, pH-triggered endosomolytic activity and enhanced gene transfer potential. J Gene Med 9:797–805

204. Meyer M, Philipp A, Oskuee R, Schmidt C, Wagner E (2008) Breathing life into polycations: functionalization with pH-responsive endosomolytic peptides and polyethylene glycol enables siRNA delivery. J Am Chem Soc 130:3272–3273
205. Philipp A, Meyer M, Zintchenko A, Wagner E (2011) Functional modification of amide-crosslinked oligoethylenimine for improved siRNA delivery. React Funct Polym 71:288–293
206. Huth S, Hoffmann F, von Gersdorff K, Laner A, Reinhardt D, Rosenecker J, Rudolph C (2006) Interaction of polyamine gene vectors with RNA leads to the dissociation of plasmid DNA-carrier complexes. J Gene Med 8:1416–1424
207. Read ML, Bremner KH, Oupicky D, Green NK, Searle PF, Seymour LW (2003) Vectors based on reducible polycations facilitate intracellular release of nucleic acids. J Gene Med 5:232–245
208. Itaka K, Harada A, Yamasaki Y, Nakamura K, Kawaguchi H, Kataoka K (2004) In situ single cell observation by fluorescence resonance energy transfer reveals fast intra-cytoplasmic delivery and easy release of plasmid DNA complexed with linear polyethylenimine. J Gene Med 6:76–84
209. Matsumoto Y, Itaka K, Yamasoba T, Kataoka K (2009) Intranuclear fluorescence resonance energy transfer analysis of plasmid DNA decondensation from nonviral gene carriers. J Gene Med 11:615–623
210. Knorr V, Russ V, Allmendinger L, Ogris M, Wagner E (2008) Acetal linked oligoethylenimines for use as pH-sensitive gene carriers. Bioconjug Chem 19:1625–1634
211. Hoon JJ, Christensen LV, Yockman JW, Zhong Z, Engbersen JF, Jong KW, Feijen J, Wan KS (2007) Reducible poly(amido ethylenimine) directed to enhance RNA interference. Biomaterials 28:1912–1917
212. Lin C, Blaauboer CJ, Timoneda MM, Lok MC, van Steenbergen M, Hennink WE, Zhong Z, Feijen J, Engbersen JF (2008) Bioreducible poly(amido amine)s with oligoamine side chains: synthesis, characterization, and structural effects on gene delivery. J Control Release 126:166–174
213. Forrest ML, Koerber JT, Pack DW (2003) A degradable polyethylenimine derivative with low toxicity for highly efficient gene delivery. Bioconjug Chem 14:934–940
214. Kloeckner J, Bruzzano S, Ogris M, Wagner E (2006) Gene carriers based on hexanediol diacrylate linked oligoethylenimine: effect of chemical structure of polymer on biological properties. Bioconjug Chem 17:1339–1345
215. Zugates GT, Peng W, Zumbuehl A, Jhunjhunwala S, Huang YH, Langer R, Sawicki JA, Anderson DG (2007) Rapid optimization of gene delivery by parallel end-modification of poly(beta-amino ester)s. Mol Ther 15:1306–1312
216. Neu M, Germershaus O, Behe M, Kissel T (2007) Bioreversibly crosslinked polyplexes of PEI and high molecular weight PEG show extended circulation times in vivo. J Control Release 124:69–80
217. Neu M, Germershaus O, Mao S, Voigt KH, Behe M, Kissel T (2007) Crosslinked nanocarriers based upon poly(ethylene imine) for systemic plasmid delivery: in vitro characterization and in vivo studies in mice. J Control Release 118:370–380
218. Russ V, Elfberg H, Thoma C, Kloeckner J, Ogris M, Wagner E (2008) Novel degradable oligoethylenimine acrylate ester-based pseudodendrimers for in vitro and in vivo gene transfer. Gene Ther 15:18–29
219. Russ V, Gunther M, Halama A, Ogris M, Wagner E (2008) Oligoethylenimine-grafted polypropylenimine dendrimers as degradable and biocompatible synthetic vectors for gene delivery. J Control Release 132:131–140
220. Klutz K, Russ V, Willhauck MJ, Wunderlich N, Zach C, Gildehaus FJ, Goke B, Wagner E, Ogris M, Spitzweg C (2009) Targeted radioiodine therapy of neuroblastoma tumors following systemic nonviral delivery of the sodium iodide symporter gene. Clin Cancer Res 15:6079–6086
221. Schreiber S, Kampgen E, Wagner E, Pirkhammer D, Trcka J, Korschan H, Lindemann A, Dorffner R, Kittler H, Kasteliz F, Kupcu Z, Sinski A, Zatloukal K, Buschle M, Schmidt W, Birnstiel M, Kempe RE, Voigt T, Weber HA, Pehamberger H, Mertelsmann R, Brocker EB, Wolff K, Stingl G (1999) Immunotherapy of metastatic malignant melanoma by a vaccine

consisting of autologous interleukin 2-transfected cancer cells: outcome of a phase I study. Hum Gene Ther 10:983–993
222. Davis PB, Cooper MJ (2007) Vectors for airway gene delivery. AAPS J 9:E11–E17
223. Fewell JG, Matar MM, Rice JS, Brunhoeber E, Slobodkin G, Pence C, Worker M, Lewis DH, Anwer K (2009) Treatment of disseminated ovarian cancer using nonviral interleukin-12 gene therapy delivered intraperitoneally. J Gene Med 11:718–728
224. Ohana P, Gofrit O, Ayesh S, Al-Sharef W, Mizrahi A, Birman T, Schneider T, Matouk I, de Groot N, Tavdy E, Sidi AA, Hochberg A (2004) Regulatory sequences of the H19 gene in DNA based therapy of bladder cancer. Gene Ther Mol Biol 8:181–192
225. Schaffert D, Kiss M, Rodl W, Shir A, Levitzki A, Ogris M, Wagner E (2011) Poly(I:C)-mediated tumor growth suppression in EGF-receptor overexpressing tumors using EGF-polyethylene glycol-linear polyethylenimine as carrier. Pharm Res 28:731–741
226. Shir A, Ogris M, Roedl W, Wagner E, Levitzki A (2011) EGFR-homing dsRNA activates cancer targeted immune response and eliminates disseminated EGFR over-expressing tumors in mice. Clin Cancer Res 17:1033–1043
227. Klutz K, Schaffert D, Willhauck MJ, Grunwald GK, Haase R, Wunderlich N, Zach C, Gildehaus FJ, Senekowitsch-Schmidtke R, Goke B, Wagner E, Ogris M, Spitzweg C (2011) Epidermal growth factor receptor-targeted (131)I-therapy of liver cancer following systemic delivery of the sodium iodide symporter gene. Mol Ther 19:676–685
228. Kircheis R, Ostermann E, Wolschek MF, Lichtenberger C, Magin-Lachmann C, Wightman L, Kursa M, Wagner E (2002) Tumor-targeted gene delivery of tumor necrosis factor-alpha induces tumor necrosis and tumor regression without systemic toxicity. Cancer Gene Ther 9:673–680
229. Wagner E, Kircheis R, Walker GF (2004) Targeted nucleic acid delivery into tumors: new avenues for cancer therapy. Biomed Pharmacother 58:152–161
230. Dufes C, Keith WN, Bilsland A, Proutski I, Uchegbu IF, Schatzlein AG (2005) Synthetic anticancer gene medicine exploits intrinsic antitumor activity of cationic vector to cure established tumors. Cancer Res 65:8079–8084
231. Bartlett DW, Davis ME (2008) Impact of tumor-specific targeting and dosing schedule on tumor growth inhibition after intravenous administration of siRNA-containing nanoparticles. Biotechnol Bioeng 99:975–985
232. Bartlett DW, Su H, Hildebrandt IJ, Weber WA, Davis ME (2007) Impact of tumor-specific targeting on the biodistribution and efficacy of siRNA nanoparticles measured by multimodality in vivo imaging. Proc Natl Acad Sci USA 104:15549–15554
233. Heidel JD, Yu Z, Liu JY, Rele SM, Liang Y, Zeidan RK, Kornbrust DJ, Davis ME (2007) Administration in non-human primates of escalating intravenous doses of targeted nanoparticles containing ribonucleotide reductase subunit M2 siRNA. Proc Natl Acad Sci USA 104:5715–5721
234. Davis ME (2009) The first targeted delivery of siRNA in humans via a self-assembling, cyclodextrin polymer-based nanoparticle: from concept to clinic. Mol Pharmaceutics 6:659–668
235. Davis ME, Zuckerman JE, Choi CH, Seligson D, Tolcher A, Alabi CA, Yen Y, Heidel JD, Ribas A (2010) Evidence of RNAi in humans from systemically administered siRNA via targeted nanoparticles. Nature 464:1067–1070
236. Hartmann L, Hafele S, Peschka-Suss R, Antonietti M, Borner HG (2008) Tailor-made poly (amidoamine)s for controlled complexation and condensation of DNA. Chemistry 14:2025–2033
237. Schaffert D, Badgujar N, Wagner E (2011) Novel Fmoc-polyamino acids for solid-phase synthesis of defined polyamidoamines. Org Lett 13:1586–1589

Adv Polym Sci (2012) 247: 31–64
DOI: 10.1007/12_2011_150

Published online: 20 October 2011

Biodegradable Nanoparticles as Vaccine Adjuvants and Delivery Systems: Regulation of Immune Responses by Nanoparticle-Based Vaccine

Takami Akagi, Masanori Baba, and Mitsuru Akashi

Abstract Polymeric nano- and microparticles have recently been shown to possess significant potential as drug delivery systems. In particular, the use of biodegradable polymeric nanoparticles with entrapped antigens such as proteins, peptides, or DNA represents an exciting approach for controlling the release of vaccine antigens and optimizing the desired immune response via selective targeting of the antigen to antigen-presenting cells (APCs). The efficient delivery of antigens to APCs, especially in dendritic cells (DCs), and the activation of APCs are some of the most important issues in the development of effective vaccines. Using nanoparticle-based vaccine delivery systems, it is possible to target delivery to DCs, activate these APCs, and control release of the antigen. Nanoparticles prepared from biodegradable and biocompatible polymers such as poly(lactide-*co*-glycolide) (PLGA), poly(amino acid)s, and polysaccharides have been shown to be effective

T. Akagi and M. Akashi (✉)
Department of Applied Chemistry, Graduate School of Engineering, Osaka University, 2-1 Yamadaoka, Suita, Osaka 565-0871, Japan

Japan Science and Technology Agency (JST), Core Research for Evolutional Science and Technology (CREST), Kawaguchi Center Building, 4-1-8 Honcho, Kawaguchi 332-0012, Japan
e-mail: akashi@chem.eng.osaka-u.ac.jp

M. Baba
Division of Antiviral Chemotherapy, Center for Chronic Viral Diseases, Graduate School of Medical and Dental Sciences, Kagoshima University, 8-35-1 Sakuragaoka, Kagoshima 890-8544, Japan

Japan Science and Technology Agency (JST), Core Research for Evolutional Science and Technology (CREST), Kawaguchi Center Building, 4-1-8 Honcho, Kawaguchi 332-0012, Japan vaccine carriers for a number of antigens. This review mainly focuses on amphiphilic poly(amino acid) and PLGA nanoparticles as vaccine delivery systems and summarizes the investigations of our research group and others on the properties of these antigen-loaded naoparticles.

Keyword Adjuvant · Biodegradable nanoparticles · Poly(γ-glutamic acid) · Protein delivery · Vaccine

Contents

Abbreviations

APCs	Antigen-presenting cells
BSA	Bovine serum albumin
CFA	Complete Freund's adjuvant
CLSM	Confocal laser scanning microscopy
CT	Chitosan
CTL	Cytotoxic T lymphocyte
DCC	*N,N*-Dicyclohexyl carbodiimide
DCs	Dendritic cells
DDS	Drug delivery system
DLS	Dynamic light scattering
FCM	Flow cytometry

HBcAg	Hepatitis B core antigen
HIV	Human immunodeficiency virus
HOBt	1-Hydroxybenzotriazole
HTLV-I	Human T-cell leukemia virus type-I
LPS	Lipopolysaccharide
MHC	Major histocompatibility complex
MPLA	Monophospholipid A
o/w	Oil-in-water
OVA	Ovalbumin
OVA-NPs	OVA encapsulating within γ-PGA-Phe nanoparticles
PCL	Poly(ε-caprolactone)
pDNA	Plasmid DNA
PEI	Polyethylenimine
PGA	Poly(glycolic acid)
PHB	Poly(hydroxybutyrate)
Phe	L-Phenylalanine
PIC	Polyion complex
PLA	Poly(lactic acid)
PLGA	Poly(lactide-*co*-glycolide)
SAXS	Small angle X-ray scattering
SEM	Scanning electron microscopy
TEM	Transmission electron microscopy
Th	T helper
TLR	Toll-like receptor
Trp	L-Tryptophan
w/o/w	Water-in-oil-in-water
γ-PGA	Poly(γ-glutamic acid)
γ-PGA-Phe	γ-PGA-*graft*-Phe copolymer
ε-PL	Poly(ε-lysine)
ε-PL-CHS	ε-PL-*graft*-cholesterol hydrogen succinate

1 Introduction

Vaccination to induce an adaptive immune response is expected for a broad range of infectious diseases and cancers. Traditional vaccines are mainly composed of live attenuated viruses, whole inactivated pathogens, or inactivated bacterial toxins. In general, these approaches have been successful for developing vaccines that can induce an immune response based on antigen-specific antibody and cytotoxic T lymphocyte (CTL) responses, which kill host cells infected with intracellular organisms (Fig. 1) [1, 2]. One of the most important current issues in vaccinology is the need for new adjuvants (immunostimulants) and delivery systems. Many of the vaccines currently in development are based on purified subunits, recombinant

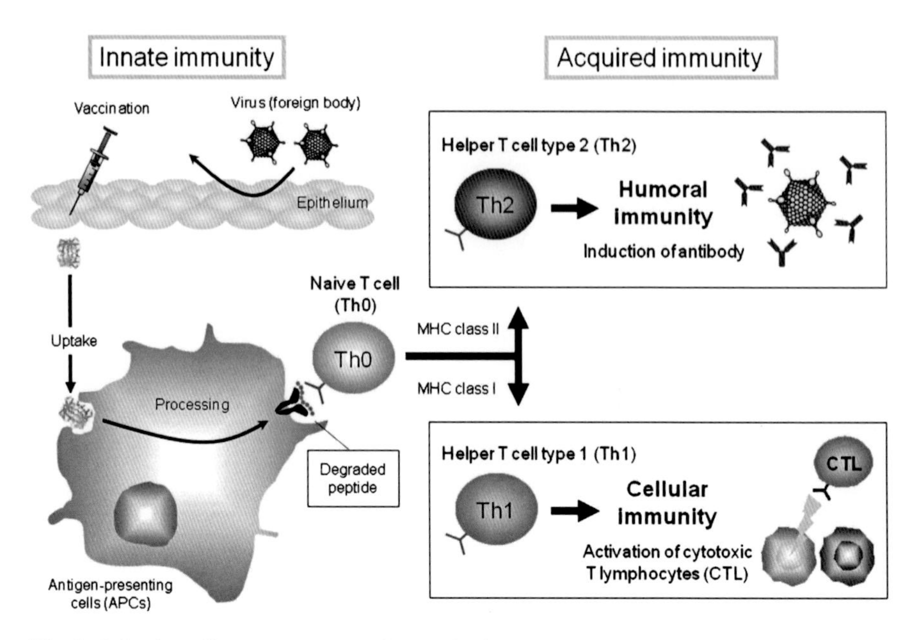

Fig. 1 Induction of immune responses by vaccination

proteins, or synthetic peptides. These new generation of vaccines are generally very safe, with well-defined components. However, these antigens are often poorly immunogenic, and thus require the use of adjuvants and delivery systems to induce optimal immune responses [3–5]. Immunological adjuvants were originally described by Ramon as "substances used in combination with a specific antigen that produced a more robust immune response than the antigen alone" [6]. Until recently, the hydroxide and phosphate salts of aluminum and calcium were the only adjuvants licensed for human use. However, the use of alum-type adjuvants for vaccination has some disadvantages [7, 8]. They are not effective for all antigens, induce local reactions, induce IgE antibody responses, and generally fail to induce cell-mediated immunity, particular CTL responses. Therefore, the development of more efficient and safe adjuvants and vaccine delivery systems to obtain high and long-lasting immune responses is of primary importance.

Polymeric nanoparticles formulated from biodegradable polymers are being widely explored as carriers for controlled delivery of different agents including proteins, peptides, plasmid DNA (pDNA), and low molecular weight compounds [9–11]. Self-assembling polymer or block/graft copolymers that can form nanostructures have been extensively investigated in the field of biotechnology and pharmaceuticals. In general, hydrophobic interactions, electrostatic forces, hydrogen bonds, van der Waal forces, or combinations of these interactions are available as the driving forces for the formation of the polymer complexes [12–16]. Numerous investigators have shown that the biological distribution of drugs, proteins, and DNA can be modified, both at the cellular and organ levels, using nano- or microparticle delivery systems [17–19]. For the development of effective vaccines, biodegradable nanoparticles show great promise as vaccine delivery systems. Controlled delivery systems consisting of nanoparticles can potentially delivery either the antigens or adjuvants to the desired location at predetermined rates and durations to generate an optimal immune response. The carrier may also protect the vaccine from degradation until it is released. Other potential advantages of the controlled delivery approach include reduced systemic side effects and the possibility of co-encapsulating multiple antigenic epitopes or both antigen and adjuvant in a single carrier. Biodegradable polymers provide sustained release of the encapsulated antigen and degrade in the body to nontoxic, low molecular weight products that are easily eliminated.

On the other hand, recent strategies for developing preventative and therapeutic vaccines have focused on the ability to deliver antigen to dendritic cells (DCs) in a targeted and prolonged manner. These strategies use nanoparticles because they can achieve longevity on intact antigen to increase the opportunity for DC uptake and processing. DCs are the most effective antigen-presenting cells (APCs), and have a crucial role in initiating T-cell-mediated immunity. DCs can control a substantial part of the adaptive immune response by internalizing and processing antigens through major histocompatibility complex (MHC) class I and class II pathways, and then presenting antigenic peptides to $CD4^+$ and $CD8^+$ T lymphocytes (Fig. 1) [20]. Therefore, targeting DCs with an antigen delivery system provides tremendous potential in developing new vaccines [21]. Antigen uptake by DCs is enhanced by the association of the antigens with polymeric nanoparticles. The adjuvant effect of particulate materials appears to largely be a consequence of their uptake into DCs. More importantly, particulate antigens have been shown to be more efficient than soluble antigens for the induction of immune responses [22, 23]. Furthermore, the submicron size of nanoparticles offers a number of distinct advantages over microparticles, and nanoparticles generally have relatively higher intracellular uptake as compared to microparticles [24, 25]. There are several factors that can affect the immune response induced by immunization with particulate antigens. Among them are particle size, the chemical structure of particles, surface hydrophobicity, zeta potential, and adjuvants used within the formulations.

This review focuses on biodegradable polymeric nanoparticles as vaccine delivery systems and immunostimulants, and summarizes the preparation of antigen-conjugated particles and the mechanism of nanoparticle-based vaccines. Using these systems, it is possible to target antigen delivery to APCs, activate these APCs, and control intracellular release and distribution of the antigen. By understanding immune activation, we can rationally design particulate adjuvant to not only deliver antigen but also to directly activate innate immune cells providing the pro-inflammatory context for antigen recognition. The generation of more potent particulate adjuvants may allow the development of prophylactic and therapeutic vaccines against cancers and chronic infectious diseases.

2 Preparation of Biodegradable Polymeric Nanoparticles

2.1 PLGA Nanoparticles

Biodegradable polymeric nanoparticles have attracted much attention for their potential in biomedical applications, such as drug, gene, and vaccine delivery systems. The biodegradation rate and the release kinetics of loaded drugs can be controlled by the composition ratio and the molecular weight of the polymer and block/graft copolymers [26–28]. Furthermore, by modulating the polymer characteristics, one can control the release of a therapeutic agent from the nanoparticles to achieve a desired therapeutic level in a target tissue for the required duration for optimal therapeutic efficacy. The commonly used biodegradable polymers are aliphatic polyesters such as poly(lactic acid) (PLA), poly(glycolic acid) (PGA), poly(ε-caprolactone) (PCL), poly(hydroxybutyrate) (PHB) and their copolymers (Fig. 2) [29]. In particular, poly(lactide-*co*-glycolide) (PLGA) has been the most extensively investigated for developing nano- and microparticles encapsulating therapeutic drugs in controlled release applications [30–32] due to their inherent advantages. The copolymers have the advantage of sustaining the release of the encapsulated therapeutic agent over a period of days to several weeks. As polyesters in nature, these polymers undergo hydrolysis upon administration into the body, forming biologically compatible and metabolizable moieties (lactic acid and glycolic acid) that are eventually removed from the body by the citric acid cycle.

Several methods have been reported for the preparation of biodegradable nanoparticles from PLGA, PLA, and PCL by dispersing preformed polymers. Emulsion solvent evaporation techniques are frequently used to prepare nano- and microparticles [33, 34]. The polymer is dissolved in an organic solvent like dichloromethane, chloroform, or ethyl acetate and then emulsified into an aqueous solution to create an oil-in-water (o/w) emulsion by using a surfactant such as poly (vinyl alcohol). After the formation of a stable emulsion, the organic solvent is evaporated by increasing the temperature under pressure (Fig. 3). The effect of this process is variable, depending on the properties of the nanoparticles. Often, surfactants are used to stabilize the nanoparticles in aqueous solution in order to prevent the aggregation and/or precipitation of water-insoluble polymers. However, adequate removal of the surfactant remains a problem, and surfactant molecules are sometimes harmful in biomedical applications.

$-(O-CH(CH_3)-C(=O))_n-$
Poly(lactic acid) (PLA)

$-(O-CH(CH_3)-C(=O))_m-(O-CH_2-C(=O))_n-$
Poly(lactic acid-*co*-glycolide) (PLGA)

$-[O-(CH_2)_5-C(=O)]_n-$
Poly(ε-caprolactone) (PCL)

$-(O-CH(CH_3)-CH_2-C(=O))_n-$
Poly(hydroxybutyrate) (PHB)

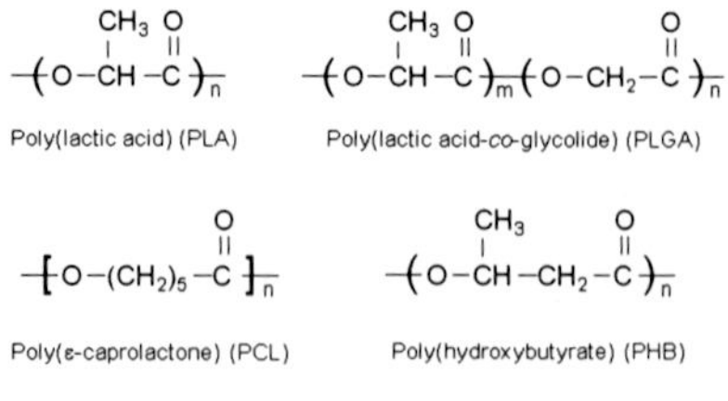

Fig. 2 Chemical structures of biodegradable polyesters used for preparation of nanoparticles

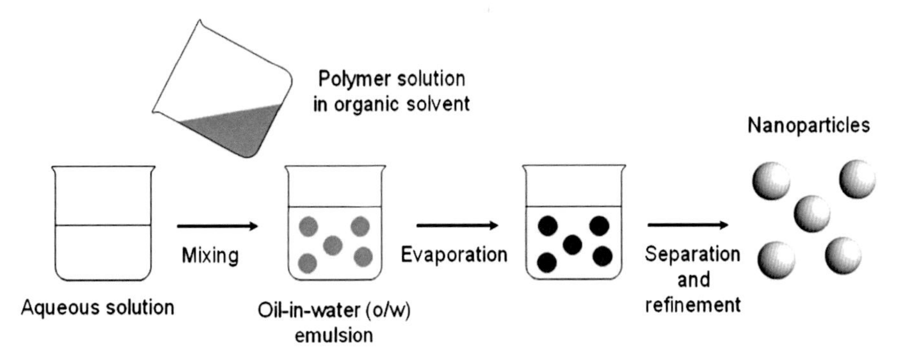

Fig. 3 Preparation of polymeric nanoparticles by emulsion solvent evaporation technique

$-(NH-CH(R)-C(=O))_n-$ Poly(α-amino acid)

$-(NH-CH(COOH)-CH_2-CH_2-C(=O))_n-$ Poly(γ-glutamic acid) (γ-PGA)

$-[NH-(CH_2)_4-CH(NH_2)-C(=O)]_n-$ Poly(ε-lysine) (ε-PL)

Fig. 4 Chemical structures of synthetic and naturally occurring poly(amino acid)s

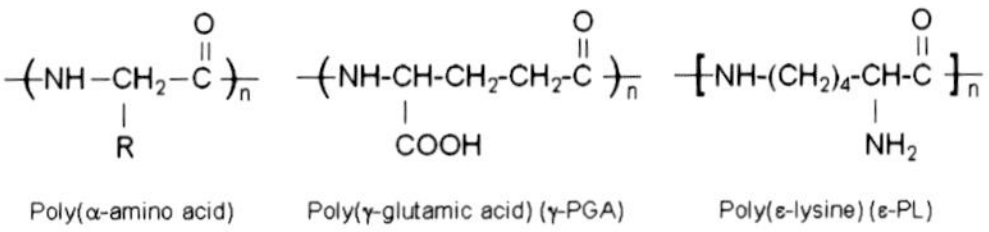

2.2 *Amphiphilic Poly(amino acid) Nanoparticles*

Recently, many studies have focused on self-assembled biodegradable nanoparticles for biomedical and pharmaceutical applications. Nanoparticles fabricated by the self-assembly of amphiphilic block copolymers or hydrophobically modified polymers have been explored as drug carrier systems. In general, these amphiphilic copolymers consisting of hydrophilic and hydrophobic segments are capable of forming polymeric structures in aqueous solutions via hydrophobic interactions. These self-assembled nanoparticles are composed of an inner core of hydrophobic moieties and an outer shell of hydrophilic groups [35, 36].

In particular, poly(amino acid)s have received considerable attention for their medical applications as potential polymeric drug carriers. Several amphiphilic block and graft copolymers based on poly(amino acid)s have been employed, such as poly(α-L-glutamic acid) [37], poly(γ-glutamic acid) [38], poly(ε-lysine) [39] (Fig. 4), poly(L-aspartic acid) [40], poly(L-lysine) [41], poly(L-arginine) [42], and poly(L-asparagine) [43] as hydrophilic segments, and poly(β-benzyl-L-aspartate) [44], poly(γ-benzyl-L-glutamate) [45], and poy(L-histidine) [46] as hydrophobic segments. In general, amphiphilic copolymers based on poly(amino acid)s form micelles through self-association in water.

Poly(γ-glutamic acid) (γ-PGA) is a naturally occurring poly(amino acid) that is synthesized by certain strains of *Bacillus* [47]. The polymer is made of D- and L-glutamic acid units linked through the α-amino and the γ-carboxylic acid groups, and its α-carboxylate side chains can be chemically modified to introduce various bioactive ligands, or to modulate the overall function of the polymer [48–52]. Unlike general poly(amino acid)s, γ-PGA has unique characteristics of enzymatic degradation and immunogenicity. It has been reported that γ-PGA has resistance

Fig. 5 Synthesis of γ-PGA hydrophobic derivatives

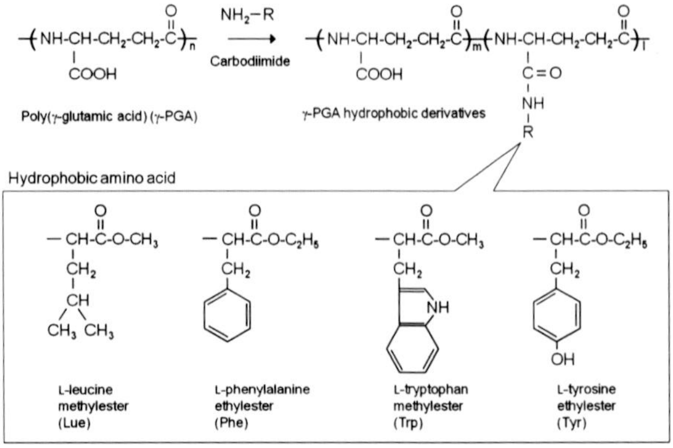

against many proteases because γ-linked glutamic acids are not easily recognized by common proteases [53, 54]. Moreover, several studies have shown that γ-PGA by itself is a poor immunogen and does not induce booster responses, probably because of its simple homopolymeric structure, similar to those of polysaccharides [55–59]. Therefore, the potential applications of γ-PGA and its derivatives have been of interest in a broad range of fields, including medicine, food, cosmetics, and water treatment [60].

Akashi et al. prepared nanoparticles composed of hydrophobically modified γ-PGA [38, 61, 62] (Fig. 5). γ-PGA (400 kDa) as the hydrophilic backbone and L-phenylalanine (Phe) as the hydrophobic segment were synthesized by grafting Phe to γ-PGA using water-soluble carbodiimide. The γ-PGA-*graft*-Phe copolymers (γ-PGA-Phe) with more than 50% grafting degree formed monodispersed nanoparticles in water due to their amphiphilic properties. To prepare nanoparticles, γ-PGA-Phe dissolved in dimethyl sulfoxide (DMSO) was added to various concentration of NaCl solution, and then the resulting solutions were dialyzed and freeze-dried. The γ-PGA-Phe formed monodispersed nanoparticles, and the particle size of the γ-PGA-Phe nanoparticles could be easily controlled (30–200 nm) by changing NaCl concentration [63]. Similarly, γ-PGA conjugated with L-tryptophan (γ-PGA-Trp) showed the same tendency (Fig. 6). The size of the nanoparticles increased with increasing NaCl concentration during formation of particles. The addition of NaCl leads to enhanced screening of the Coulomb interactions between the carboxyl groups of γ-PGA-Phe. Therefore, according to the increase in NaCl concentration, a larger number of graft copolymers was involved in the formation nanoparticles. The nanoparticles showed a highly negative zeta potential (−25 mV) due to the ionization of the carboxyl groups of γ-PGA located near the surfaces. The specific self-assembly behavior of γ-PGA-Phe in aqueous solution was due to multiple stacking of phenyl groups. Beside the particle formation of γ-PGA by using hydrophobic interaction, nanoparticles formed by complexation of γ-PGA with a bivalent metal ion complex [64] or by chemical crosslinking of carboxyl groups of γ-PGA [65] have been reported.

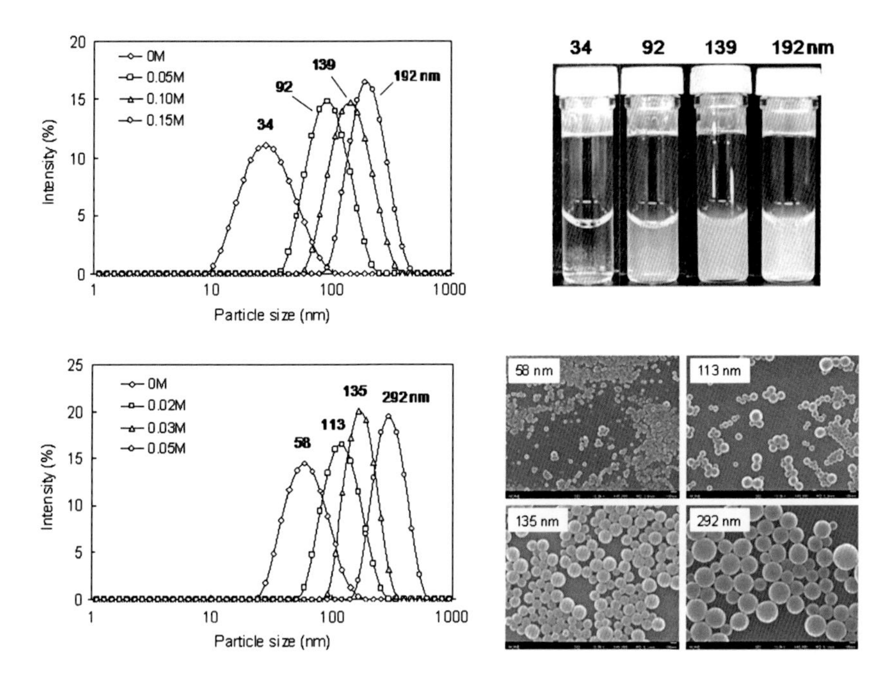

Fig. 6 Size changes of (**a**) γ-PGA-Phe and (**c**) γ-PGA-Trp nanoparticles prepared at various NaCl concentrations. The size of nanoparticles was measured by DLS. (**b**) Photographs of γ-PGA-Phe nanoparticles (2.5 mg/mL) dispersed in water. (**d**) Scanning electron microscope (SEM) images of γ-PGA-Trp nanoparticles prepared at various NaCl concentrations

Poly(ε-lysine) (ε-PL) is produced by a *Streptomyces albulus* strain, and has been used as a food additive due to its antimicrobial activities [66, 67]. ε-PL is water soluble and biodegradable and has a molecular weight of approximately 5,000. ε-PL is an L-lysine homopolymer (25–30 residues) with a linkage between the carboxyl group and the ε-amino group (Fig. 4). Matsusaki et al. reported the nanoparticle formation of amphiphilic ε-PL-*graft*-cholesterol hydrogen succinate (ε-PL-CHS) in water. ε-PL was hydrophobically modified by CHS in the presence of *N*,*N*-dicyclohexyl carbodiimide (DCC) and 1-hydroxybenzotriazole (HOBt) in *N*,*N*-dimethylformamide (DMF) (Fig. 7) [39]. ε-PL-CHS nanoparticles were prepared by the solvent (tetrahydrofuran) exchange method. ε-PL-CHS could form stable nanoparticles in water following the hydrophobic interactions of its CHS groups. The size of the ε-PL-CHS nanoparticles was approximately 150–200 nm. For the purposes of nonviral gene delivery, cationic polymers such as poly(L-lysine) and polyethylenimine (PEI) have been used as carriers for complexing gene vectors into polyplexes [68–70]. A polyplex can be easily formed when the oppositely charged DNA and polycation are mixed in aqueous solution and interact via electrostatic interactions. These polyplexes result in an increased net positive charge of the complexes, and promote cellular uptake and transfection efficiency. However, the in vivo applications of polyplexes are limited by low gene expression and toxicity due to their cationic nature [71–73]. ε-PL is

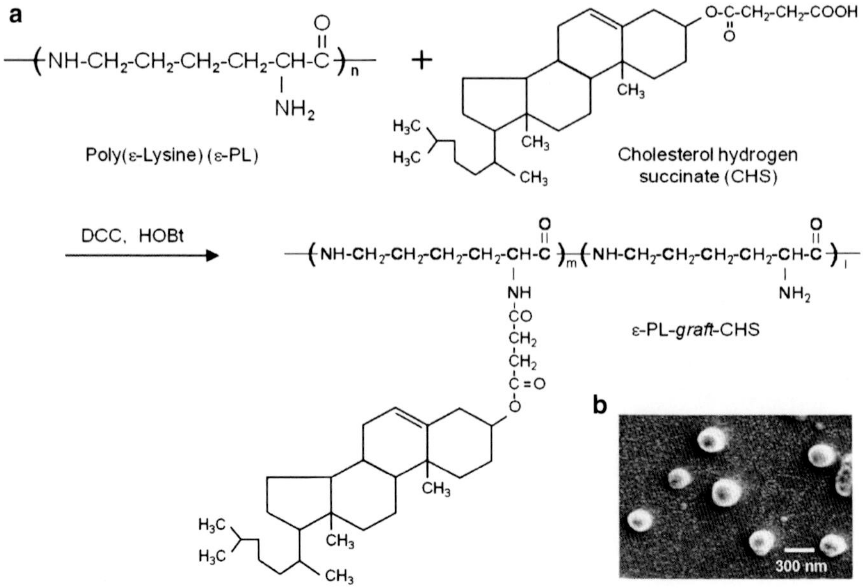

Fig. 7 (**a**) Synthesis of amphiphilic ε-PL-*graft*-cholesterol hydrogen succinate (ε-PL-CHS). (**b**) SEM image of nanoparticles prepared from ε-PL-CHS

a very safe material for use in humans. Therefore, the nanoparticles fabricated from ε-PL may be useful for DNA vaccine delivery and adjuvants.

2.3 Amphiphilic Polysaccharide Nanoparticles

Polysaccharidic hydrogel particles have been often used for designing protein-loaded systems for therapeutic applications. Polysaccharides are very hydrophilic polymers, and their hydrogels thus exhibit a good biocompatibility. Various type of hydrophobized polysaccharides, such as pullulan [74, 75], curdlan [76], dextran [77], alginic acid [78], and chitosan [79], have been used for preparation of nanoparticles. Akiyoshi et al. reported that self-aggregated hydrogel nanoparticles could be formed from cholesterol-bearing pullulan by an intra- and/or intermolecular association in diluted aqueous solutions [80]. Recently, much attention has been paid to chitosan as a drug or gene carrier because of its biocompatibility and biodegradability. Chitosan is a polysaccharide constituted of *N*-glucosamine and *N*-acetyl-glucosamine units, in which the number of *N*-glucosamine units exceeds 50%. Chitosan can be degraded into nontoxic products in vivo, and thus has been widely used in various biomedical applications [81, 82]. Chitosan has cationic characters even in neutral conditions to form complexes with negatively charged pDNA. Jeong et al. prepared nanosized self-aggregates composed of

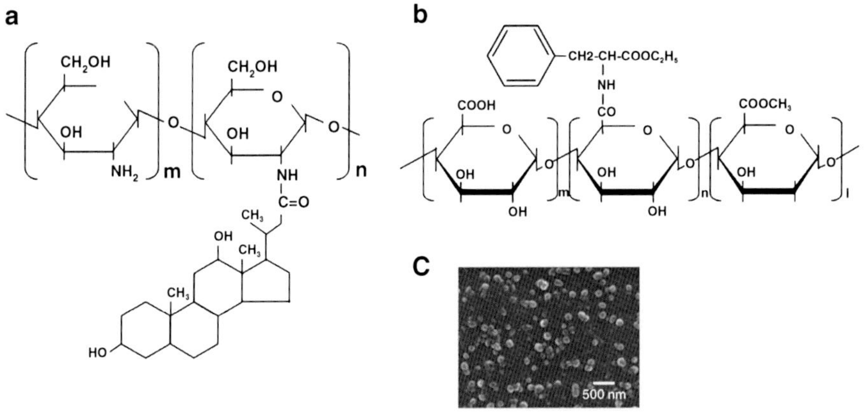

Fig. 8 Preparation of amphiphilic polysaccharide. Chemical structures of deoxycholic acid-modified chitosan (**a**) and Phe-modified pectin (pectin-*graft*-Phe) (**b**). SEM image of nanoparticles prepared from pectin-*graft*-Phe (**c**)

hydrophobically modified chitosans with deoxycholic acids (Fig. 8a) [83, 84]. The size of self-aggregates varied in the range of 130–300 nm in diameter, and their structures were found to depend strongly on the molecular weight of chitosan. To explore the potential applications of self-aggregates as a gene delivery carrier, complexes between chitosan self-aggregates and pDNA were prepared. The complex formation had a strong dependency on the size and structure of chitosan self-aggregates and significantly influenced the transfection efficiency of cells. It is expected that these approaches to control the size and structure of chitosan-derived self-aggregates will have a wide range of applications in gene delivery. Also, Kida et al. reported that novel polysaccharide-based nanoparticles were successfully prepared by the self-assembly of amphiphilic pectins, which were easily synthesized by the reaction of pectins with Phe as hydrophobic group (Fig. 8b) [85]. Pectin is a polymer of D-galacturonic acid. The galacturonic acid molecule has a carboxyl group on C5, some of which are esterified to form methyl esters. The pectin-*graft*-Phe could form about 200 nm-sized nanoparticles (Fig. 8c), and were able to retain entrapped protein in the nanoparticles for one week without any significant leakage.

2.4 Polyion Complex Nanoparticles

Polymer complexes associated with two or more complementary polymers are widely used in potential applications in the form of particles, hydrogels, films, and membranes. In particular, a polyion complex (PIC) can be easily formed when oppositely charged polyelectrolytes are mixed in aqueous solution and interact via electrostatic (coulombic) interactions. Nanoscaled structural materials (e.g., nanoparticles, micelles, nanogels, and hollow nanospheres) composed of PIC are prepared by tuning the preparation conditions, such as the charge ratio of the anionic-to-cationic polymers, temperature, concentration, and type of polyelectrolyte [12, 86, 87].

PIC containing γ-PGA and chitosan (CT) as a cationic polymer has been used for preparation of nanoparticles, hydrogels, and films for biomedical applications. Sung et al. investigated the PIC particle formation of γ-PGA and CT by self-assembly in aqueous media [88]. Nanoparticles were obtained upon addition of a γ-PGA (160 kDa) aqueous solution (pH 7.4) into a low molecular weight CT (50 kDa) aqueous solution (pH 6.0). It was found that the particle size and the zeta potential of the prepared nanoparticles were mainly determined by the relative amount of the local concentration of γ-PGA in the added solution to the surrounding concentration of CT. The size (80–400 nm) and surface charge (from −35 to +25 mV) of γ-PGA-CT nanoparticles could be easily controlled by changing the mixing ratio of two polymers. Hajdu et al. also prepared γ-PGA (1,200 kDa)–CT (320 kDa) nanoparticles [89]. The size and size distribution of the nanoparticles depended on the concentrations of γ-PGA and CT solutions and their ratio as well as on the pH of the mixture and the order of addition. The particle size was in the range of 20–285 nm, as measured by transmission electron microscopy (TEM), and the average hydrodynamic diameters were between 150 and 330 nm.

The stability and characteristics of prepared PIC are influenced by various factors involving their chemical compositions and their surrounding environment. In particular, for PIC micelles or nanoparticles, the ionic strength and pH of the solution is a key parameter for stability because of the shielding effect of the ionic species on the electrostatic interactions [90]. Therefore, destabilization of PIC under physiological conditions limits their applications as a drug carrier. For the development of stable PIC nanoparticles under physiological conditions, Akagi et al. focused on a novel approach for the stabilization of PIC nanoparticles by hydrophobic interactions. Amphiphilic γ-PGA-Phe as the biodegradable anionic polymer, and ε-PL as the cationic polymer were used for preparation of PIC nanoparticles (Fig. 9) [91]. The PIC nanoparticles were prepared by mixing

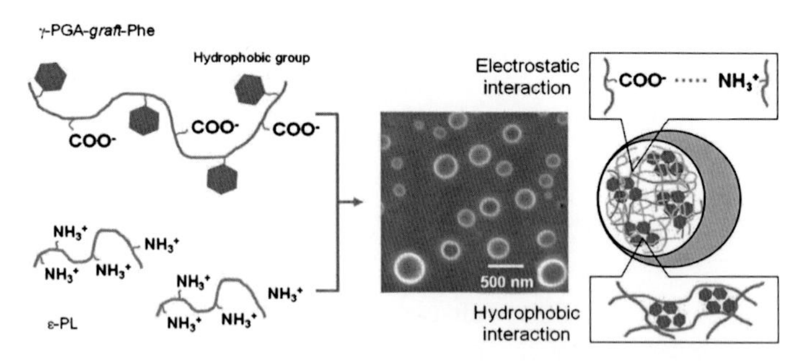

Fig. 9 Stabilization of polyion complex nanoparticles composed of poly(amino acid)s using hydrophobic interactions

γ-PGA-Phe (water soluble) with ε-PL in phosphate-buffered saline (PBS). The formation and stability of the PIC nanoparticles was investigated by dynamic light scattering (DLS) measurements. Monomodal anionic PIC nanoparticles were obtained using nonstoichiometric mixing ratios. When unmodified γ-PGA was mixed with ε-PL in PBS, the formation of PIC nanoparticles was observed. However, within a few hours after the preparation, the PIC nanoparticles dissolved in the PBS. In contrast, γ-PGA-Phe/ε-PL nanoparticles showed high stability for a prolonged period of time in PBS, and over a wide range of pH values. The stability and size of the PIC nanoparticles depended on the γ-PGA-Phe/ε-PL mixing ratio and the hydrophobicity of the γ-PGA. The improved stability of the PIC nanoparticles was attributed to the formation of hydrophobic domains in the core of the nanoparticles. The fabrication of PIC nanoparticles using hydrophobic interactions was very useful for the stabilization of the nanoparticles.

3 Polymeric Nanoparticles for Antigen Delivery and Adjuvant

3.1 Preparation of Antigen-Loaded Nanoparticles

Nanoparticles containing encapsulated, surface-immobilized or surface-adsorbed antigens are being investigated as vaccine delivery systems as alternatives to the currently used alum, with an objective to develop better vaccine systems and minimize the frequency of immunization. The encapsulation of antigenic proteins or peptides into PLGA nanoparticle carrier system can be carried out through mainly three methods: the water-in-oil-in-water (w/o/w) emulsion technique, the phase separation method, and spray drying. The w/o/w double emulsion process is popularly used to load proteins into nanoparticles (Fig. 10) [92, 93]. In this process, an antigen is first dissolved in an aqueous solution, which is then emulsified in an organic solvent to make a primary water-in-oil emulsion. This initial emulsion is further mixed in an emulsifier-containing aqueous solution to make a w/o/w double emulsion. The ensuing removal of the solvent leaves nano- and microparticles in the aqueous continuous phase, making it possible to collect them by filtration or centrifugation. However, the possible denaturation of the proteins at the oil–water interface limits the usage of this method. It has been reported that this interface causes conformational changes in bovine serum albumin (BSA) [94, 95]. Moreover, it has a disadvantage in that the entrapment efficiency is very low. The prevention of protein denaturation and degradation, as well as high entrapment efficiency, would be of particular importance in the preparation of nanoparticles containing water-soluble drugs such as a protein. Improved protein integrity has been achieved by the addition of stabilizers such as carrier proteins (e.g., albumin), surfactants during the primary emulsion phase, or molecules such as trehalose and mannitol to

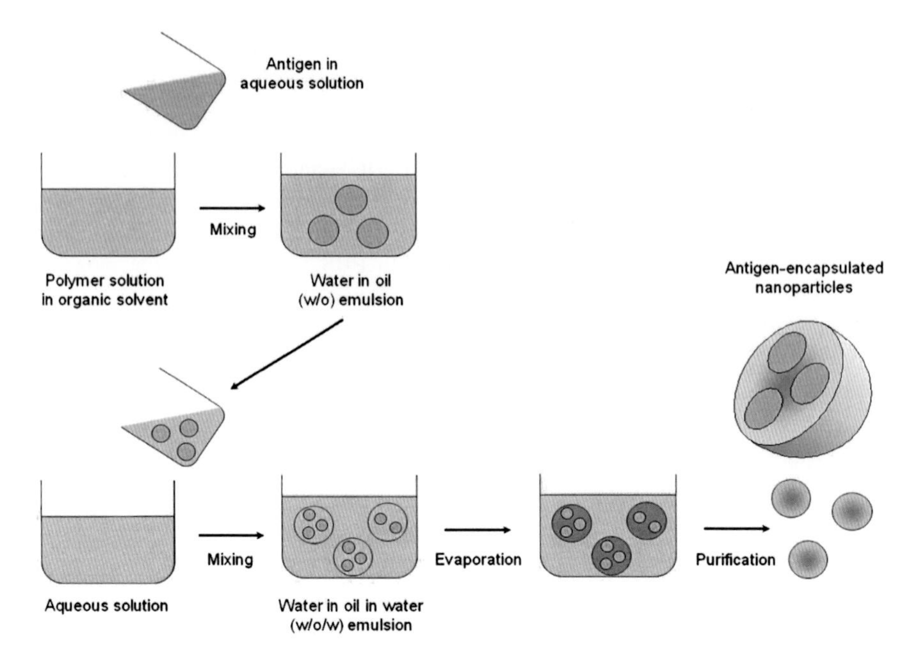

Fig. 10 Preparation of antigen-encapsulating nanoparticles by w/o/w emulsion method

the protein phase. Protein stability may also be enhanced if the protein is encapsulated as a solid rather than in solution.

We have recently found that nanoparticles consisting of amphiphilic poly(amino acid)s can efficiently and stably encapsulate various types of protein into the nanoparticles. Protein-loaded γ-PGA-Phe nanoparticles were prepared by encapsulation, covalent immobilization, or physical adsorption methods in order to study their potential applications as protein carriers [96, 97]. To prepare the protein-encapsulating γ-PGA-Phe nanoparticles, proteins with various molecular weights and isoelectric points were dissolved in saline, and the γ-PGA-Phe dissolved in DMSO was added to the protein solutions. The resulting solutions were then centrifuged and repeatedly rinsed (Fig. 11). The encapsulation of proteins into the nanoparticles was successfully achieved. All proteins used in this experiment were successfully encapsulated into the nanoparticles. The encapsulation efficiency was found to be in the range of 30–60% for most samples. For all samples tested, it was observed that the encapsulation efficiency for a given protein was not markedly influenced by the physical properties of that protein. Ovalbumin (OVA) encapsulated into the nanoparticles was not released (less than 10%) over the pH range of 4–8, even after 10 days. Moreover, it was found that the γ-PGA-Phe nanoparticles have some excellent properties. The enzyme-encapsulating nanoparticles showed high enzymatic activity. In the case of protein-encapsulating nanoparticles prepared by the self-assembly of γ-PGA-Phe, the encapsulated protein may be more stable than via the emulsion method. Proteins encapsulated into the nanoparticles appear to be adequate in terms

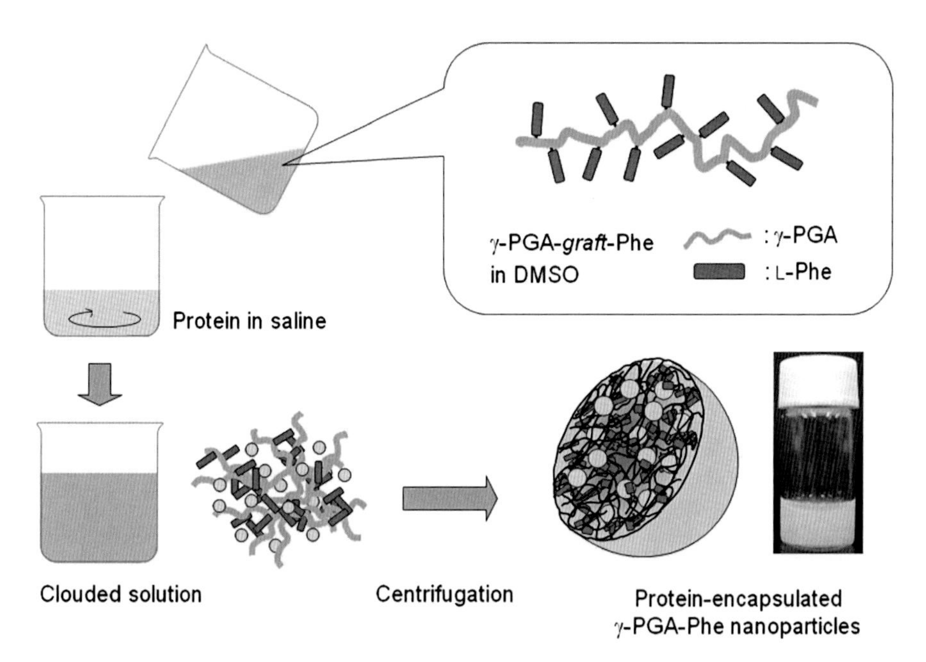

Fig. 11 Preparation of protein-encapsulating γ-PGA-Phe nanoparticles

of the preservation of the protein structure. The γ-PGA-Phe nanoparticles and protein-encapsulating nanoparticles could be preserved by freeze-drying. The results of cytotoxicity tests showed that the nanoparticles did not cause any relevant cell damage. Therefore, it is expected that the γ-PGA-Phe nanoparticles will have great potential as multifunctional carriers in pharmaceutical and biomedical applications, such as drug and vaccine delivery systems. Also, Portilla-Arias et al. reported preparation of nanoparticles made of alkyl esters of γ-PGA and described their potential application as drug and protein carriers [98].

3.2 Delivery of Antigens Using Nanoparticles

Antigen-loaded polymeric nanoparticles represent an exciting approach to the enhancement of antigen-specific humoral and cellular immune responses via selective targeting of the antigen to APCs [99, 100]. DCs are considered to be initiators and modulators of immune responses and are capable of processing antigens through both major histocompatibility complex (MHC) class I and II pathways. Immature DCs encounter pathogens (e.g., virus or bacteria), antigens, or particulate materials at the injection site and, after phagocytosis, the foreign bodies taken up into the DCs present antigens on MHC class II molecules or even on MHC class I molecules by cross-priming [101]. Therefore, the antigen delivery to DCs is of key importance in the development of effective vaccines (Fig. 12).

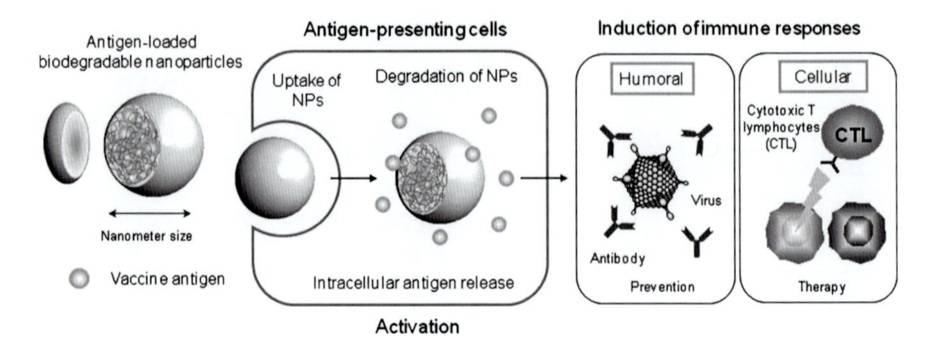

Fig. 12 Induction of immune responses by nanoparticle-based vaccine

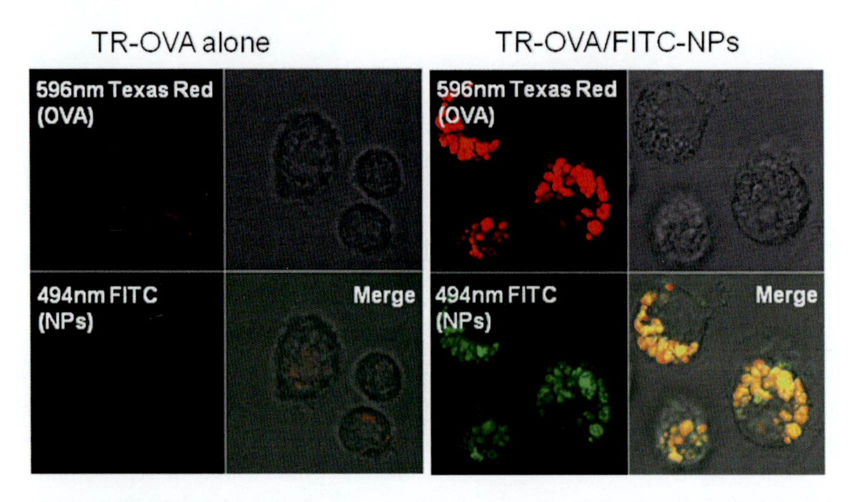

Fig. 13 Uptake of OVA-encapsulating γ-PGA-Phe nanoparticles by DCs. DCs were incubated with Texas Red-labeled OVA (*TR-OVA*) alone (**a**) or TR-OVA encapsulated within fluorescein-labeled nanoparticles (*TR-OVA/FITC-NPs*) (**b**). The intracellular localization of OVA (*red*) and NPs (*green*) was observed by confocal laser scanning microscopy

Akagi et al. demonstrated the use of nanoparticles composed of amphiphilic poly (amino acid) derivatives as vaccine delivery and adjuvants [62, 102–104]. To evaluate the uptake of OVA encapsulated within γ-PGA-Phe nanoparticles (OVA-NPs) by DCs, murine bone marrow-derived DCs were incubated with 250 nm-sized OVA-NPs for 30 min at 37 °C. The cells were then analyzed by flow cytometry (FCM) and confocal laser scanning microscopy (CLSM). OVA-NPs were efficiently taken up into DCs, whereas the uptake of OVA alone was barely detectable at the same concentration of OVA (Fig. 13). OVA-NPs were more efficiently taken up than OVA alone by DCs, and the uptake of OVA-NPs was inhibited at 4 °C. These results suggest that OVA-NPs were phagocytosed mainly via endocytosis by the DCs. In the case of OVA alone, an approximately 30-fold higher concentration was required to elicit a similar amount of intracellular OVA as compared to OVA-NPs. Likewise, it has been reported that PLGA nanoparticles or liposomes are efficiently phagocytosed by the DCs in culture, resulting in their intracellular localization [105–107]. Foged et al. investigated DC uptake of model fluorescent polystyrene particles with a broad size range (0.04–15 μm). The results showed that DCs internalized particles in the tested size range with different efficiencies. The optimal particle size for DC uptake was 500 nm and below. In the smaller the particle size, a higher percentage of the DCs interacted with the polystyrene spheres [24]. Kanchan et al. also reported that PLA nanoparticles (200–600 nm) were efficiently taken up by macrophages in comparison to microparticles (2–8 μm) [25]. In contrast, in hydrogel particles composed of polyacrylamide, there was no difference in uptake by APCs of particles sized between 3.5 μm and 35 nm [108]. This disparity in uptake may be related to fundamental differences in the material properties of those carriers.

Particle shape and surface charge are equally important particulate physicochemical factors and play crucial roles in the interaction between particles and APCs. In general, cationic particles induced high phagocytosis activity of APCs, because of the anionic nature of cell membranes [24]. Recently, particle shape has been identified as having a significant effect on the ability of macrophages to internalize particles via actin-driven movement of the macrophage membrane. Champion et al. observed that the cellular uptake of particles strongly depends on the shape of particles. The worm-like particles with very high aspect ratios exhibited negligible phagocytosis when compared to traditional spherical particles [109, 110]. These result suggest that uptake of particles by APCs strongly depends on the local geometry at the interface between particles and cells.

3.3 Activation of Dendritic Cells by Nanoparticles

Research on biomaterial adjuvant potential has been focused largely on determining the degree of DC maturation induced by exposure to polymeric nanoparticles or liposomes [111–113]. The maturation of DCs is associated with increased expression of several cell surface markers, including the co-stimulatory molecules CD40, CD80, CD83, CD86, MHC class I, and MHC class II. It is well known that DC maturation can be induced by inflammatory factors such as lipopolysaccharide (LPS), bacterial DNA, or inflammatory cytokines such as TNF-α, and the process is highly importance for the initiation of acquired and innate immune responses by these cells [114, 115]. Therefore, in addition to the antigen delivery to DCs, the control of DC maturation is deeply involved in the development of effective vaccines.

In vitro studies have shown that γ-PGA-Phe or PLGA nanoparticle-pulsed DCs result in DC maturation by upregulation of co-stimulatory molecule expression and cytokine production (Fig. 14). To determine whether the uptake of γ-PGA-Phe

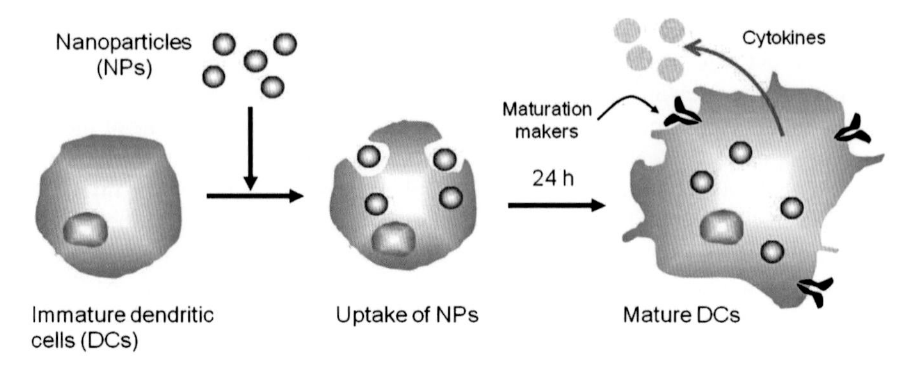

Fig. 14 Maturation and activation of DCs by nanoparticles

nanoparticles mediates the phenotypic maturation of DCs, the DCs were incubated with γ-PGA-Phe nanoparticles for 24 or 48 h, and the expression of surface molecules was measured by fluorescence confocal microscopy (FCM). Upon exposure of these DCs to the nanoparticles, the expression of co-stimulatory molecules (maturation markers) was increased in a dose-dependent manner. The expression levels of co-stimulatory molecules in nanoparticle-pulsed DCs were similar to those of LPS-pulsed DCs. These results suggest that γ-PGA-Phe nanoparticles have great potential as adjuvant for DC maturation [62, 102, 103]. The mechanisms responsible for DC maturation by γ-PGA-Phe nanoparticles are still unclear. However, it is hypothesized that not only the uptake of nanoparticles but also the characteristics of the polymers forming the nanoparticles are important for the induction of DC maturation. The DC uptake of 30 nm-sized nanoparticles was lower than for 200 nm-sized nanoparticles, but the effect of DC activation by the nanoparticles was high for the small sizes [116, 117]. Thus, it is considered that the surface interactions between the nanoparticles and DCs predominately affect DC maturation. In addition, soluble γ-PGA-induced innate immune responses in a Toll-like receptor 4 (TLR4)-dependent manner in DCs have been reported [118, 119]. TLRs are abundantly expressed on professional APCs. TLRs play a major role in pathogen recognition, and in the initiation of the inflammatory and immune responses. The stimulation of TLRs by TLR ligands induces the surface expression of co-stimulatory molecules, and this phenotypic modulation is a typical feature of DC maturation. Treatment with high molecular weight γ-PGA (2,000 kDa), but not low molecular weight γ-PGA (10 kDa) induced a significant upregulation of CD40, CD80, and CD86 expression in wild-type DCs. The stimulatory capacity of γ-PGA was not significantly affected by pretreatment with Polymyxin B (PmB). In contrast, DCs from TLR4-defective mice did not show an enhanced expression of maturation markers in response to the 2,000 kDa γ-PGA treatment. It is suggested that the γ-PGA-Phe nanoparticles also induce DC maturation in a TLR4-dependent manner using the same 2,000 kDa γ-PGA, because γ-PGA is located near the nanoparticle surface.

Tomayo et al. have also reported that poly(anhydride) nanoparticles act as agonists of various TLRs. The nanoparticles were useful as Th1 adjuvants in immunoprophylaxis and immunotherapy through TLR exploitation [120].

Similar results have been obtained with PLGA nano- and microparticles [121, 122], liposomes [107], cationic polystyrene microparticles [123], polystyrene nanoparticles [124], and acid-degradable cationic nanoparticles [125]. Elamanchili et al. examined DC maturation by PLGA nanoparticles. The results showed that after PLGA nanoparticle pulsing, DCs exhibited a modest increase in the expression of MHC class II and CD86 compared to untreated controls. In addition, DCs pulsed with PLGA nanoparticles containing an immunomodulator, monophosphoryl lipid A (MPLA), induced further DC maturation [106]. The PLGA-based nanoparticulate system offers the flexibility for incorporation of broad range of TLR ligands. Copland et al. investigated whether formulation of antigen in mannosylated liposomes enhanced uptake and DC maturation. Exposure to liposomes containing OVA resulted in enhanced expression of maturation markers when compared to exposure to antigen in solution. Expression was highest following exposure to mannosylated liposomes [107]. These particulate systems hold promise as a vaccine delivery system and immunostimulant. However, it has also been reported that PLGA particles failed to mature DCs in vitro [126, 127]. These differences may be attributed to particle size, particle concentration in DCs, presence or absence of antigen, and experimental conditions.

3.4 Gene Delivery by Polyion Complex Nanoparticles

Gene delivery has great potential for the treatment of many different diseases. The basic idea of gene therapy involves delivery of an exogenous gene into the cells to express the encoded protein, which may be insufficiently or aberrantly expressed naturally [128]. DNA delivery is, however, a difficult process and a suitable vector is required for efficient protection as well as release. Both viral and nonviral vectors have been used for gene delivery. Nonviral gene delivery relies on DNA condensation induced by cationic agents. Cationic polymers have been widely chosen to condense DNA through electrostatic interactions between negatively charged DNA and the positively charged cationic sites [129]. PIC nanoparticles composed of γ-PGA and chitosan (CT) have been used as a DNA delivery system. CT/DNA complex nanoparticles have been considered as a vector for gene delivery. Although advantageous for DNA packing and protection from enzymatic degradation, CT-based complexes may lead to difficulties in DNA release at the site of action. To improve the transfection efficiency of CT/DNA complexes, γ-PGA/CT/DNA conjugated nanoparticles were prepared by an ionic-gelation method for transdermal DNA delivery using a low-pressure gene gun [130]. pDNA was mixed with aqueous γ-PGA (20 kDa). Nanoparticles were obtained upon addition of the mixed solution to aqueous CT (80 kDa). The prepared γ-PGA/CT/DNA nanoparticles were pH-sensitive and had a more compact internal structure with a greater density than the conventional CT/DNA. Analysis using small angle X-ray scattering (SAXS) indicated that incorporating γ-PGA caused the formation of compounded nanoparticles whose internal structure might facilitate the dissociation of CT and DNA. As compared with CT/DNA, γ-PGA/CT/DNA nanoparticles improved their penetration depth into mouse skin and enhanced gene expression. Moreover, in addition to improving the release of DNA intracellularly, the incorporation of γ-PGA in nanoparticles markedly increased their cellular internalization [131]. Taken together, the results show that γ-PGA significantly enhanced the transfection efficiency of this developed gene delivery system. The results indicated that γ-PGA played multiple important roles in enhancing the cellular uptake and transfection efficiency of γ-PGA/CT/DNA nanoparticles. This delivery system may be useful for DNA vaccine development.

Kurosaki et al. also developed a vector coated by γ-PGA for effective and safe gene delivery [132]. To develop a useful nonviral vector, PIC constructed with pDNA, PEI, and various polyanions, such as polyadenylic acid, polyinosinic–polycytidylic acid, α-polyaspartic acid, and γ-PGA were prepared. The pDNA/PEI complex had a strong cationic surface charge and showed extremely high transgene efficiency although it agglutinated with erythrocytes and had extremely high cytotoxicity. The γ-PGA could electrostatically coat the pDNA/PEI complex to form stable anionic particles. The coating of γ-PGA dramatically decreased the toxicities of pDNA/PEI complex. Moreover, the pDNA/PEI/γ-PGA complex was highly taken up by the cells via a γ-PGA-specific receptor-mediated pathway and showed extremely high transgene efficiencies. Further studies are necessary to examine the detailed uptake mechanism and clinical safety as gene delivery vector.

4 Control of Intracellular Distribution of Nanoparticles

4.1 pH-Responsive Nanoparticles

In general, particulate materials can be easily internalized into the cells via endocytosis, depending on their size, shape, and surface charge. However, the internalized materials are mostly trafficked from acidic endosomes to lysosomes, where degradation may occur. Thus, degraded exogenous antigens are presented by the MHC class II presentation pathway, and a part of the pathway involves antibody-mediated immune responses. In contrast, antigens within the cytosol are processed into proteasomes and presented by the MHC class I pathway, a pathway involved in the cytotoxic T-lymphocyte (CTL) response [99–101]. Therefore, the induction of antigen-specific cellular immunity by exogenous antigens is needed for the regulation of intracellular distribution of antigens. The escape of internalized antigens from endosomes to the cytoplasm is an important subject relating to control of the antigen processing/presentation pathways.

The release of biomolecules from acidic endosomes requires a membrane-disruptive agent, which can release the internalized compounds into the cytoplasm. Approaches include the use of membrane-penetrating peptides, pathogen-derived pore-forming proteins, and "endosome escaping" polymers or lipids that disrupt the endosomal membrane in response to the pH reduction that occurs in these compartments. Thus, in recent years, there has been significant interest in developing pH-sensitive nanoparticles that can enhance the cytoplasmic delivery of various biomolecules [133–136]. Standley et al. reported an acid-degradable particle composed of acrylamide and acid-degradable crosslinker for protein-based vaccines. These particles released encapsulated protein in a pH-dependent manner. They were stable at the physiological pH of 7.4 but degraded quickly in the pH 5.0 environment of endosomes. The degradation of particles led to the endosome escape of encapsulated proteins. The colloid osmotic mechanism generates a quick degradation of the particles into many molecules, thus increasing the osmotic pressure within the endosomes, leading to a rapid influx of water across the membrane, resulting in its disruption. In fact, the MHC class I presentation levels achieved with these particles were vastly enhanced as a result of their ability to deliver more protein into the cytoplasm of APCs. In a mouse immunization study, these acid-sensitive particles could induce antigen-specific CTL responses and showed antitumor activity [137, 138]. Hu et al. also reported the endosome escape of pH-responsive core–shell nanoparticles. pH-sensitive poly (diethylamino ethyl methacrylate) (PDEAEMA)- core/poly(ethylene glycol) dimethacrylate (PAEMA)-shell nanoparticles were capable of efficient cytosolic delivery of membrane-impermeable molecules such as calcein and OVA to DCs. These particles effectively disrupted endosomes and delivered molecules to the cytosol of cells without cytotoxicity, and enhanced priming of $CD8^+$ T cells by DCs pulsed with OVA/PDEAEMA-core nanoparticles [139]. Polycations that absorb protons in response to the acidification of endosomes can disrupt these vesicles via the proton sponge effect. The proton sponge effect arises from a large number of weak conjugate bases with buffering capacities between 7.2 and 5.0, such as PEI, leading to proton absorption in acid organelles and an osmotic pressure buildup across the organelle membrane. This osmotic pressure causes swelling and/or rupture of the acidic endosomes and a release of the internalized molecules into the cytoplasm [140].

4.2 Amphiphilic Polymers for Cytosolic Delivery

Synthetic poly(alkylacrylic acid) [141, 142] and poly(alkylacrylic acid-*co*-alkyl acrylate) [143, 144] also have pH-dependent, membrane-disruptive properties.

These polymers contain a combination of carboxyl groups and hydrophobic alkyl groups, and are protonated at the endosomal pH range. Upon a decrease in pH, they increase their hydrophobicity, and penetrate into the endosomal membranes and disrupt them. The hydrophobicity of the polymers is important for disrupting lipid membranes. Foster et al. have applied these amphiphilic polymers to nanoparticle delivery systems [145]. pH-responsive nanoparticles (180 nm) incorporating OVA-conjugated poly(propylacrylic acid) (PPAA) (PPAA-OVA) were evaluated to test whether improved cytosolic delivery of a protein antigen could enhance $CD8^+$ CTL and prophylactic tumor vaccine responses. Nanoparticles containing PPAA-OVA were formed by ionic complexation of cationic poly(dimethylaminoethyl methacrylate) (PDMAEMA) with the anionic PPAA-OVA conjugate (PPAA-OVA/PDMAEMA). The PPAA-OVA/PDMAEMA nanoparticles were stably internalized and could access the MHC class I pathway in the cytosol by triggering endosome escape. In an EG.7-OVA mouse tumor protection model, PPAA-OVA/PDMAEMA-immunized mice delayed tumor growth for nearly 5 weeks, whereas control mice injected with PBS and free OVA developed tumors in less than 10 days. This response was attributed to the eightfold increase in production of OVA-specific $CD8^+$ T-lymphocytes and an 11-fold increase in production of anti-OVA IgG. However, these vinyl polymers are not biodegradable and, thus, their molecular weight presents a limitation for medical applications.

Recently, our group developed novel biodegradable nanoparticles composed of hydrophobically modified γ-PGA (γ-PGA-Phe). The nanoparticles showed a highly negative zeta potential (−25 mV) due to the ionization of the carboxyl groups of γ-PGA located near the surfaces. Protein-encapsulating γ-PGA-Phe nanoparticles efficiently delivered proteins from the endosomes to the cytoplasm in DCs [146]. To evaluate their potential applications as membrane disruptive nanoparticles, the nanoparticles were characterized with respect to their hemolytic activity against erythrocytes as a function of pH. The nanoparticles showed hemolytic activity with decreasing pH from 7 to 5.5, and were membrane-inactive at physiological pH. As the pH decreased, the hemolytic activity of the nanoparticles gradually increased, reaching a peak at pH 5.5. This activity was dependent on the hydrophobicity of γ-PGA. The mechanism responsible for the pH-dependent hemolysis by the nanoparticles involved a conformational change of γ-PGA-Phe and corresponding increase in the surface hydrophobicity. Increased polymer hydrophobicity resulted in increased membrane disruption. The γ-PGA-Phe has carboxyl side chain groups, so the p*K*a of the proton of the carboxyl groups is also a very important factor for the pH sensitivity of the γ-PGA-Phe [104].

It has also been reported that antigen delivery to DCs via PLGA particles increased the amount of protein that escaped from endosomes into the cytoplasm. How proteins or peptides encapsulated within PLGA particles become accessible to the cytoplasm is still not clear. It is suggested that the gradual acidification of endosomes leads to protonation of the PLGA polymer, resulting in enhanced hydrophobicity and attachment and rupture of the endosomal membrane [147].

5 Regulation of Immune Responses by Nanoparticle-Based Vaccines

5.1 *Induction of Immune Responses Using Amphiphilic Poly(amino acid) Nanoparticles*

Induction and regulation of an adaptive immune response by vaccination is possible for a broad range of infectious diseases or cancers. Vaccine delivery and adjuvant that can induce antigen-specific humoral and cellular immunity are useful for development of effective vaccine systems. Cellular immunity is required to remove intracellular pathogens, while humoral immunity plays a central role in neutralizing extracellular microorganisms. The efficacy of antigen-loaded γ-PGA-Phe nanoparticles on the induction of antigen-specific humoral and cellular immune responses was examined using OVA as a model antigen [102, 103, 148, 149]. The immune responses were investigated in mice after subcutaneous immunization with OVA-NPs. The OVA-specific CTL responses were not observed in the spleen cells obtained from the control (PBS) and OVA-alone-immunized mice. In contrast, the spleen cells obtained from the mice immunized with OVA-NPs showed a more potent CTL response than those obtained from mice immunized with OVA plus complete Freund's adjuvant (OVA + CFA) (Fig. 15). When anti-OVA antibody responses were examined and compared among the groups after immunization, both OVA-NP- and OVA + CFA-immunized mice showed significantly higher levels of OVA-specific total IgG, IgG1, and IgG2a antibodies than OVA-alone-immunized mice. These results indicate that the γ-PGA-Phe nanoparticles have the ability to prime cellular and humoral immunity by vaccination. It has been

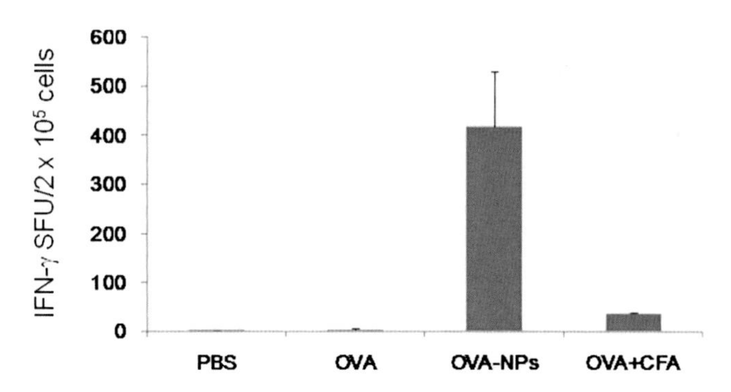

Fig. 15 Induction of cellular immunity by subcutaneous immunization with OVA-encapsulating γ-PGA-Phe nanoparticles. Mice were subcutaneously immunized one time with OVA alone (10 μg), 10 μg of OVA and 100 μg of NPs (*OVA-NPs*), 10 μg of OVA and 100 μL of complete Freund's adjuvant (*OVA + CFA*), or PBS (control). Splenocytes were obtained from the immunized mice on day 10 after the immunization and stimulated with the OVA peptide. The number of IFN-γ-producing cells was measured by enzyme-linked immunospot assay. *SFU* spot forming units

demonstrated that the γ-PGA-Phe nanoparticles are also effective for vaccines against human immunodeficiency virus (HIV) [150, 151], influenza virus [152, 153], Japanese encephalitis virus [154], human T-cell leukemia virus type-I (HTLV-I) [155], or cancers [146, 156, 157]. The antigen-loaded γ-PGA-Phe nanoparticles can provide a safe antigen delivery and adjuvant system for vaccination against viral infections or tumors because of their biocompatibility and biodegradability [158–161].

5.2 Vaccination Using Antigen-Loaded PLGA Nanoparticles

PLGA or PLA nano- and microparticles are suitable vehicles for the delivery of recombinant proteins, peptides, and pDNA to generate immune responses in vivo. Several studies have shown that PLGA nanoparticles can be used to modulate immune responses against encapsulated antigens due to their ability to efficiently target APCs and to facilitate appropriate processing and presenting of antigens to T cells [93, 162–170]. Gutierro et al. investigated the immune response to BSA-loaded PLGA nanoparticles after subcutaneous, oral, and intranasal administration to evaluate parameters that can affect the immune response [171]. These parameters include size, the internal structure of nanoparticles, surface hydrophobicity, zeta potential, and co-encapsulated surfactants, adjuvants or excipients during formulation, which are known to influence targeting strategies.

Many different vaccine antigens encapsulated into PLGA nanoparticles were shown to induce broad and potent immune responses. For example, hepatitis B therapeutic vaccines were designed and formulated by loading the hepatitis B core antigen (HBcAg) into PLGA nanoparticles (300 nm) with or without monophospholipid A (MPLA) adjuvant [172]. A single immunization with HBcAg-encapsulating PLGA nanoparticles containing MPLA induced a stronger cellular immune response than those induced by HBcAg alone or by HBcAg mixed with MPLA in a murine model. More importantly, the level of HBcAg-specific immune responses could be significantly increased further by a booster immunization with the PLGA nanoparticles. These results suggested that co-delivery of HBcAg and MPLA in PLGA nanoparticles promoted HBcAg-specific cellular immune responses. These findings suggest that appropriate design of the vaccine formulation and careful planning of the immunization schedule are important for the successful development of effective therapeutic vaccines for hepatitis B virus.

5.3 Effect of Particle Size on Nanoparticle-Based Vaccines

To design optimal drug carriers, polymeric nanoparticles have the advantage of being able to regulate their physicochemical properties, such as particle size, shape, surface charge, polymer composition, hydrophobicity, and biodegradability. In particular, a method for regulating the size of polymeric nanoparticles is essential for effective vaccine delivery, and to elicit a specific immune response.

Pluronic-stabilized polypropylene sulfide nanoparticles of 20, 45, and 100 nm diameter were prepared to compare the effective targeting of DCs in lymph nodes [173, 174]. Among the three different sizes of nanoparticles, 20 nm-sized nanoparticles were the most readily taken up into the lymphatics via interstitial flow, and activated lymph node-residing DCs more efficiently than the other sizes of nanoparticles. Different sized antigen-conjugated carboxylated polystyrene nanoparticles were also investigated for their size-dependent immunogenicity in vivo. The optimal size for an effective immune response was narrowly defined at 40–50 nm, in the viral range [175]. Furthermore, it has been reported that the size of antigen-loaded PLA particles modulated the immune response [25]. Immunization with PLA nanoparticles (200–600 nm) was associated with higher levels of IFN-γ production related to the T helper 1 (Th1)-type immune response. In contrast, immunization with PLA microparticles (2–8 μm) promoted IL-4 secretion related to the Th2-type immune response. Gutierro et al. also demonstrated that the vaccination of 1,000 nm-sized BSA-loaded PLGA particles generally elicited a higher serum IgG response than that obtained with the vaccination of 500 or 200 nm-sized particles, the immune response for 500 nm particles being similar than that obtained with 200 nm by the subcutaneous and the oral route, and higher by the intranasal route [171]. The vaccination of 1,000 nm particles generally elicited a higher serum IgG response than that obtained with the vaccination of 500 or 200 nm-sized particles, the immune response for 500 nm particles being similar than that obtained with 200 nm by the subcutaneous and the oral route, and higher by the intranasal route. These results suggest that the biodistribution of nano- and microparticles and the particle-related immune response can be regulated by controlling the size of the particles. Consequently, the size of the particulate delivery system is an important factor for modulating immune responses via differential interactions with APCs.

6 Concluding Remarks and Future Perspectives

Biodegradable nanoparticles with entrapped vaccine antigens, such as proteins, peptides and DNA, have recently been shown to possess significant potential as vaccine delivery systems. There are three primary mechanisms of adjuvant function: (1) stabilization of antigen, (2) delivery of antigen, and (3) activation of innate immunity. The duration of delivery is likely to affect immunity. Delivery of antigen is particularly important in cases where the vaccine is intended to act through DCs, as is the case for new vaccine applications requiring cell-mediated immunity. Nanoparticles are extremely flexible delivery systems capable of encapsulating a wide range of antigens. Improving delivery to DCs by nanoparticles will improve vaccine efficiency. Nanoparticle-based vaccine systems will also reduce the vaccine dosage frequency and will increase patient compliance. In the near future, these vaccine systems can be used for treating many infectious diseases or cancers.

γ-PGA is a very promising biodegradable polymer that is produced by various strains of *Bacillus*. Potential applications of γ-PGA as thickener, cytoprotectant, humectant, biological adhesive, fluoculant, or heavy metal absorbent, etc. have been reported. This review describes the preparation of polymeric nanoparticles composed of γ-PGA and their pharmaceutical and biomedical applications. The production of γ-PGA has already been established on the industrial scale because it can be produced easily and extracellularly in high yield by culture of bacteria in a fermenter. Moreover, various molecular weights of γ-PGA can be obtained commercially. γ-PGA by itself is shown to be weakly or non-immunogenic and safe. Amphiphilic γ-PGA nanoparticles have potential use as a new adjuvant instead of alum, and the nanoparticles are very suitable for use as vaccine delivery systems. These systems are expected to be introduced into clinical studies in the near future.

There is a growing interest in identifying the relationship between the size of nanoparticles and their adjuvant activities, but the results from recent studies remain controversial. Many investigators are in agreement that the size of the particles is crucial to their adjuvant activities. Some factors that may affect the conflicting findings include: (1) the polymeric materials used to form the nanoparticles, (2) the nature of the antigen used, (3) the methods of antigen conjugation, and (4) the immunization route. To clarify the influence of nanoparticles on adjuvant activity, there is a need to more comprehensively compare immune responses induced by precisely size-controlled nanoparticles prepared with the same materials and loaded with the same antigens by the proper method.

References

1. Zhao Z, Leong KW (1996) Controlled delivery of antigens and adjuvants in vaccine development. J Pharm Sci 85:1261–1270
2. Singh M, O'Hagan DT (2002) Recent advances in vaccine adjuvants. Pharm Res 19:715–728
3. Singh M, O'Hagan DT (1999) Advances in vaccine adjuvants. Nat Biotechnol 17:1075–1081
4. O'Hagan DT, Rappuoli R (2004) Novel approaches to vaccine delivery. Pharm Res 21:1519–1530
5. Peek LJ, Middaugh CR, Berkland C (2008) Nanotechnology in vaccine delivery. Adv Drug Deliv Rev 60:915–928
6. Ramon G (1924) Sur la toxine et surranatoxine diphtheriques. Ann Inst Pasteur 38:1–10
7. Gupta RK (1998) Aluminum compounds as vaccine adjuvants. Adv Drug Deliv Rev 32:155–172
8. Brewer JM (2006) (How) do aluminium adjuvants work? Immunol Lett 102:10–15
9. Soppimath KS, Aminabhavi TM, Kulkarni AR et al (2001) Biodegradable polymeric nanoparticles as drug delivery devices. J Control Release 70:1–20
10. Hans ML, Lowman AM (2002) Biodegradable nanoparticles for drug delivery and targeting. Curr Opin Solid State Mater Sci 6:319–327
11. Greenland JR, Letvin NL (2007) Chemical adjuvants for plasmid DNA vaccines. Vaccine 25:3731–3741
12. Kakizawa Y, Kataoka K (2002) Block copolymer micelles for delivery of gene and related compounds. Adv Drug Deliv Rev 54:203–222

13. Zhang L, Eisenberg A (1995) Multiple morphologies of crew-cut aggregates of polystyrene-*b*-poly(acrylic acid) block copolymers. Science 1268:1728–1731
14. Dou H, Jiang M, Peng H et al (2003) pH-dependent self-assembly: micellization and micelle-hollow-sphere transition of cellulose-based copolymers. Angew Chem Int Ed 42:1516–1519
15. Reihs T, Muller M, Lunkwitz K (2004) Preparation and adsorption of refined polyelectrolyte complex nanoparticles. J Colloid Interface Sci 271:69–79
16. Kang N, Perron ME, Prudhomme RE et al (2005) Stereocomplex block copolymer micelles: core-shell nanostructures with enhanced stability. Nano Lett 5:315–319
17. Panyam J, Labhasetwar V (2003) Biodegradable nanoparticles for drug and gene delivery to cells and tissue. Adv Drug Deliv Rev 55:329–347
18. Torchilin VP (2006) Multifunctional nanocarriers. Adv Drug Deliv Rev 58:1532–1555
19. Vasir JK, Labhasetwar V (2007) Biodegradable nanoparticles for cytosolic delivery of therapeutics. Adv Drug Deliv Rev 59:718–728
20. Banchereau J, Steinman RM (1998) Dendritic cells and the control of immunity. Nature 392:245–252
21. Gamvrellis A, Leong D, Hanley JC et al (2004) Vaccines that facilitate antigen entry into dendritic cells. Immunol Cell Biol 82:506–516
22. Harding CV, Song R (1994) Phagocytic processing of exogenous particulate antigens by macrophages for presentation by class I MHC molecules. J Immunol 153:4925–4933
23. Wang X, Akagi T, Akashi M et al (2007) Development of core-corona type polymeric nanoparticles as an anti-HIV-1 vaccine. Mini-Rev Org Chem 4:281–290
24. Foged C, Brodin B, Frokjaer S et al (2005) Particle size and surface charge affect particle uptake by human dendritic cells in an in vitro model. Int J Pharm 298:315–322
25. Kanchan V, Panda AK (2007) Interactions of antigen-loaded polylactide particles with macrophages and their correlation with the immune response. Biomaterials 28:5344–5357
26. O'Hagan DT, Jeffery H, Davis SS (1994) The preparation and characterization ofpoly (lactide-co-glycolide) microparticles: III. Microparticle/polymer degradation rates and the in vitro release of a model protein. Int J Pharm 103:37–45
27. Li X, Deng X, Yuan M et al (2000) In vitro degradation and release profiles of poly-DL-lactide-poly(ethylene glycol) microspheres with entrapped proteins. J Appl Polym Sci 78:140–148
28. Liggins RT, Burt HM (2001) Paclitaxel loaded poly(L-lactic acid) microspheres: properties of microspheres made with low molecular weight polymers. Int J Pharm 222:19–33
29. Lemoine D, Francois C, Kedzierewicz F et al (1996) Stability study of nanoparticles of poly (ε-caprolactone), poly(D, L-lactide) and poly(D, L-lactide-co-glycolide). Biomaterials 17:2191–2197
30. Jiang W, Gupta RK, Deshpande MC et al (2005) Biodegradable poly(lactic-co-glycolic acid) microparticles for injectable delivery of vaccine antigens. Adv Drug Deliv Rev 57:391–410
31. Mohamed F, van der Walle CF (2008) Engineering biodegradable polyester particles with specific drug targeting and drug release properties. J Pharm Sci 97:71–87
32. Kumari A, Yadav SK, Yadav SC (2009) Biodegradable polymeric nanoparticles based drug delivery systems. Colloids Surf B 75:1–18
33. O'Donnell PB, McGinity JW (1997) Preparation of microspheres by the solvent evaporation technique. Adv Drug Deliv Rev 28:25–42
34. Lü JM, Wang X, Marin-Muller C et al (2009) Current advances in research and clinical applications of PLGA-based nanotechnology. Expert Rev Mol Diagn 9:325–341
35. Gaucher G, Dufresne MH, Sant VP et al (2005) Block copolymer micelles: preparation, characterization and application in drug delivery. J Control Release 109:169–188
36. Letchford K, Burt H (2007) A review of the formation and classification of amphiphilic block copolymer nanoparticulate structures: micelles, nanospheres, nanocapsules and polymersomes. Eur J Pharm Biopharm 65:259–269
37. Holowka EP, Pochan DJ, Deming TJ (2005) Charged polypeptide vesicles with controllable diameter. J Am Chem Soc 127:12423–12428

38. Matsusaki M, Hiwatari K, Higashi M et al (2004) Stably-dispersed and surface-functional bionanoparticles prepared by self-assembling amphipathic polymers of hydrophilic poly (γ-glutamic acid) bearing hydrophobic amino acids. Chem Lett 33:398–399
39. Matsusaki M, Fuchida T, Kaneko T et al (2005) Self-assembling bionanoparticles of poly (ε-lysine) bearing cholesterol as a biomesogen. Biomacromolecules 6:2374–2379
40. Arimura H, Ohya Y, Ouchi T (2005) Formation of core-shell type biodegradable polymeric micelles from amphiphilic poly(aspartic acid)-*block*-polylactide diblock copolymer. Biomacromolecules 6:720–725
41. Akiyoshi K, Ueminami A, Kurumada S et al (2000) Self-association of cholesteryl-bearing poly(L-lysine) in water and control of its secondary structure by host – guest interaction with cyclodextrin. Macromolecules 33:6752–6756
42. Holowka EP, Sun VZ, Kamei DT et al (2007) Polyarginine segments in block copolypeptides drive both vesicular assembly and intracellular delivery. Nat Mater 6:52–57
43. Jeong JH, Kang HS, Yang SR et al (2003) Polymer micelle-like aggregates of novel amphiphilic biodegradable poly(asparagine) grafted with poly(caprolactone). Polymer 44:583–591
44. Kataoka K, Matsumoto T, Yokoyama M et al (2000) Doxorubicin-loaded poly(ethylene glycol)-poly(β-benzyl-L-aspartate) copolymer micelles: their pharmaceutical characteristics and biological significance. J Control Release 64:143–153
45. Lin J, Zhang S, Chen T et al (2007) Micelle formation and drug release behavior of polypeptide graft copolymer and its mixture with polypeptide block copolymer. Int J Pharm 336:49–57
46. Lee ES, Shin HJ, Na K et al (2003) Poly(L-histidine)-PEG block copolymer micelles and pH-induced destabilization. J Control Release 90:363–374
47. Kubota H, Matsunobu T, Uotani K et al (1993) Production of poly(γ-glutamic acid) by *Bacillus subtilis* F-2-01. Biosci Biotechnol Biochem 57:1212–1213
48. King EC, Watkins WJ, Blacker AJ et al (1998) Covalent modification in aqueous solution of poly-γ-D-glutamic acid from *Bacillus licheniformis*. J Polym Sci A: Polym Chem 36:1995–1999
49. Morillo M, Martinez de Ilarduya A, Munoz-Guerra S (2001) Comblike alkyl esters of biosynthetic poly(γ-glutamic acid). 1. Synthesis and characterization. Macromolecules 34:7868–7875
50. Prodhomme EJF, Tutt AL, Glennie MJ et al (2003) Multivalent conjugates of poly-γ-D-glutamic acid from *Bacillus licheniformis* with antibody F(ab') and glycopeptide ligands. Bioconjug Chem 14:1148–1155
51. Tachibana Y, Kurisawa M, Uyama H et al (2003) Thermo- and pH-responsive biodegradable poly(α-*N*-substituted γ-glutamine)s. Biomacromolecules 4:1132–1134
52. Shimokuri T, Kaneko T, Serizawa T et al (2004) Preparation and thermosensitivity of naturally occurring polypeptide poly(γ-glutamic acid) derivatives modified by alkyl groups. Macromol Biosci 4:407–411
53. Oppermann FB, Fickaitz S, Steinbiichel A (1998) Biodegradation of polyamides. Polym Degrad Stab 59:337–344
54. Obst M, Steinbuchel A (2004) Microbial degradation of poly(amino acid)s. Biomacromolecules 5:1166–1176
55. Schneerson R, Kubler-Kielb J, Liu TY et al (2003) Poly(γ-D-glutamic acid) protein conjugates induce IgG antibodies in mice to the capsule of *Bacillus anthracis*: a potential addition to the anthrax vaccine. Proc Natl Acad Sci USA 100:8945–8950
56. Rhie GE, Roehrl MH, Mourez M et al (2003) A dually active anthrax vaccine that confers protection against both bacilli and toxins. Proc Natl Acad Sci USA 100:10925–10930
57. Wang TT, Fellows PF, Leighton TJ et al (2004) Induction of opsonic antibodies to the gamma-D-glutamic acid capsule of *Bacillus anthracis* by immunization with a synthetic peptide-carrier protein conjugate. FEMS Immunol Med Microbiol 40:231–237

58. Joyce J, Cook J, Chabot D et al (2006) Immunogenicity and protective efficacy of *Bacillus anthracis* poly-γ-D-glutamic acid capsule covalently coupled to a protein carrier using a novel triazine-based conjugation strategy. J Biol Chem 281:4831–4843
59. Kubler-Kielb J, Liu TY, Mocca C et al (2006) Additional conjugation methods and immunogenicity of *Bacillus anthracis* poly-γ-D-glutamic acid-protein conjugates. Infect Immun 74:4744–4749
60. Shih IL, Van YT (2001) The production of poly(γ-glutamic acid) from microorganisms and its various application. Bioresource Technol 79:207–225
61. Kaneko T, Higashi M, Matsusaki M et al (2005) Self-assembled soft nanofibrils of amphipathic polypeptides and their morphological transformation. Chem Mater 17:2484–2486
62. Akagi T, Wang X, Uto T et al (2007) Protein direct delivery to dendritic cells using nanoparticles based on amphiphilic poly(amino acid) derivatives. Biomaterials 28:3427–3436
63. Kim H, Akagi T, Akashi M (2009) Preparation of size tunable amphiphilic poly(amino acid) nanoparticles. Macromol Biosci 9:842–848
64. Bodnar M, Kjoniksen AL, Molnar RM et al (2008) Nanoparticles formed by complexation of poly-γ-glutamic acid with lead ions. J Hazard Mater 153:1185–1192
65. Radu JEF, Novak L, Hartmann JF et al (2008) Structural and dynamical characterization of poly-γ-glutamic acid-based cross-linked nanoparticles. Colloid Polym Sci 286:365–376
66. Shih IL, Shen MH, Van YT (2006) Microbial synthesis of poly(ε-lysine) and its various applications. Bioresour Technol 97:1148–1159
67. Saimura M, Takehara M, Mizukami S et al (2008) Biosynthesis of nearly monodispersed poly (ε-L-lysine) in *Streptomyces* species. Biotechnol Lett 30:377–385
68. Harada A, Kawamura M, Matsuo T et al (2006) Synthesis and characterization of head-tail type polycation block copolymer as non-viral gene vector. Bioconjug Chem 17:3–5
69. Wagner E, Kloeckner J (2006) Gene delivery using polymer therapeutics. Adv Polym Sci 192:135–173
70. Nguyen DN, Green JJ, Chan JM et al (2009) Polymeric materials for gene delivery and DNA vaccination. Adv Mater 21:847–867
71. Tousignan JD, Gates AL, Ingram LA et al (2002) Comprehensive analysis of the acute toxicities induced by systemic administration of cationic lipid: Plasmid DNA complexes in mice. Hum Gene Ther 11:2493–2513
72. Chollet P, Favrot MC, Hurbin A et al (2002) Side-effects of a systemic injection of linear polyethylenimine-DNA complexes. J Gene Med 4:84–91
73. Nishiyama N, Kataoka K (2006) Nanostructured devices based on block copolymer assemblies for drug delivery: designing structures for enhanced drug function. Adv Polym Sci 193:67–101
74. Akiyoshi K, Kobayashi S, Shichibe S et al (1998) Self-assembled hydrogel nanoparticle of cholesterol-bearing pullulan as a carrier of protein drugs: Complexation and stabilization of insulin. J Control Release 54:313–320
75. Jung SW, Jeong Y, Kim SH (2003) Characterization of hydrophobized pullulan with various hydrophobicities. Int J Pharm 254:109–121
76. Na K, Park KH, Kim SW et al (2000) Self-assembled hydrogel nanoparticles from curdlan derivatives: characterization, anti-cancer drug release and interaction with a hepatoma cell line (HepG2). J Control Release 69:225–236
77. Gref R, Rodrigues J, Couvreur P (2002) Polysaccharides grafted with polyesters: novel amphiphilic copolymers for biomedical applications. Macromolecules 35:9861–9867
78. Leonard M, Boisseson MRD, Hubert P et al (2004) Hydrophobically modified alginate hydrogels as protein carriers with specific controlled release properties. J Control Release 98:395–405
79. Park JH, Kwona S, Nam JO et al (2004) Self-assembled nanoparticles based on glycol chitosan bearing 5h-cholanic acid for RGD peptide delivery. J Control Release 95:579–588

80. Akiyoshi K, Deguchi S, Moriguchi N et al (1993) Self-aggregates of hydrophobized polysaccharides in water. Formation and characteristics of nanoparticles. Macromolecules 26:3062–3068
81. Hsieh CY, Tsai SP, Wang DM et al (2005) Preparation of γ-PGA/chitosan composite tissue engineering matrices. Biomaterials 26:5617–5623
82. Kang HS, Park SH, Lee YG et al (2007) Polyelectrolyte complex hydrogel composed of chitosan and poly(γ-glutamic acid) for biological application: Preparation, physical properties, and cytocompatibility. J Appl Polym Sci 103:386–394
83. Kim YH, Gihm SH, Park CR et al (2001) Structural characteristics of size-controlled self-aggregates of deoxycholic acid-modified chitosan and their application as a DNA delivery carrier. Bioconjug Chem 12:932–938
84. Lee KY, Jo WH, Kwon IC et al (1998) Structural determination and interior polarity of self-aggregates prepared from deoxycholic acid-modified chitosan in water. Macromolecules 31:378–383
85. Kida T, Inoue K, Akagi T et al (2007) Preparation of novel polysaccharide nanoparticles by the self-assembly of amphiphilic pectins and their protein-encapsulation ability. Chem Lett 36:940–941
86. Muller M, Reihs T, Ouyang W (2005) Needlelike and spherical polyelectrolyte complex nanoparticles of poly(L-lysine) and copolymers of maleic acid. Langmuir 21:465–469
87. Hartig SM, Greene RR, DasGupta J et al (2007) Multifunctional nanoparticulate polyelectrolyte complexes. Pharm Res 24:2353–2369
88. Lin YH, Chung CK, Chen CT et al (2005) Preparation of nanoparticles composed of chitosan/poly-γ-glutamic acid and evaluation of their permeability through Caco-2 cells. Biomacromolecules 6:1104–1112
89. Hajdu I, Bodnar M, Filipcsei G et al (2009) Nanoparticles prepared by self-assembly of chitosan and poly-γ-glutamic acid. Colloid Polym Sci 286:343–350
90. Lin YH, Sonaje K, Lin KM et al (2008) Multi-ion-crosslinked nanoparticles with pH-responsive characteristics for oral delivery of protein drugs. J Control Release 132:141–149
91. Akagi T, Watanabe K, Kim H et al (2010) Stabilization of polyion complex nanoparticles composed of poly(amino acid) using hydrophobic interactions. Langmuir 26:2406–2413
92. Tamber H, Johansen P, Merkle HP et al (2005) Formulation aspects of biodegradable polymeric microspheres for antigen delivery. Adv Drug Deliv Rev 57:357–376
93. Mundargi RC, Babu VR, Rangaswamy V et al (2008) Nano/micro technologies for delivering macromolecular therapeutics using poly(D, L-lactide-co-glycolide) and its derivatives. J Control Release 125:193–209
94. Sah H (1999) Stabilization of proteins against methylene chloride/water interface induced denaturation and aggregation. J Control Release 58:143–151
95. Panyam J, Dali MM, Sahoo SK et al (2003) Polymer degradation and in vitro release of a model protein from poly(D, L-lactide-co-glycolide) nano- and microparticles. J Control Release 92:173–187
96. Akagi T, Kaneko T, Kida T et al (2005) Preparation and characterization of biodegradable nanoparticles based on poly(γ-glutamic acid) with L-phenylalanine as a protein carrier. J Control Release 108:226–236
97. Akagi T, Kaneko T, Kida T et al (2006) Multifunctional conjugation of proteins on/into core-shell type nanoparticles prepared by amphiphilic poly(γ-glutamic acid). J Biomater Sci Polym Ed 17:875–892
98. Portilla-Arias JA, Camargo B, Garcia-Alvarez M et al (2009) Nanoparticles made of microbial poly(γ-glutamate)s for encapsulation and delivery of drugs and proteins. J Biomater Sci Polym Ed 20:1065–1079
99. O'Hagan DT (1998) Recent advances in immunological adjuvants: the development of particulate antigen delivery systems. Exp Opin Invest Drugs 7:349–359
100. Storni T, Kundig TM, Senti G et al (2005) Immunity in response to particulate antigen-delivery systems. Adv Drug Deliv Rev 57:333–355

101. Shen H, Ackerman AL, Cody V et al (2006) Enhanced and prolonged cross-presentation following endosomal escape of exogenous antigens encapsulated in biodegradable nanoparticles. Immunology 117:78–88
102. Uto T, Wang X, Sato K et al (2007) Targeting of antigen to dendritic cells with poly (γ-glutamic acid) nanoparticles induce antigen-specific humoral and cellular immunity. J Immunol 178:2979–2986
103. Uto T, Akagi T, Hamasaki T et al (2009) Modulation of innate and adaptive immunity by biodegradable nanoparticles. Immunol Lett 125:46–52
104. Akagi T, Kim H, Akashi M (2010) pH-dependent disruption of erythrocyte membrane by amphiphilic poly(amino acid) nanoparticles. J Biomater Sci Polym Edn 21:315–328
105. Lutsiak ME, Robinson DR, Coester C et al (2002) Analysis of poly(D, L-lactic-co-glycolic acid) nanosphere uptake by human dendritic cells and macrophages in vitro. Pharm Res 19:1480–1487
106. Elamanchili P, Diwan M, Cao M et al (2004) Characterization of poly(D, L-lactic-co-glycolic acid) based nanoparticulate system for enhanced delivery of antigens to dendritic cells. Vaccine 22:2406–2412
107. Copland MJ, Baird MA, Rades T et al (2003) Liposomal delivery of antigen to human dendritic cells. Vaccine 21:883–890
108. Cohen JA, Beaudette TT, Tseng WW et al (2009) T-cell activation by antigen-loaded pH-sensitive hydrogel particles in vivo: the effect of particle size. Bioconjug Chem 20:111–119
109. Champion JA, Mitragotri S (2006) Role of target geometry in phagocytosis. Proc Natl Acad Sci USA 103:4930–4934
110. Champion JA, Mitragotri S (2009) Shape induced inhibition of phagocytosis of polymer particles. Pharm Res 26:244–249
111. Reddy ST, Swartz MA, Hubbell JA (2006) Targeting dendritic cells with biomaterials: developing the next generation of vaccines. Trends Immunol 27:573–579
112. Jones KS (2008) Biomaterials as vaccine adjuvants. Biotechnol Prog 24:807–814
113. Babensee JE (2007) Interaction of dendritic cells with biomaterials. Semin Immunol 20:101–108
114. Jilek S, Merkle HP, Walter E (2005) DNA-loaded biodegradable microparticles as vaccine delivery systems and their interaction with dendritic cells. Adv Drug Deliv Rev 57:377–390
115. Black M, Trent A, Tirrell M et al (2010) Advances in the design and delivery of peptide subunit vaccines with a focus on toll-like receptor agonists. Expert Rev Vaccines 9:157–173
116. Kim H, Uto T, Akag T et al (2010) Amphiphilic poly(amino acid) nanoparticles induce size-dependent dendritic cell maturation. Adv Funct Mater 20:3925–3931
117. Akagi T, Shima F, Akashi M (2011) Intracellular degradation and distribution of protein-encapsulated amphiphilic poly(amino acid) nanoparticles. Biomaterials 32:4959–4967
118. Kim TW, Lee TY, Bae HC et al (2007) Oral administration of high molecular mass poly-γ-glutamate induces NK cell-mediated antitumor immunity. J Immunol 179:775–780
119. Lee TY, Kim YH, Yoon SW et al (2009) Oral administration of poly-γ-glutamate induces TLR4- and dendritic cell-dependent antitumor effect. Cancer Immunol Immunother 58:1781–1794
120. Tamayo I, Irache JM, Mansilla C et al (2010) Poly(anhydride) nanoparticles act as active Th1 adjuvants through Toll-like receptor exploitation. Clin Vaccine Immunol 17:1356–1362
121. Yoshida M, Babensee JE (2004) Poly(lactic-*co*-glycolic acid) enhances maturation of human monocyte-derived dendritic cells. J Biomed Mater Res 71:45–54
122. Jilek S, Ulrich M, Merkle HP et al (2004) Composition and surface charge of DNA-loaded microparticles determine maturation and cytokine secretion in human dendritic cells. Pharm Res 21:1240–1247
123. Thiele L, Rothen-Rutishauser B, Jilek S et al (2001) Evaluation of particle uptake in human blood monocyte-derived cells in vitro. Does phagocytosis activity of dendritic cells measure up with macrophages? J Control Release 76:59–71
124. Matsusaki M, Larsson K, Akagi T et al (2005) Nanosphere induced gene expression in human dendritic cells. Nano Lett 5:2168–2173

125. Kwon YJ, Standley SM, Goh SL et al (2005) Enhanced antigen presentation and immunostimulation of dendritic cells using acid-degradable cationic nanoparticles. J Control Release 105:199–212
126. Sun H, Pollock KG, Brewer JM (2003) Analysis of the role of vaccine adjuvants in modulating dendritic cell activation and antigen presentation in vitro. Vaccine 21:849–855
127. Waeckerle-Men Y, Allmen EU, Gander B et al (2006) Encapsulation of proteins and peptides into biodegradable poly(D, L-lactide-co-glycolide) microspheres prolongs and enhances antigen presentation by human dendritic cells. Vaccine 24:1847–1857
128. Li SD, Huang L (2006) Gene therapy progress and prospects: non-viral gene therapy by systemic delivery. Gene Ther 13:1313–1319
129. Mann A, Richa R, Ganguli M (2008) DNA condensation by poly-L-lysine at the single molecule level: role of DNA concentration and polymer length. J Control Release 125:252–262
130. Lee PW, Peng SF, Su CJ et al (2008) The use of biodegradable polymeric nanoparticles in combination with a low-pressure gene gun for transdermal DNA delivery. Biomaterials 29:742–751
131. Peng SF, Yang MJ, Su CJ et al (2009) Effects of incorporation of poly(γ-glutamic acid) in chitosan/DNA complex nanoparticles on cellular uptake and transfection efficiency. Biomaterials 30:1797–1808
132. Kurosaki T, Kitahara T, Fumoto S et al (2009) Ternary complexes of pDNA, polyethylenimine, and gamma-polyglutamic acid for gene delivery systems. Biomaterials 30:2846–2853
133. Plank C, Zauner W, Wagner E (1998) Application of membrane-active peptides for drug and gene delivery across cellular membranes. Adv Drug Deliv Rev 34:21–35
134. Shai Y (1999) Mechanism of the binding, insertion and destabilization of phospholipid bilayer membranes by α-helical antimicrobial and cell non-selective membrane-lytic peptides. Biochim Biophys Acta 1462:55–70
135. Yessine MA, Leroux JC (2004) Membrane-destabilizing polyanions: interaction with lipid bilayers and endosomal escape of biomacromolecules. Adv Drug Deliv Rev 56:999–1021
136. Chen R, Yue Z, Eccleston ME et al (2005) Modulation of cell membrane disruption by pH-responsive pseudo-peptides through grafting with hydrophilic side chains. J Control Release 108:63–72
137. Murthy N, Xu M, Schuck S et al (2003) A macromolecular delivery vehicle for protein-based vaccines: Acid-degradable protein-loaded microgels. Proc Natl Acad Sci USA 29:4995–5000
138. Standley SM, Kwon TJ, Murthy N et al (2004) Acid-degradable particles for protein-based vaccines: Enhanced survival rate for tumor-challenged mice using ovalbumin model. Bioconjug Chem 15:1281–1288
139. Hu Y, Litwin T, Nagaraja AR et al (2007) Cytosolic delivery of membrane-impermeable molecules in dendritic cells using pH-responsive core-shell nanoparticles. Nano Lett 7:3056–3064
140. Boussif O, Lezoualc'h F, Zanta MA et al (1995) A versatile vector for gene and oligonucleotide transfer into cells in culture and in vivo. Proc Natl Acad Sci USA 92:7297–7301
141. Murthy N, Robichaud JR, Tirrell DA et al (1999) The design and synthesis of polymers for eukaryotic membrane disruption. J Control Release 61:137–143
142. Jones RA, Cheung CY, Black FE et al (2003) Poly(2-alkylacrylic acid) polymers deliver molecules to the cytosol by pH-sensitive disruption of endosomal vesicles. Biochem J 372:65–75
143. Kusonwiriyawong C, van de Wetering P, Hubbell JA et al (2003) Evaluation of pH-dependent membrane-disruptive properties of poly(acrylic acid) derived polymers. Eur J Pharm Biopharm 56:237–246
144. Yessine MA, Meier C, Petereit HU et al (2006) On the role of methacrylic acid copolymers in the intracellular delivery of antisense oligonucleotides. Eur J Pharm Biopharm 63:1–10

145. Foster S, Duvall CL, Crownover EF et al (2010) Intracellular delivery of a protein antigen with an endosomal-releasing polymer enhances CD8 T-cell production and prophylactic vaccine efficacy. Bioconjug Chem 21:2205–2212
146. Yoshikawa T, Okada N, Oda A et al (2008) Development of amphiphilic γ-PGA-nanoparticle based tumor vaccine: potential of the nanoparticulate cytosolic protein delivery carrier. Biochem Biophys Res Commun 366:408–413
147. Panyam J, Zhou WZ, Prabha S et al (2002) Rapid endo-lysosomal escape of poly(dl-lactide-co-glycolide) nanoparticles: implications for drug and gene delivery. FASEB J 16:1217–1226
148. Uto T, Wang X, Akagi T et al (2009) Improvement of adaptive immunity by antigen-carrying biodegradable nanoparticles. Biochem Biophys Res Commun 379:600–604
149. Hamasaki T, Uto TA et al (2010) Modulation of gene expression related to Toll-like receptor signaling in dendritic cells by poly(γ-glutamic acid) nanoparticles. Clin Vaccine Immunol 17:748–756
150. Wang X, Uto T, Akagi T et al (2007) Induction of potent $CD8^+$ T-cell responses by novel biodegradable nanoparticles carrying human immunodeficiency virus type 1 gp120. J Virol 81:10009–10016
151. Wang X, Uto T, Akagi T et al (2008) Poly(γ-glutamic Acid) nanoparticles as an efficient antigen delivery and adjuvant system: potential for an anti-AIDS vaccine. J Med Virol 80:11–19
152. Okamoto S, Yoshii H, Akagi T et al (2007) Influenza hemagglutinin vaccine with poly (γ-glutamic acid) nanoparticles enhances the protection against influenza virus infection through both humoral and cell-mediated immunity. Vaccine 25:8270–8278
153. Okamoto S, Matsuura M, Akagi T et al (2009) Poly(γ-glutamic acid) nano-particles combined with mucosal influenza virus hemagglutinin vaccine protects against influenza virus infection in mice. Vaccine 27:5896–5905
154. Okamoto S, Yoshii H, Ishikawa T et al (2008) Single dose of inactivated Japanese encephalitis vaccine with poly(γ-glutamic acid) nanoparticles provides effective protection from Japanese encephalitis virus. Vaccine 26:589–594
155. Matsuo K, Yoshikawa T, Oda A et al (2007) Efficient generation of antigen-specific cellular immunity by vaccination with poly(γ-glutamic acid) nanoparticles entrapping endoplasmic reticulum-targeted peptides. Biochem Biophys Res Commun 362:1069–1072
156. Yoshikawa T, Okada N, Oda A et al (2008) Nanoparticles built by self-assembly of am phiphilic poly(γ-glutamic acid) can deliver antigens to antigen-presenting cells with high efficiency: A new tumor-vaccine carrier for eliciting effector T cells. Vaccine 26:1303–1313
157. Yamaguchi S, Tatsumi T, Takehara T et al (2010) EphA2-derived peptide vaccine with amphiphilic poly(gamma-glutamic acid) nanoparticles elicits an anti-tumor effect against mouse liver tumor. Cancer Immunol Immunother 59:759–767
158. Akagi T, Higashi M, Kaneko T et al (2005) In vitro enzymatic degradation of nanoparticles prepared from hydrophobically-modified poly(γ-glutamic acid). Macromol Biosci 5:598–602
159. Akagi T, Higashi M, Kaneko T et al (2006) Hydrolytic and enzymatic degradation of nanoparticles based on amphiphilic poly(γ-glutamic acid)-*graft*-L-phenylalanine copolymer. Biomacromolecules 7:297–303
160. Akagi T, Baba M, Akashi M (2007) Preparation of nanoparticles by the self-organization of polymers consisting of hydrophobic and hydrophilic segments: potential applications. Polymer 48:6729–6747
161. Kim H, Akagi T, Akashi M (2010) Preparation of CpG ODN-encapsulated anionic poly (amino acid) nanoparticles for gene delivery. Chem Lett 39:278–279
162. Raghuvanshi RS, Katare YK, Lalwani K et al (2002) Improved immune response from biodegradable polymer particles entrapping tetanus toxoid by use of different immunization protocol and adjuvants. Int J Pharm 245:109–121
163. Ataman-Onal Y, Munier S, Ganée A et al (2006) Surfactant-free anionic PLA nanoparticles coated with HIV-1 p24 protein induced enhanced cellular and humoral immune responses in various animal models. J Control Release 112:175–185

164. Hamdy S, Elamanchili P, Alshamsan A et al (2007) Enhanced antigen-specific primary CD4+ and CD8+ responses by codelivery of ovalbumin and toll-like receptor ligand monophosphoryl lipid A in poly(D, L-lactic-co-glycolic acid) nanoparticles. J Biomed Mater Res A 81:652–662
165. Solbrig CM, Saucier-Sawyer JK, Cody V et al (2007) Polymer nanoparticles for immunotherapy from encapsulated tumor-associated antigens and whole tumor cells. Mol Pharm 4:47–57
166. Wendorf J, Chesko J, Kazzaz J et al (2008) A comparison of anionic nanoparticles and microparticles as vaccine delivery systems. Hum Vaccin 4:44–49
167. Nayak B, Panda AK, Ray P et al (2009) Formulation, characterization and evaluation of rotavirus encapsulated PLA and PLGA particles for oral vaccination. J Microencapsul 26:154–165
168. Hamdy S, Molavi O, Ma Z et al (2008) Co-delivery of cancer-associated antigen and Toll-like receptor 4 ligand in PLGA nanoparticles induces potent $CD8^{+}$ T cell-mediated anti-tumor immunity. Vaccine 26:5046–5057
169. Caputo A, Sparnacci K, Ensoli B et al (2008) Functional polymeric nano/microparticles for surface adsorption and delivery of protein and DNA vaccines. Curr Drug Deliv 5:230–242
170. Slütter B, Plapied L, Fievez V et al (2009) Mechanistic study of the adjuvant effect of biodegradable nanoparticles in mucosal vaccination. J Control Release 138:113–121
171. Gutierro I, Hernández RM, Igartua M et al (2002) Size dependent immune response after subcutaneous, oral and intranasal administration of BSA loaded nanospheres. Vaccine 21:67–77
172. Chong CS, Cao M, Wong WW et al (2005) Enhancement of T helper type 1 immune responses against hepatitis B virus core antigen by PLGA nanoparticle vaccine delivery. J Control Release 102:85–99
173. Reddy ST, Rehor A, Schmoekel HG (2006) In vivo targeting of dendritic cells in lymph nodes with poly(propylene sulfide) nanoparticles. J Control Release 112:26–34
174. Reddy ST, Van Der Vlies AJ, Simeoni E et al (2007) Exploiting lymphatic transport and complement activation in nanoparticle vaccines. Nat Biotechnol 25:1159–1164
175. Fifis T, Gamvrellis A, Crimeen-Irwin B et al (2004) Size-dependent immunogenicity: therapeutic and protective properties of nano-vaccines against tumors. J Immunol 173:3148–3154

Adv Polym Sci (2012) 247: 65–114
DOI: 10.1007/12_2011_160

Published online: 5 November 2011

Biodegradable Polymeric Assemblies for Biomedical Materials

Yuichi Ohya, Akihiro Takahashi, and Koji Nagahama

Abstract Recently, self-assembled systems using biodegradable polymers at the nanometer scale, such as microspheres, nanospheres, polymer micelles, nanogels, and polymersomes, have attracted much attention especially in biomedical fields. To construct such self-assembled systems, it is extremely important to have precise control of intermolecular noncovalent interactions, such as hydrophobic interactions based on their amphiphilic molecular structures. Biodegradable polymers, especially aliphatic polyesters such as polylactide, polyglycolide, poly(ε-caplolactone) and their copolymers, have been used as biomedical materials for a long time. This chapter is mainly focused on aliphatic polyesters and related polymers, and reviews the synthetic methods for amphiphilic biodegradable polymers containing aliphatic polyesters as components. Moreover, the application of various types of self-assembly systems using amphiphilic biodegradable copolymers such as micro- or nanosized particles (microspheres, nanospheres, polymer micelles, nanogels, polymersomes), supramolecular physically interlocked systems, and stimuli-responsive systems for biomedical use such as drug delivery systems are also reviewed.

Y. Ohya (✉)
Department of Chemistry and Materials Engineering, Faculty of Chemistry, Materials and Bioengineering, Kansai University, 3-3-35 Yamate-cho, Suita, Osaka 564-8680, Japan

Organization for Research and Development of Innovative Science and Technology (ORDIST), Kansai University, 3-3-35 Yamate-cho, Suita, Osaka 564-8680, Japan
e-mail: yohya@kansai-u.ac.jp

A. Takahashi
Organization for Research and Development of Innovative Science and Technology (ORDIST), Kansai University, 3-3-35 Yamate-cho, Suita, Osaka 564-8680, Japan

K. Nagahama
Department of Nanobiochemistry, Frontiers of Innovative Research in Science and Technology (FIRST), Konan University, 7-1-20 Minatojima-Minamimachi, Chuo-ku, Kobe 650-0047, Japan

Keywords Aliphatic polyesters · Amphiphilic polymers · Polymer micelles · Self-assembly · Smart materials

Contents

Abbreviations

5FU	5-Fluorouracil
Ala	Alanine
Arg	Arginine
Asp	Aspartic acid
ATRP	Atom transfer radical polymerization
BSA	Bovine serum albumin
CD	Cyclodextrin
cDP	Cyclodepsipeptide
CHESG	Cholesterol group-modified enzymatically synthesized glycogen
CHP	Cholesterol–pullulan
CL	ε-Caprolactone
CMC	Critical micelle concentration
Cys	Cysteine
DDS	Drug delivery system
Dex	Dextran
DLA	D-Lactide
DMAE	Dimethylaminoethyl
DTT	Dithiothreitol
DXO	1,4-Dioxepan-5-one
DXR	Doxorubicin
EP	Ethyl ethylene phosphate
EPR	Enhanced permeability and retention

FDA	Food and Drug Administration
FITC	Fluorescein isothiocyanate
GA	Glycolide
Glc	Glycolic acid
Gly	Glycine
HA	Hyaluronic acid
HMDI	Hexamethylene diisocyanate
HPEI	Heparin–polyethyleneimine
IP	Injectable polymer
IPP	Isopropyl ethylene phosphate
LA	Lactide
LCST	Lower critical solution temperature
Leu	Leucine
LH-RH	Luteinizing hormone-releasing hormone
LLA	L-Lactide
LSEC	Liver sinusoidal endothelial cell
Lys	Lysine
MA	Malic acid
MDBE	Malide dibenzyl ester
MeO	Methoxy
ML	Mevalonolactone
mRNA	Messenger RNA
MS	Microsphere
NCA	N-carboxy anhydride
NIRF	Near-infrared fluorescence
NMR	Nuclear magnetic resonance
NS	Nanosphere
o/w	Oil in water
OCL	Oligocaprolactone
Oct	Octylate
P2VP	Poly(2-vinylpyridine)
PAA	Poly(acrylic acid)
PBD	Poly(butadiene)
PBS	Phosphate-buffered saline
PCL	Poly(ε-caplolactone)
PDLA	Poly(D-lactide)
PDP	Polydepsipeptide
PEE	Poly(ethyl ethylene)
PEG	Poly(ethylene glycol)
PEGMA	Poly(ethylene glycol) methacrylate
PEO	Poly(ethyle oxide)
PGA	Poly(glycolic acid), polyglycolide
PIC	Polyion complex
PLA	Poly(lactic acid), polylactide

PLGA	Poly(lactide-*co*-glycolide)
PLLA	Poly(L-lactide)
PMA	Poly(malic acid)
PMPC	Poly(2-methacryloyloxyethyl phosphorylcholine)
PNIPAAm	Poly(*N*-isopropylacrylamide)
pPRX	Pseudopolyrotaxane
PRX	Polyrotaxane
PS	Polystyrene
PTMC	Poly(trimethylene carbonate)
PTX	Paclitaxel
PVA	Poly(vinyl alcohol)
RES	Reticuloendothelial system
RGD	Arginine–glycine–aspartic acid
ROP	Ring-opening polymerization
SC	Stereocomplex
Ser	Serine
siRNA	Small interfering RNA
SMP	Shape-memory polymers
TCA	Tricarboxylic acid
T_g	Glass transition temperature
THF	Tetrahydrofuran
Thr	Threonine
T_m	Melting temperature
TMC	Trimethylene carbonate
TMS	Trimethylsilyl
Tyr	Tyrosine
w/o/w	Water in oil in water

1 Introduction

In biomedical fields, biodegradable polymers can be defined as polymers that would be degraded into low molecular weight compounds under physiological conditions or in the body within a significantly shorter period than a (usually human) lifetime. The importance of biodegradable polymers has increased more and more in biomedical fields [1–19] because biodegradable polymers can provide the following advantages compared with nondegradable polymers: (1) It is not necessary to remove the polymers from the body after their roles have been achieved. (2) The low molecular weight degradation products are expected to be metabolized or excreted, and not to cause long-term toxicity. (3) Degradation itself can offer some significant functions, for example, sustained release of drugs from biodegradable matrices can be achieved by degradation-dependent release mechanisms.

Biodegradable polymers can be categorized into natural and synthetic polymers. The typical examples of natural biodegradable polymers are proteins, polysaccharides, and nucleic acids. They are hydrophilic, and most of them are water-soluble. Proteins are polypeptides, whose amide (peptide) bonds are highly stable against spontaneous (nonenzymatic) hydrolysis under neutral conditions but can be degraded by enzymatic hydrolysis. For example, collagen, gelatin, and fibrin have been used as biodegradable medical material [20–22]. However, proteins of nonhuman origin may cause immunogenetic problems. In addition, both human-origin and animal-origin proteins may cause infection problems, such as bovine spongiform encephalopathy (BSE) and human immunodeficiency virus (HIV). Polysaccharides are highly hydrophilic polymers having many hydroxyl groups. Some polysaccharides have other functional groups, such as carboxylic acid, sulfate, amino, and acetamide groups. The glucoside (ether) bonds of polysaccharides are also highly stable against spontaneous (nonenzymatic) hydrolysis under neutral conditions but can be degraded by enzymatic hydrolysis. Polysaccharides are generally nontoxic and have no or low immunogeneticity, and are also used as biomedical materials. Nucleic acids (DNA and RNA) are highly water-soluble anionic polymers and have not been used very often as biomedical materials. The phosphorotriester linkage of DNA is relatively stable under physiological conditions, but is very sensitive to enzymatic (nuclease) degradation. These natural polymers are hydrophilic and not suitable for solid-state (bulk) materials having firm physical properties.

On the other hand, many synthetic biodegradable polymers have been developed for biomedical materials. Some of them are semicrystalline or noncrystalline polymers having strong physical properties, and can be used as biodegradable plastics. Typical examples of synthetic (artificial) biodegradable polymers are polyamides (including synthetic polypeptides), polyesters [23, 24], polyanhydrides [25–28], polycarbonates [29], poly(ortho ester)s [30–32], polyacetals [33, 34], polyphophazenes [35, 36], and polyphosphoesters [37–40]. They have various degradation rates and physical properties based on their molecular structures. In the design of biodegradable biomaterials, many important properties must be considered [5]. These materials must (1) not evoke a sustained inflammatory response; (2) possess a degradation time coinciding with their function; (3) have appropriate mechanical properties for their intended use; (4) produce nontoxic degradation products that can be readily resorbed or excreted; and (5) include appropriate permeability and processability for designed application.

Among these synthetic biodegradable polymers, aliphatic polyester [poly (hydroxyl acid)s] such as poly(lactic acid) (polylactide, PLA); poly(glycolic acid) (polyglycolide, PGA); poly(ε-caplolactone) (PCL); and their copolymers have been used often as implantable biomaterials (e.g., absorbable sutures, bone fixation materials, and drug delivery devices) because these aliphatic polyesters can provide favorable degradation rates, high mechanical properties, low- or nontoxic metabolizable degradation products, and are FDA-approved for clinical use [8–11].

Recently, biodegradable polymers have been used to fabricate macro- and nanometer scale self-assembled systems such as microspheres (MSs), nanospheres (NSs), polymer micelles, nanogels, and polymersomes (Fig. 1). These have attracted growing interest because of their potential utility for drug delivery systems (DDS), tissue engineering, and other applications. To construct these self-assembled systems

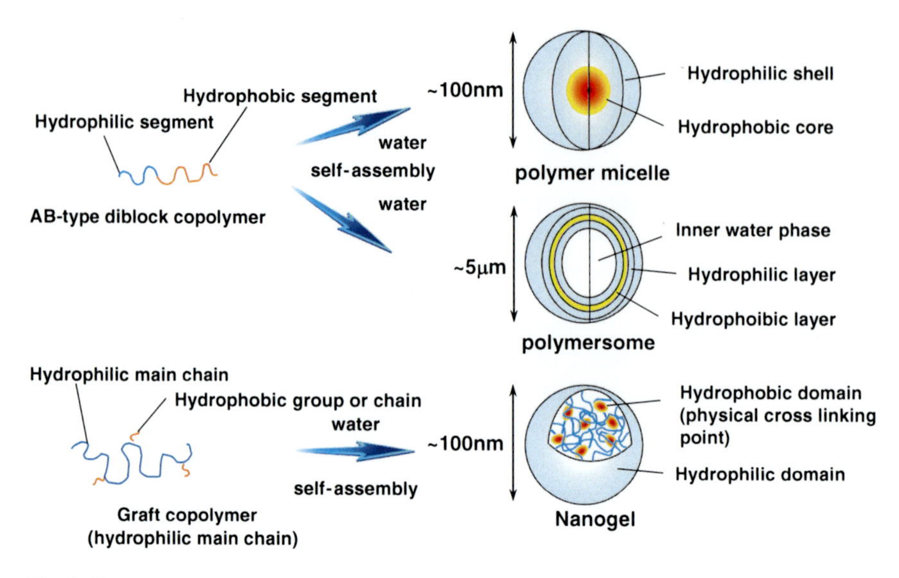

Fig. 1 Typical examples of nanometer-scale polymeric assemblies: polymer micelles, polymersomes, and nanogels

it is extremely important to have precise control of intermolecular noncovalent interactions such as hydrophobic interactions based on their molecular structures. In fact, most such noncovalent assembly systems are made of amphiphilic copolymers. Aliphatic polyesters are basically hydrophobic semicrystalline polymers having no reactive functional groups. Many methods for adding hydrophilicity and functionality to the aliphatic polyesters have been carried out by copolymerization with functional monomers or hybridization with other functional hydrophilic polymers [13–19].

This chapter focuses on biodegradable polymers, mainly aliphatic polyesters, and reviews the synthesis of amphiphilic biodegradable copolymers containing aliphatic polyesters as components. Moreover, the application of various types of self-assembled systems using amphiphilic biodegradable copolymers, such as micro- or nanosized particles (MSs, NSs, polymer micelles, nanogels, and polymersomes), supramolecular physically interlocked systems, and stimuli-responsive systems including physically crosslinked hydrogels for biomedical use such as DDS are also reviewed.

2 Synthesis of Biodegradable Amphiphilic Polyesters

2.1 Homopolymers and Random Copolymers

2.1.1 Aliphatic Polyesters

So far, many studies have focused on the development and application of aliphatic polyesters such as PLA [1–3], PGA [41, 42], and PCL [43, 44]. Figure 2 shows the structures of their monomers: lactides (LAs), glycolide (GA), ε-caprolactone (CL), and some typical comonomers.

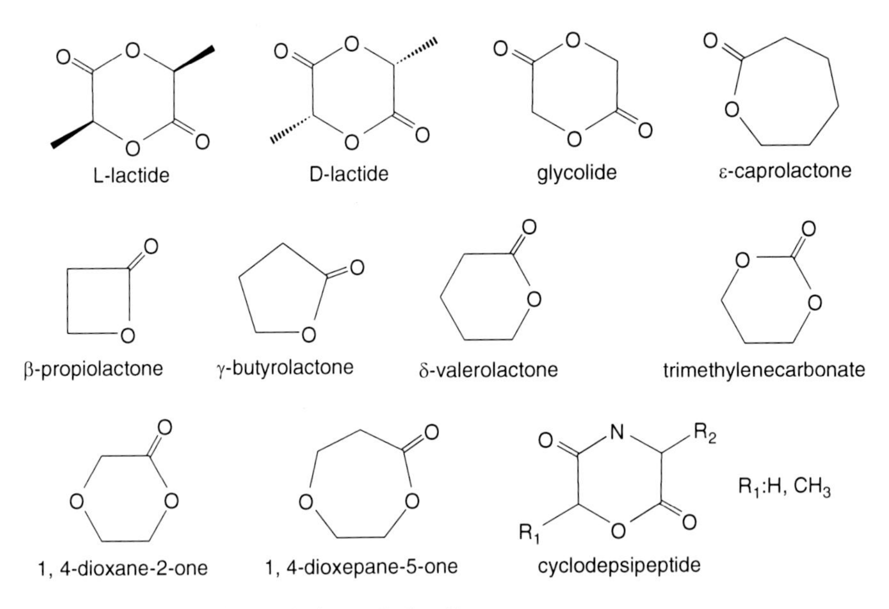

Fig. 2 Lactides, glycolide, and other typical cyclic monomers

PLA is one of the most popular aliphatic polyesters, and is obtained by ring-opening polymerization (ROP) of LA (the cyclic dimer of lactic acid, 2,6-dimethyl-1,4-dioxane-2,5-dione) using a catalyst or by direct polycondensation of lactic acid. The most popular catalyst for ROP of LA is organic Sn, typically $Sn(Oct)_2$. Most commercially available PLAs are usually prepared from the L,L-isomer of lactide (LLA) to give poly(L-lactide) (PLLA), because the naturally occurring lactic acid is the L-isomer produced from glucose by microorganisms. PLLA is a semicrystalline polymer with relatively high melting temperature (T_m), glass transition temperature (T_g), and mechanical strength. Because PLLA is FDA-approved, and can be degraded under physiological conditions to give L-lactic acid, which can be metabolized in the human body, it has been applied for implantable biomedical materials.

Since PLLA is a degradable polymer having relatively good biocompatibility, low toxicity and immunogenicity, and excellent mechanical properties, it has been used in a variety of applications in the pharmaceutical and biomedical fields, and as a degradable plastic for disposable consumer products. Recently, biodegradable cellular scaffolds for tissue regeneration using PLLA and other aliphatic polyesters have also been extensively studied. However, PLLA also has some disadvantages for biomedical use. It is a brittle and hard polymer with very low elongation to break and low compatibility with soft tissues. It is a hydrophobic polymer without reactive functional groups (besides its termini). The degradation of bulk semicrystalline PLLA takes a long time, and the control of degradation is not so easy. The properties of PLLA-based polymers can be modified by copolymerization (random, block, and graft), hybridization with other polymers, change in molecular architecture (branched, star-shaped, or dendrimers), and functionalization (end group and pendant group) with reactive and hydrophilic groups (hydroxyl, amino, carboxyl, thiol and so on). The physical properties such as crystallinity, T_m, T_g, hydrophobicity, and mechanical properties can be affected by such modifications to give somewhat controllable degradation rates.

PGA was one of the very first degradable polymers ever investigated for biomedical use. PGA found favor as a degradable suture, and has been actively used since 1970 [45–47]. Because PGA is poorly soluble in many common solvents, limited research has been conducted with PGA-based drug delivery devices. Instead, most recent research has focused on short-term tissue engineering scaffolds. PGA is often fabricated into a mesh network and has been used as a scaffold for bone [48–51], cartilage [52–54], tendon [55, 56], and tooth [57].

PCL is a semicrystalline polyester with great solubility in common organic solvents, a low T_m (55–60 °C), and low T_g (−54 °C) [58]. Because of PCL's very low in vivo degradation rate and high drug permeability, it has found favor as a long-term implant delivery device. PCL has low tensile strength (< 23 MPa), but very high elongation at breakage (4,700%), making it a very good elastic biomaterial [59]. PCL and PCL composites have also been used as tissue engineering scaffolds for the regeneration of bone [60–62], cartilage [63], skin [64], nerve [65], and other tissues.

These representative aliphatic polyesters are often used in copolymerized form in various combinations, for example, poly(lactide-*co*-glycolide) (PLGA) [66–68] and poly(lactide-*co*-caprolactone) [69–73], to improve degradation rates, mechanical properties, processability, and solubility by reducing crystallinity. Other monomers such as 1,4-dioxepan-5-one (DXO) [74–76], 1,4-dioxane-2-one [77], and trimethylene carbonate (TMC) [28] (Fig. 2) have also been used as comonomers to improve the hydrophobicity of the aliphatic polyesters as well as their degradability and mechanical properties.

2.1.2 Polyesters Having Reactive (Hydrophilic) Side-Chain Groups

To design amphiphilic and/or reactive copolymers containing aliphatic polyesters, one of the most promising approaches is copolymerization with functional monomers having protected reactive side-chain groups. Some kinds of monomers having reactive (hydrophilic) side-chain groups have been reported (Fig. 3). Recently, the synthesis of various types of functional polyesters has been reviewed [15–19].

Malic acid (MA) has two carboxylic acid groups and one hydroxyl group, and can be metabolized by the tricarboxylic acid (TCA) cycle. There are three different types of poly(malic acid)s (PMAs), poly(α-malic acid) (α-PMA), poly(β-malic acid) (β-PMA), and poly(α,β-malic acid) (α,β-PMA). α-PMA can be synthesized by ROP of the protected cyclic dimer malide dibenzyl ester (MDBE), and subsequent deprotection [78]. Although the molecular weight of the α-PMA homopolymer is less than 3,800 Da because of its low reactivity, copolymerization of MDBE with lactide was possible to give a higher molecular weight copolymer having reactive carboxylic acid side-chain groups [78]. The use of α-PMA as a water-soluble carrier

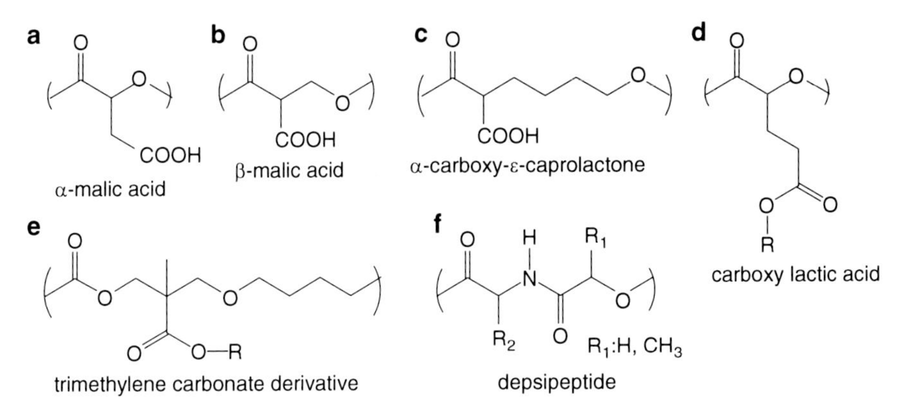

Fig. 3 Examples of monomer units having reactive side-chain groups, which can be copolymerized with polyesters: (**a**) α-malic acid, (**b**) β-malic acid, (**c**) α-carboxyl-ε-caprolactone, (**d**) carboxy lactic acid, (**e**) trimethylene carbonate derivative, and (**f**) depsipeptide

polymer for macromolecular prodrugs for anticancer drugs was achieved by attachment of anticancer drugs, 5-fluorouracil (5FU) and doxorubicin (DXR), utilizing the reactivity of side-chain groups [79, 80]. β-PMA was also synthesized by ROP of cyclic monomer [81] and its use as a drug carrier was also reported [82]. Kimura reported the synthesis of poly(malic acid-*co*-lactide) and poly(malic acid-*co*-glycolic acid) by ROP of 1:1 cyclic dimer of malic acid benzyl ester and lactic acid or glycolic acid (Glc) [83, 84]. They later reported the utility of the copolymer as a biodegradable scaffold for tissue engineering, exhibiting high cell-adhesive properties by immobilizing the cell-adhesive peptide arginine–glycine–aspartic acid (RGD) [85].

Lavasanifar and coworkers reported PCL-like copolymers, poly(α-carboxyl-ε-caprolactone), having reactive carboxylic acid side-chain groups [86, 87]. They reported the synthesis of block copolymers with poly(ethylene glycol) (PEG) and the use of the copolymers as micelle-forming reactive polymers for immobilizing hydrophobic anticancer drugs [88]. Nottelet reported the synthesis of LA-like and β-propiolactone-like cyclic monomers from glutamic acid (Glu) and aspartic acid (Asp), respectively, and the block copolymerization of them with LA [89]. Hedrick et al. reported poly(trimethylene carbonate) (PTMC)-like polymers bearing functional carboxylate side chains [90–92]. They reported the preparation of PTMC block copolymers immobilizing sugar units for DDS.

2.1.3 Polydepsipeptides

Copolymers of α-hydroxy acids and α-amino acids are one type of poly(ester-amide)s and are called polydepsipeptides (PDPs) [17]. Since some of natural occurring α-amino acids, typically Asp, Glu, lysine (Lys), cysteine (Cys), serine (Ser), and threonine (Thr), possess reactive (hydrophilic) side-chain groups, PDPs are highly valuable for providing functional amphiphilic biodegradable polyester-based materials. In other words, PDPs have both the functionality of polypeptides and the degradability of polyesters. These polymers contain both ester and amide groups in the chain, so their biodegradation behavior is different to that of the homopolymers.

Initially PDPs were synthesized by stepwise polycondensation of linear activated depsipeptide [93]. In 1985, Helder, Feijen and coworkers reported the synthesis of PDPs by ROP of a morpholine-2,5-dione derivative (cyclic dimer of α-hydroxy- and α-amino acid; cyclodepsipeptide, cDP) [94, 95]. The ROP method gives an alternative type of PDP by homopolymerization and also allows the copolymerization with other monomers (lactones and cyclic diesters) including LA, GA, and CL to give a wide variety of functional biodegradable materials. The synthesis of PDPs as functional biomaterials has been recently reviewed [17].

Several groups have tried to polymerize 3-alkyl-substituted morpholine-2,5-dione derivatives (3-alkyl substituted cDPs, i.e.,combinations of amino acids with alkyl side chains with glycolic acid or lactic acid) to synthesize aliphatic PDPs [96–99]. After these studies, the syntheses of PDPs and poly(DP-*co*-LA) s with reactive (hydrophilic) functional groups using amino acids (Asp, Glu, Lys, Cys, Ser) with protected reactive side-chain groups (–COOH, $-NH_2$, –SH, –OH), and subsequent deprotection, were reported by some groups including one of the authors (Fig. 4) [100–107]. Using these methods, aliphatic polyesters having reactive side-chain groups can be produced and utilized in various applications such as DDS and tissue engineering using the reactivity and hydrophilicity of the side-chain groups. Langer et al. reported copolymerization of cDP containing Lys as amino acid to give poly(DP-*co*-LA) [poly(Lys-LA)], and immobilization of RGD peptide on the PLA-based materials for a biodegradable cell-adhesive scaffold for tissue

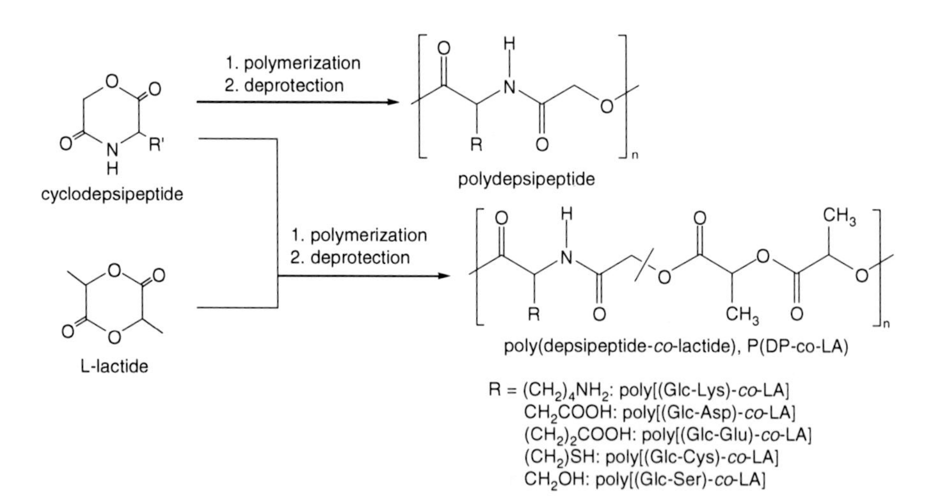

Fig. 4 Synthesis of polydepsipeptides and poly(depsipeptide-*co*-lactide)s having reactive side-chain groups

engineering [103–105]. We evaluated the poly(DP-*co*-LA)s having carboxylic acid and amino groups as scaffold for tissue engineering with controllable degradation rates, various physicochemical properties, and chemical modification abilities [108–111].

2.2 Block Copolymers

The ROP of cyclic esters, including LA, GA, CL, and cDPs, can be initiated with alcoholic hydroxyl groups so that biodegradable polyesters having terminal alcoholic residues can be obtained. Using this principle, block copolymers containing aliphatic polyester segments can easily be obtained by polymerization of cyclic esters with polymers having terminal hydroxyl groups as macro-initiators. Amphiphilic block copolymers can also be synthesized by a coupling reaction of hydrophobic polymers and hydrophilic polymers with reactive termini. Some amphiphilic block copolymers were also obtained using polyesters having terminal initiating groups. For example, alkylbromide can be used as initiating group for atom transfer radical polymerization (ATRP) to synthesize vinyl-type block copolymers, and primary amino groups can be used for ROP of amino acid *N*-carboxy anhydride (NCA) to synthesized polypeptide-*block*-polyesters. Aliphatic polyesters are usually hydrophobic and combine with hydrophilic polymers to give amphiphilic block copolymers. Such amphiphilic block copolymers have been used in biomedical fields as polymeric micelles, temperature-responsive materials and so on (see Sects. 4 and 8). The preparation and application of PLA-based amphiphilic block copolymers were reviewed recently [16].

The most popular hydrophilic polymers used in conjugation with aliphatic polyesters to prepare amphiphilic block copolymers are polyethers, especially PEG, which can also be called polyoxyethylene or poly(ethylene oxide) (PEO). PEG is nonionic, nonimmunogenetic, nontoxic and biocompatible, soluble in water and common organic solvents, and FDA-approved for clinical use. PEG can provide a prolonged blood circulation and diminished reticuloendothelial system (RES) uptake [112]. Although PEG is not biodegradable, low molecular weight (below ca. 30,000 Da) PEG can be excreted mainly from the kidneys. Because of such favorable properties, PEG has been used in the biomedical field. There are many reports on the preparation of amphiphilic block copolymers of aliphatic polyesters and polyethers that include PEG [113–134]. Most of the research on amphiphilic block copolymers of aliphatic polyesters and polyethers has been carried out using AB-type diblock copolymers and ABA (hydrophobic–hydrophilic–hydrophobic)-type or BAB (hydrophilic–hydrophobic–hydrophilic)-type triblock copolymers. In addition, multiblock copolymers of polyesters and polyethers have also been synthesized, mainly by polycondensation of the hydrophilic and hydrophobic segments, and investigated as biomedical materials such as antiadhesive membranes and temperature-responsive injectable polymers [135–137].

AB-type diblock copolymers of PDP and polyester can be synthesized by two-step polymerization using a basically similar strategy to that outlined above. After first step polymerizaiton to obtain a polymer with terminal hydroxyl group, the hydroxyl group on the terminal can be used as initiating group for second step polymerization to give an AB-type diblock copolymer. Ouchi et al. synthesized amphiphilic AB-type diblock copolymers of PLA and PDP with reactive side-chain groups [138–140]. They also reported the preparation of polymer micelles using the PDP-*b*-PLA [poly(Glc–Lys)-*b*-PLA] [138, 139].

Several polypeptides possess pendent functional groups, including carboxylic acid (–COOH), thiol (–SH), and amine ($-NH_2$). These functional groups are further utilized for bioconjugation and shell-crosslinking of polymer micelles. A facile method for preparation of polypeptide-*b*-polyester is ROP of NCAs using primary amino group-terminated polyester. We reported on the preparation of AB-type polypeptide-*b*-PLA containing Asp residues with a reactive carboxylic acid group using amino-terminated PLA (NH_2-PLA) [141] and the subsequent formation of polymer micelles with a positively charged shell [142, 143].

Polysaccharides are hydrophilic natural polymers that can be degraded enzymatically. Block copolymers containing polysaccharide as a block were reviewed recently [144]. The synthesis of block copolymers of polysaccharides and aliphatic polyesters has also been tried. But, many successful results were not reported because the reactivity of many hydroxyl groups on polysaccharides was an obstacle to the ROP of cyclic polyester or coupling reactions using terminal-activated polysaccharides. Li and Zhang reported the synthesis of maltoheptaose-*b*-PCL copolymers by ROP [145]. Even though the short oligosaccharide segment made of seven units may not be considered as a true polymer chain, the chemistry devised by them should be easily applicable to longer saccharidic chains. Liu and Zhang used the Michael reaction for coupling a dextran (Dex) with an amino-functionalized terminal and acrylolyl end-capped PCL [146]. Sun et al. also synthesized Dex-*b*-PCL by disulfide bond formation [147].

Some other degradable (i.e., nonvinyl-type) polymers have been reported as components for amphiphilic block copolymers. For example, Hsiue reported the synthesis of a block copolymer of poly(2-ethyl oxazoline) and PLA by ROP. They reported the use of ABA-type triblock copolymers as pH-responsive polymer micelles [148].

Many kinds of nonbiodegradable vinyl-type hydrophilic polymers were also used in combination with aliphatic polyesters to prepare amphiphilic block copolymers. Two typical examples of the vinyl-polymers used are poly(*N*-isopropylacrylamide) (PNIPAAm) [149–152] and poly(2-methacryloyloxyethyl phosphorylcholine) (PMPC) [153]. PNIPAAm is well known as a temperature-responsive polymer and has been used in biomedicine to provide smart materials. Temperature-responsive nanoparticles or polymer micelles could be prepared using PNIPAAm-*b*-PLA block copolymers [149–152]. PMPC is also a well-known biocompatible polymer that suppresses protein adsorption and platelet adhesion, and has been used as the hydrophilic outer shell of polymer micelles consisting of a block copolymer of PMPC-*co*-PLA [153]. Many other vinyl-type polymers used for PLA-based amphiphilic block copolymers were also introduced in a recent review [16].

2.3 *Graft Copolymers and Branched Polymers*

2.3.1 Graft Copolymers

Graft copolymers have three different parameters: length of main chain, length of graft chain, and number of graft chains. So, precise control of properties such as hydrophilic/hydrophobic balance and crystallinity may be achieved by changing the molecular structure. Graft copolymers having a aliphatic polyester as either main chain or graft chain have been reported.

To synthesize graft copolymers having aliphatic polyesters as graft chains, any of three general methods, "grafting through," "grafting from" and "grafting to," can be used. "Grafting through" can be achieved by copolymerization of a macro-monomer (an aliphatic polyester having a terminal vinyl group) with other monomers to give graft copolymers having vinyl-type main chains. "Grafting from" can be applied by ROP of cyclic esters using polymers having hydroxyl side-chain groups as multifunctional macro-initiators. "Grafting to" is a coupling reaction of aliphatic polyesters having reactive termini onto main chain polymers having reactive side chains.

Vinyl-type graft copolymers having aliphatic polyester graft chains were commonly synthesized by the "grafting through" method using aliphatic polyesters having terminal acryloyl or mathacryloyl groups. PNIPAAm-*g*-PLA and PMPC-*g*-PLA can be synthesized by this method [154, 155]. Many other examples of such graft copolymers were introduced in recent reviews [13, 16].

Aliphatic polyester-grafted polysaccharides can be synthesized by both "grafting from" and "grafting to" methods. We reported preparation of PLA-grafted polysaccharides, pullulan, amylose, and Dex, by "grafting from" using protected polysaccharides as macro-initiators (Fig. 5) [156–161]. Polysaccharides are highly hydrophilic polymers and are not soluble in either common organic solvents or melted cyclic ester monomers. So, the grafting polymerization must be heterogeneous using polysaccharide as a macro-initiator. In addition, since polysaccharides have too many hydroxyl groups, the control of grafting numbers and sites on the polysaccharide is very difficult. We employed trimethylsilyl (TMS) as a protecting group, which can be easily removed after graft polymerization. Most hydroxyl groups were protected by TMS groups, and the obtained TMS–polysaccharide was soluble in common organic solvent such as THF. Graft ROP of LA in THF solution proceeded using alkali metal as a catalyst, and the residual (nonreacted) hydroxyl groups of the polysaccharides were used as initiating groups. After removal of the TMS groups under mild conditions, PLA-grafted polysaccharides were obtained. The obtained amphiphilic PLA-grafted polysaccharides showed interesting properties such as tenacious mechanical properties, microphase separated structures, accelerated degradation behavior, and anti-cell-adhesive properties, depending on the polysaccharide content [156–161]. PLA-grafted polysaccharides having relatively short graft chains can also be synthesized by coupling using a "grafting to" method [162–167]. The obtained oligolactide-grafted polysaccharides

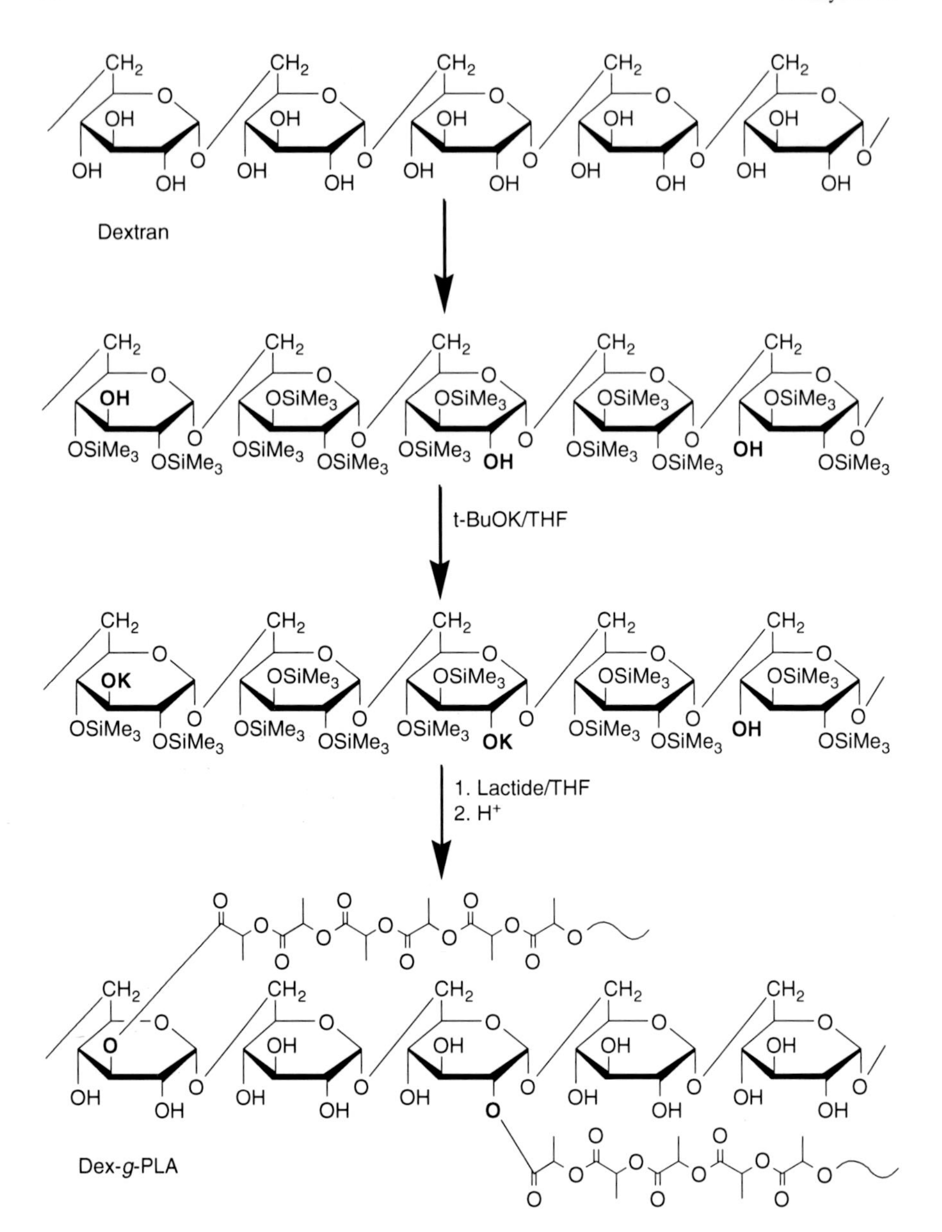

Fig. 5 Synthesis of polylactide-grafted polysaccharide (Dex-*g*-PLA) by trimethylsilyl-protection method

can be utilized as biodegradable nanogels in aqueous solution (see Sect. 6) or as hydrogel-forming biodegradable materials.

Polyglycidol is a polymer having a PEG-like backbone and hydroxyl side chains. We also synthesized PLA-grafted polyglycidol (polyglycidol-*g*-PLA) using a "grafting from" method without any protecting groups to give highly grafted

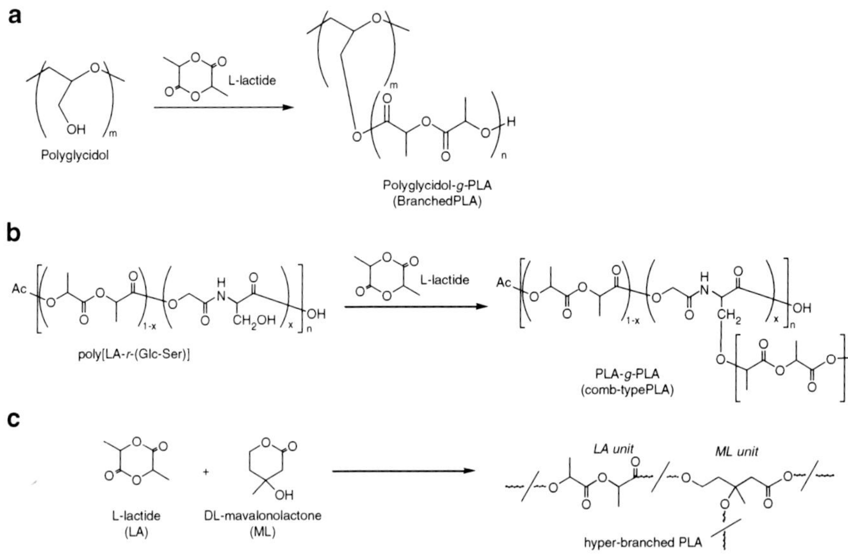

Fig. 6 Synthesis of branched PLAs: (**a**) polyglycidol-*g*-PLA, (**b**) PLA-*g*-PLA (comb-type PLA), and (**c**) hyperbranched PLA

polymer as a kind of branched PLA (Fig. 6a) [168]. The obtained branched PLA showed modified degradability and physical properties.

Graft copolymers having aliphatic polyester main chains can be synthesized by "grafting to" methods. For example, PLA-*g*-PEG having hydrophobic PLA-based main chain and hydrophilic PEG graft chains could be obtained by a coupling reaction of poly(DP-*co*-LA) having carboxylic acid side chains [poly(Glc–Asp)-*co*-LA] with one terminal reactive PEG (MeO-PEG-OH) [111]. The obtained amphiphilic graft copolymers showed temperature-responsive sol–gel transition behavior.

2.3.2 Branched Polymers

Introduction of branched structures is an effective method for modifying the physical properties of polymers by changing crystallinity, hydrodynamic diameter, and entanglement of polymer chains. Branched PLA, a graft polymer where both the main chain and graft chains are PLA (i.e., PLA-*g*-PLA or comb-type PLA), could be synthesized by a "grafting from" method using poly(DP-*co*-LA) having hydroxyl side chains [poly(Glc–Ser)-*co*-LA] [107, 169] (Fig. 6b). The obtained branched PLAs exhibited controllable degradability and physical properties based on their branched structures. Branched aliphatic polyesters can be synthesized by polyol compounds such as penta-erythritol [170] as multifunctional initiators to give star-shaped polyesters. Multiarmed PEG (four-arm, eight-arm) [171–173] and cyclodextrins (CDs) [174] can also be used as multifunctional initiators to give star-shaped block copolymers of PEG and PLA and star-shaped PLAs having a CD core. Mevalonolactone (ML) is a bifunctional monomer having a cyclic lactone structure and hydroxyl groups, and can act as both monomer and initiating group. The author tried to synthesize hyperbranched PLA by copolymerization of ML with LA (Fig. 6c) [175]. Some branched PLA could be obtained, but the molecular weights of the polymers were not so high because of low activity of ML.

3 Microspheres and Nanospheres

Micro- or nanosized polymer particles are generally called microspheres (MSs) or nanospheres (NSs), respectively, and have been used for DDS. The term "nanoparticle" is more general and includes polymer micelles and nanogels, which are described in Sects. 4–6. Although polymer micelles and nanogels have sufficient surface hydrated layers for dispersion or solubilizaton in aqueous media, MSs and NSs are basically spherical particles of hydrophobic polymers without enough hydrated layers.

MSs and NSs of aliphatic polyesters are generally prepared by emulsion–solvent evaporation methods. MSs containing lipophilic drugs are prepared by oil in water (o/w) emulsion methods. The polymer solution containing lipophilic drugs in organic solvent (which is not miscible with water and has relatively low boiling point, typically dichloromethane) was poured into a larger amount of water or buffer solution with vigorous stirring and/or sonication. The obtained emulsion was evaporated to remove organic solvent. After centrifugation, washing and drying, the MSs containing drugs would be obtained. MSs containing hydrophilic (water-soluble) drugs can be prepared by a water in oil in water (w/o/w) double emulsion method (Fig. 7). A small amount of aqueous solution containing water-soluble drugs was added to organic solvent containing polymer with vigorous stirring and/or sonication to give the primary w/o emulsion. The obtained primary emulsion was

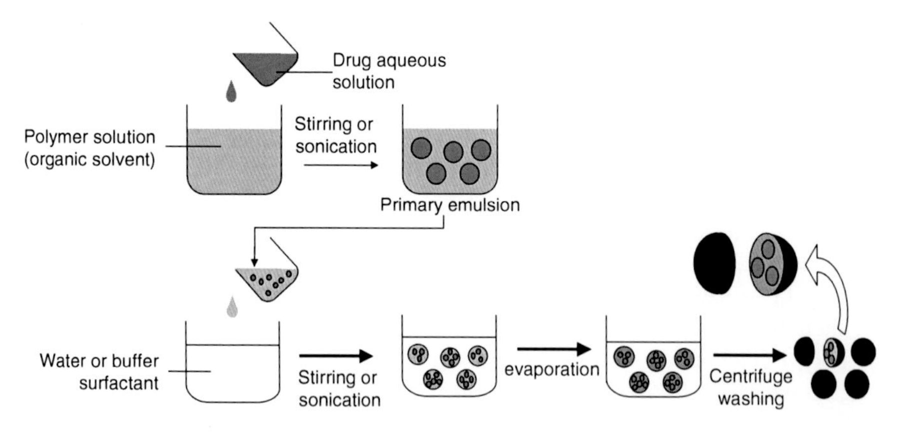

Fig. 7 Preparation of water-soluble drug-loaded microspheres using w/o/w emulsion methods

then poured into a larger amount of water to give a w/o/w emulsion. After evaporation, centrifugation, washing and drying, the MSs containing hydrophilic drugs in the internal aqueous phase would be obtained.

Biodegradable aliphatic polyesters MSs have been used for drug delivery devices that release drug in sustained manner. The most famous product is Lupron Depot, which consists of 20-μm MSs of PLGA (LA:GA = 75:25) prepared by a w/o/w emulsion method and releases luteinizing hormone-releasing hormone (LH-RH) superagonist (leuprorelin acetate) for endocrine diseases, including prostate cancer [176–179].

We could prepare MSs having reactive surfaces using poly(DP-*co*-LA)s, i.e., poly[(Glc–Lys)-*co*-LA] and poly[(Glc–Asp)-*co*-LA], having reactive amino or carboxylic acid groups [180]. In an o/w emulsion system, the hydrophilic amino and carboxylic acid groups of the polymers were condensed at the interface of aqueous and organic phases to give surface functionality to the MSs. We then reported the introduction of saccharide (galactose) residues on the MSs to give cell-specific recognition ability using the reactive functional groups on the surfaces [180].

In o/w and w/o/w emulsion methods, amphiphilic molecules or polymers, typically poly(vinyl alcohol) (PVA), are usually added to the aqueous phase as emulsion stabilizers (surfactants), to stabilize aqueous and oil phase droplets. To obtain NSs having smaller (nanometer order) diameters compared with MSs, the conditions of stirring and sonication, and the addition of surfactants are very important. Amphiphilic biodegradable polymers can be used as biodegradable surfactants instead of nonbiodegradable PVA. We reported the use of PDP-*b*-PLAs having hydrophilic PDP segments as biodegradable surfactants to obtain NSs and MSs with functionalized surfaces [181, 182]. Using PDP-*b*-PLAs having hydrophilic amino or carboxylic acid groups (i.e., poly(Glc–Lys)-*b*-PLA and poly (Glc–Asp)-*b*-PLA) as biodegradable surfactants, NSs of aliphatic polyester could be obtained without PVA by an o/w emulsion method [181]. In addition, fine and uniform distribution of internal aqueous phase containing water-soluble drugs could be achieved by the addition of the PDP-*b*-PLAs as biodegradable surfactants to the w/o/w emulsion. The fine and uniform distribution of aqueous phase containing bovine serum albumin (BSA) as model protein drug led to sustained release of the protein from the obtained MSs (Fig. 8) [182]. We also reported the use of amphiphilic PLA-grafted polysaccharides to prepare MSs containing hydrophilic drugs. We prepared MSs containing fluorescein isothiocyanate(FITC)-labeled BSA as a model drug using Dex-*g*-PLLA, and investigated the distribution of FITC-BSA in the MSs and its release behavior. The efficient entrapment and uniform distribution of FITC-BSA in the MSs compared with PLLA could be achieved [183]. In addition, the application of MSs prepared from block copolymers of PEG and PLA for a vaccine delivery system has been described in a short review [184].

The addition of amphiphilic polymer in o/w or w/o/w emulsion systems is also useful to give surface functionality to MSs and NSs. Liang et al. reported preparation of NSs from poly(γ-glutamic acid)-*b*-PLA, immobilizing galactose residues on the surface by an o/w emulsion method, and in vivo specific delivery of aniticancer

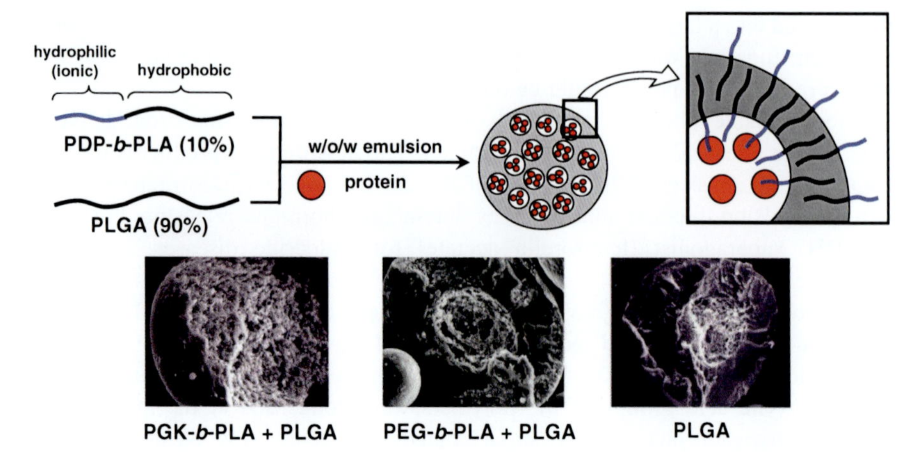

Fig. 8 Preparation of biodegradable microspheres entrapping proteins using amphiphilic PDP-*b*-PLA block copolymers as biodegradable polymeric surfactants, and SEM images of their cross-sections. Reprinted from [182] with permission

drug (paclitaxel; PTX) to liver [185]. Saeed et al. reported the ATRP of poly (ethylene glycol) methacrylate (PEGMA) using PLGA having a bromide terminal to give PLGA-based amphiphilic block copolymer. They then prepared NSs from the block copolymer and PLGA, immobilizing a folic acid derivative as a homing device by click chemistry, and reported that the obtained multifunctional NSs was useful for gene delivery [186]. Liu et al. reported preparation of MSs from PNIPAAm-*b*-PLA block copolymer containing BSA as model drug, and showed that the release rate of BSA from the MSs could be accelerated by temperature change [187].

4 Polymer Micelles

Polymer micelles are nanometer sized (usually several tens of nanometers) self-assembled particles having a hydrophobic core and hydrophilic outer shell composed of amphiphilic AB- or ABA-type block copolymers, and are utilized as drug delivery vehicles. The first polymer micelle-type drug delivery vehicle was made of PEG-*b*-poly(aspartic acid) (PEG-*b*-PAsp), immobilizing the hydrophobic anticancer drug DXR [188–191]. After this achievement by Kataoka et al., a great amount of research on polymer micelles has been carried out, and there are several reviews available on the subject [192–194].

As described above, the most common hydrophilic polymer combined with aliphatic polyesters to prepare polymer micelles is PEG. Although there have been many reports on the polymer micelles of PEG-*b*-aliphatic polyesters, only few recent examples are introduced in this review. Shin et al. reported the therapeutic potential of PEG-*b*-PLA micelles entrapping multiple anticancer drugs of poor solubility in water [195]. Mikhail and Allen prepared PEG-*b*-PCL covalently attached to docetaxel and investigated the morphology of the self-assembled structure. They reported the release behavior of the drug from the micelles, comparing with block copolymer micelles physically entrapping docetaxel [196]. Ding et al. synthesized PEG-*b*-PLA having protoporphyrin IX residues at its terminal, and investigated the micelle formation and its application to photodynamic therapy [197]. Hedrick et al. reported a block copolymer of PEG and PTMC having polar side-chain groups and investigated the effect of intramicelle hydrogen bonding on the stability of polymer micelles, their drug loading efficiency, and the cytotoxic activity of the drug-loaded polymer micelles [198].

Of course, many other hydrophilic polymers have been used as outer shell-forming segments in polymer micelle systems. This review focuses on polymer micelle systems using biodegradable polymers for both core-forming and shell-forming segments. Researches on the polymer micelles formed by combination of aliphatic polyesters and nonbiodegradable hydrophilic polymers have been reviewed in the literature [16]. Sun et al. synthesized Dex-*b*-PCL by disulfide bond formation [146]. They reported the micelle formation of the block copolymer and efficient intracellular drug release by cleavage of the disulfide bond under the reductive conditions of the cytosol. Nottelet et al. reported the preparation of fully biodegradable polymer micelles of block copolymers of carboxylic acid-functionalized PLA and PLA [89]. Wang et al. reported the micelle and vesicle formation of polyphosphate-*b*-PCL and their potential utility as cellular delivery vehicles for anticancer drugs [199]. Liu et al. also reported a polymer micelle system of star-shaped polyphosphate-*b*-PLA having disulfide linkages, and its efficient cellular delivery of DXR [200]. Ouchi et al. reported use of PDP having amino or carboxylic acid groups as hydrophilic segments for formation of biodegradable micelles [138, 139]. They reported the entrapment and release behavior of DXR from polymer micelles composed of PDP-*b*-PLA. We reported the preparation of negatively charged biodegradable polymeric micelles consisting of polypeptide-*b*-PLA (PAsp-*b*-PLLA) [142, 143].

The advantages of polymer micelles are their small size and core–shell structure, which protects bioactive agents entrapped in the core by a hydrophilic polymer shell. Such polymeric micelles can escape rapid renal excretion, and display long circulation times after administration in the body. However, all physically assembled polymeric micelles have a drawback of easy dissociation in the body fluids because of instability under extremely diluted conditions below the critical micelle concentration (CMC). Such a dissociation behavior leads to unfavorably rapid release of the bioactive agents and interferes with site-specific transport of the micelles to a target site. We reported the preparation of polyanion-coated biodegradable polymeric micelles by coating positively charged polymeric micelles consisting of poly(L-lysine)-*block*-poly(L-lactide) (PLys-*b*-PLLA) AB diblock copolymers with anionic hyaluronic acid (HA) by polyion complex (PIC) formation. The obtained HA-coated micelles showed significantly higher stability in aqueous solution (Fig. 9) [201]. The HA-coated micelles showed sustained release of model drugs and low cytotoxicity. It is known that there are receptors for HA on liver sinusoidal endothelial cells (LSECs). Specific interactions of

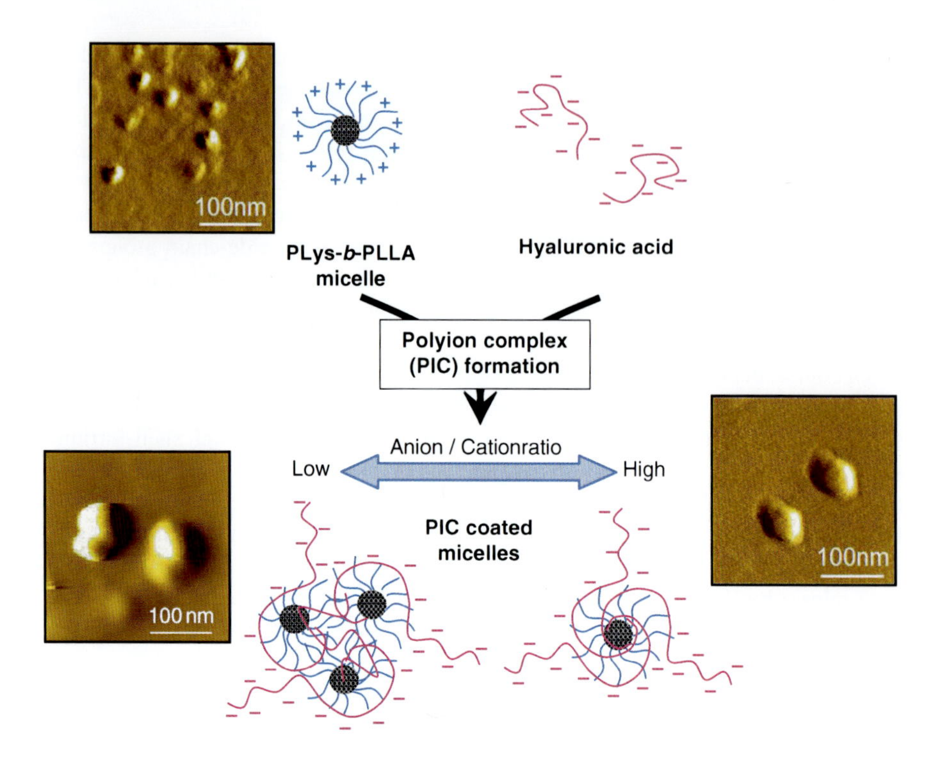

Fig. 9 Preparation of PIC-coated polymeric micelles and their atomic force microscope (AFM) images. Reprinted from [201] with permission

HA-coated micelles with LSECs and Kupffer cells were investigated. The HA-coated micelles were taken up only into LSECs. These results suggest the potential utility of the HA-coated micelles as highly stable drug delivery vehicles exhibiting specific accumulation into LSECs [202].

5 Polymersomes

Hollow capsules made of polymer-based [203, 204] and/or lipid-based [205] amphiphiles are of great interest because of their tremendous potential applications in medicine, pharmacy, and biotechnology. Generally, such capsules have fluid-filled membranes that consist of their hydrophobic part, separating the core from the outside medium. Together with micelles, they are the most common and stable structures of amphiphiles in water. However, unlike micelles, which mostly load hydrophobic molecules in the hydrophobic core, hollow capsules can encapsulate hydrophilic molecules such as dyes, proteins, and nucleic acids in the aqueous interior and also integrate hydrophobic molecules in the hydrophobic membrane. Indeed, various utilities of hollow capsules as drug- and gene-delivery carriers

[206–208], artificial cells [209], bioreactors [210, 211], and bioimaging tools [212, 213] have been reported.

The most versatile method to prepare such hollow capsules is self-assembly [203–205, 214, 215]. Owing to their amphiphilic nature and molecular geometry, lipid-based amphiphiles can aggregate into spherical closed bilayer structures in water: so-called liposomes. It is quite reasonable that the hollow sphere structure of liposomes makes them suitable as precursors for the preparation of more functional capsules via modification of the surfaces with polymers and ligand molecules [205, 216, 217]. Indeed, numerous studies based on liposomes in this context have been performed [205, 209, 213].

On the other hand, polymer-based amphiphiles, in particular amphiphilic block copolymers composed of a hydrophobic block and a hydrophilic block with optimized lengths, can be self-assembled into bilayer structures in aqueous solution: so-called polymersomes [203]. Preparation of polymersomes was first reported by Discher and coworkers in 1999 [218]. Polymersome preparation is similar to that of liposomes, usually using a film hydration technique or simple direct dissolution technique as described in the literatures [218, 219]. Since the size of polymersomes is mainly governed by the volume fraction, which is defined as the relative hydrodynamic volume ratio of hydrophilic block to the total copolymer chain, polymersomes can be tuned to sizes ranging from nano- to micrometers by modifying the polymer structures [207, 220]. Compared with liposomes, polymersomes possess several advantageous properties. The membrane thickness of polymersomes determines their properties such as elasticity, permeability, and mechanical stability [218]. Owing to the higher molecular weight of the polymers as compared to lipids, the membrane of polymersomes is generally thicker and tougher, and the membrane makes them more stable and less permeable than conventional liposomes [221]. These characteristics enhance the benefit of polymersomes, especially in drug and gene delivery systems for in vivo use, because they result in stable blood circulating properties and a decreased rate of drug release.

Polymersomes can be prepared from various amphiphilic block copolymers, for example, poly(ethylene glycol)-*b*-poly(ethyl ethylene) (PEG-*b*-PEE) [218], poly(acrylic acid)-*b*-polystyrene (PAA-*b*-PS) [222], poly(ethylene glycol)-*b*-poly (2-vinylpyridine) (PEG-*b*-P2VP) [223], poly(glutamic acid)-*b*-poly(butadiene) (PGlu-*b*-PBD) [224], and poly(ethylene glycol)-*b*-poly(*N*-isopropylacrylamide) (PEG-*b*-PNIPAAm) [225]. PEG is a common choice as hydrophilic block of copolymers that self-assemble into polymersomes, because PEG is noted for its biocompatibility and resistance to both protein adsorption and cellular adhesion, resulting in a prolonged blood circulation time for such PEG-based polymersomes.

For therapeutic applications, polymersomes preferably should be biodegradable as well as biocompatible. Biodegradable polymersomes offer several advantageous properties compared to nonbiodegradable polymersomes, such as facilitation of the sustained release of encapsulated molecules and improved safeness through removal of empty vehicles after the release of drugs. Accordingly, biodegradable polymersomes have been prepared using block copolymers of PEG as hydrophilic

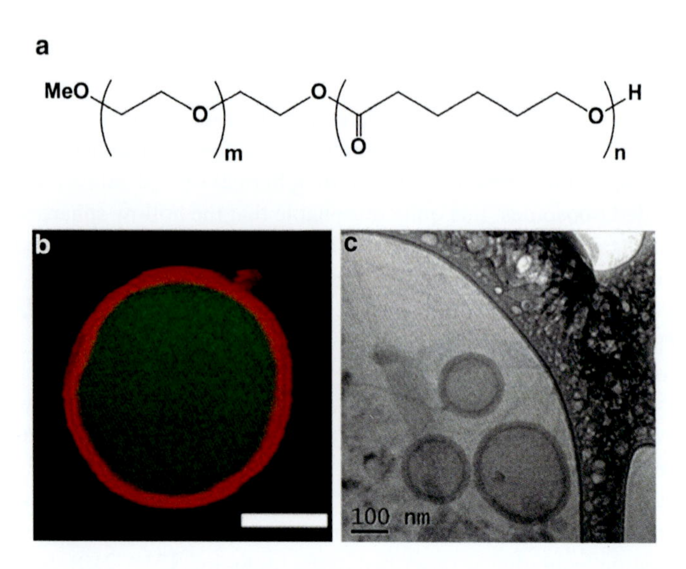

Fig. 10 (**a**) Chemical structure of PEG-*b*-PCL copolymer. (**b**) CLSM image of PEG-*b*-PCL polymersomes containing membrane-encapsulated Nile Red (2 mol%) and aqueous entrapped Calcein dyes. Scale bar: 5 μm. (**c**) Cryo-TEM image of PEG-*b*-PCL polymersomes. Scale bar: 100 nm. Reprinted from [228] with permission

block and aliphatic polyesters as hydrophobic block. PEG-*b*-PLA [226, 227] and PEG-*b*-PCL (Fig. 10a) [228] were developed by the groups of Feijen and Hammer, respectively. As described above, the block copolymers can be synthesized by ROP of cyclic esters in the presence of mono-hydroxyl-terminated PEG as a macro-initiator. The molecular weight of polyester blocks can be tuned by control of the feed molar ratio of cyclic esters to PEG in the polymerization process. Thus, the volume fraction of block copolymers can be tailored to self-assemble into polymersomes. In addition to regulation of volume fraction of the copolymers, the size distributions of the polymersomes can be roughly controlled with standard techniques such as sonication, freeze/thaw extraction, and extraction at above the T_g of polyester blocks to give monodispersed vesicular structure with diameters ranging from nano- to micrometers (Fig. 10b,c), which are useful for in vivo applications [228].

Hammer and coworkers prepared PEG-*b*-PCL polymersomes entrapping DXR (Fig. 11a). The release of DXR from the polymersomes was in a sustained manner over 14 days at 37 °C in PBS via drug permeation through the PCL membrane, and hydrolytic degradation of the PCL membrane [228]. The release rate of encapsulated molecules from polymersomes can be tuned by blending with another type of block copolymer [229]. Indeed, the release rate of encapsulated DXR from polymersomes prepared from mixtures of PEG-*b*-PLA with PEG-*b*-PBD copolymers increased linearly with the molar ratio of PEG-*b*-PLA in acidic media (Fig. 11b). Under acidic conditions, the PLA first underwent hydrolysis and, hours later, pores formed in the membrane followed by final membrane

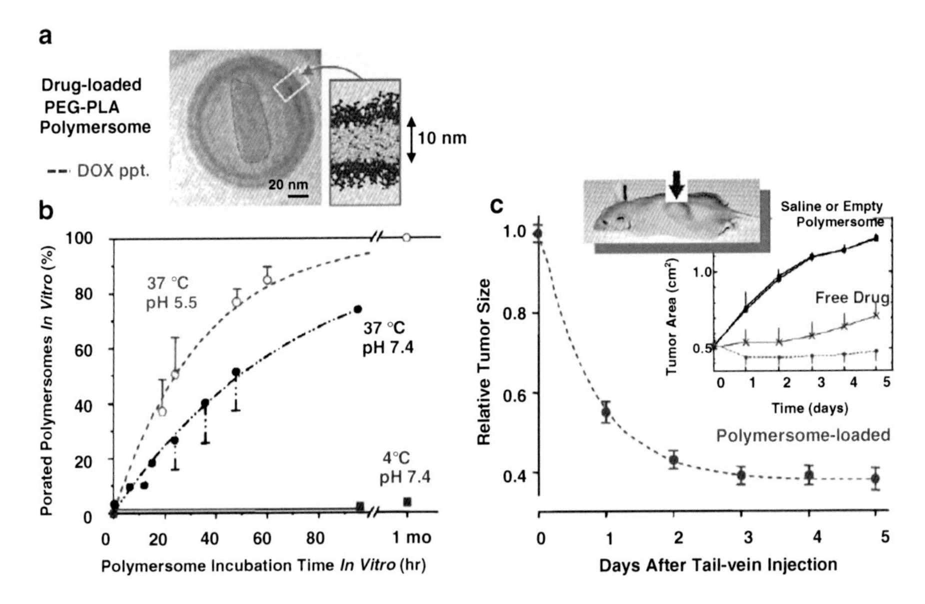

Fig. 11 Drug loading, release, and antitumor activity of biodegradable polymersomes. (**a**) Cryo-TEM image of DXR- and PTX-loaded PEG-*b*-PLA/PEG-*b*-PBD polymersomes. (**b**) PEG-*b*-PLA/PEG-*b*-PBD polymersomes visibly porate and release encapsulates in isotonic PBS, pH 7.4 at 37 °C and even faster in isotonic HEPES, pH 5.5 at 37 °C, but the polymersomes are stable in PBS at 4 °C. (**c**) Solid tumors shrink after a single injection of (DXR + PTX)-loaded polymersomes. Reprinted from [230] with permission

disintegration. Cellular uptake studies of the blend polymersomes showed that these polymersomes were hydrolyzed within the endolysosomal compartments and released their contents [230]. Furthermore, in vivo studies demonstrated growth arrest and shrinkage of rapidly growing tumors after intravenous injection of the polymersomes (Fig. 11c). Combination therapy with DXR- and PTX-loaded polymersomes triggered apoptosis in the tumors. Apoptosis was enhanced twofold with polymersome-delivered drug compare with free drug.

In addition to low molecular weight drugs, PEG-*b*-polyester polymersomes can encapsulate water-soluble macromolecular drugs, e.g. proteins and nucleic acids, into the hydrophilic interior space. Indeed, PEG-*b*-PLA and PEG-*b*-PCL polymersomes have been reported as potential oxygen nanocarriers by Palmer and coworkers [231]. Hemoglobin-loading efficiencies of up to 20% were obtained for polymersomes prepared from PEG-*b*-PLA and PEG-*b*-PCL copolymers with shorter hydrophobic polyester blocks. Furthermore, oxygen affinity, cooperativity coefficient, and methemoglobin level of the polymersomes were consistent with values required for efficient oxygen delivery in the systemic circulation. Discher and coworkers have demonstrated loading and functional delivery of siRNA with polymersomes of nonionic PEG-*b*-PLA copolymers [232]. The biodegradable polymersomes are taken up passively by cultured cells, after which the vesicles transform into micelles that allow endolysosomal escape and delivery of siRNA into cytosol for mRNA knockdown. Polymersome-mediated knockdown appears as efficient as common cationic-lipid transfection reagents such as Lipofectamine 2000, and about half as effective as Lenti-virus after sustained selection.

Recently, we have also prepared nanosized polymersomes through self-assembly of star-shaped PEG-*b*-PLLA block copolymers (eight-arm PEG-*b*-PLLA) using a film hydration technique [233]. The polymersomes can encapsulate FITC-labeled Dex, as model of a water-soluble macromolecular drug, into the hydrophilic interior space. The eight-arm PEG-*b*-PLLA polymersomes showed relatively high stability compared to that of polymersomes of linear PEG-*b*-PLLA copolymers with the equal volume fraction. Furthermore, we have developed a novel type of polymersome of amphiphilic polyrotaxane (PRX) composed of PLLA-*b*-PEG-*b*-PLLA triblock copolymer and α-cyclodextrin (α-CD) [234]. These polymersomes possess unique structures: the surface is covered by PRX structures with multiple α-CDs threaded onto the PEG chain. Since the α-CDs are not covalently bound to the PEG chain, they can slide and rotate along the PEG chain, which forms the outer shell of the polymersomes [235, 236]. Thus, the polymersomes could be a novel functional biomedical nanomaterial having a dynamic surface.

Recently, a biomimicking self-assembly approach using polypeptides has emerged for the preparation of functional polymersomes for in vivo use. The formation of fully polypeptide-based polymersome was first demonstrated using poly N^{ε}-2-[2-(2-methoxyethoxy)ethoxy] acetyl-L-lysine-*b*-poly(L-leucine) amphiphilic copolymers by Deming and coworkers [237]. For these polymersomes, the size and structure are dictated primarily by the ordered conformations of the polymer blocks, in a manner similar to viral capsid assembly. Owing to the rigid ordered structure of the hydrophobic membrane, the polymersomes showed great stability and no release of entrapped molecules over a few weeks. Kimura et al. have also developed fully polypeptide-based but non-naturally-occurring polymersomes ("peptosomes") of polysarcosine-*b*-poly(γ-methyl L-glutamate) synthesized by a NCA polymerization method (Fig. 12a) [238]. The peptosomes were found to possess stable hydrophobic membrane composed of poly(γ-methyl L-glutamate) chains having α-helical structure. Hence, they utilized the peptosomes as nanocarriers for in vivo imaging probes. The peptosomes labeled with a near-infrared fluorescence (NIRF) probe showed a relatively long half-life time in the rat blood stream, which was comparable to that of PEGylated liposomes. Moreover, NIRF imaging of a small murine cancer was performed using the peptosome as a nanocarrier (Fig. 12b, c) [239].

In addition to ordered chain conformations, polypeptides contain the abundant chemical functionality of amino acids. Deming and coworkers have developed fully naturally occurring polypeptide-based polymersomes composed of polyarginine (PArg) and polyleucine (PLeu) as a hydrophilic block and hydrophobic block, respectively [214]. The polymersomes are quite stable in physiological conditions and can encapsulate water-soluble molecules. The remarkable feature of this material is that the PArg block directs self-assembly for the polymersome formation and simultaneously provides functionality for efficient intracellular delivery to the polymersomes. This unique synergy between nanoscale self-assembly and

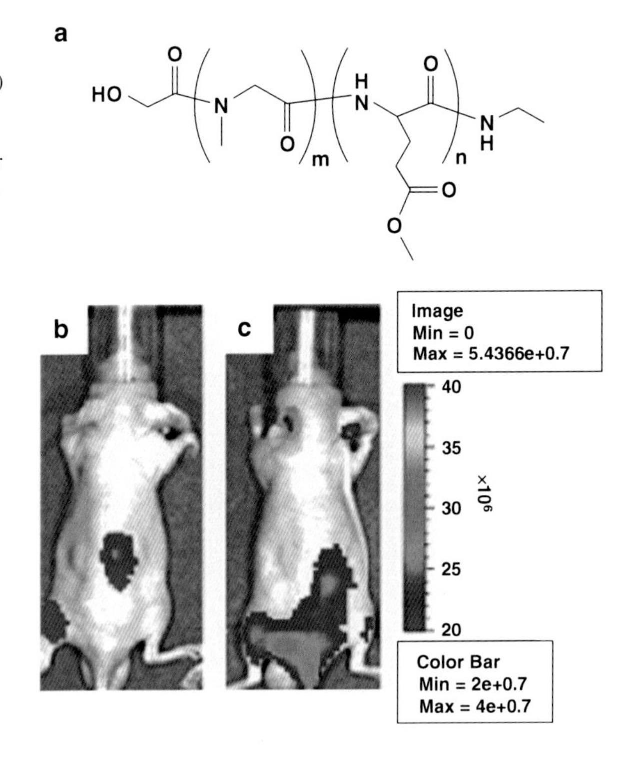

Fig. 12 (**a**) Chemical structure of poly(sarcosine)-*b*-poly(γ-methyl L-glutamate) copolymer. (**b**) Bioluminescence image of tumor-bearing mouse after administration of D-luciferin. (**c**) In vivo cancer imaging using NIRF-labeled peptosome. Fluorescence image after 1 day of administration of the labeled peptosome. Reprinted from [239] with permission

inherent peptide functionality provides a new approach for design of multifunctional polymersomes for DDS.

On the other hand, Kataoka and coworkers have developed a novel type of polypeptide-based polymersome composed of PEG-*b*-PAsp and PEG-*b*-poly [(5-aminopentyl)-α,β-aspartamide] with a nonionic PEG block and anionic or cationic polypeptide blocks [210, 240]. These block copolymers do not rely on amphiphilicity for self-assembly, but rather simple mixing of a pair of opposite and equally charged block copolymers results in a stable PIC membrane that forms polymersomes ("PICsomes"). Since PICsome membranes do not have a hydrophobic core, they have an increased permeability to hydrophilic compounds with molecular weights below 450 Da. Furthermore, PICsomes with crosslinked membrane through covalent bonds showed tuning of permeability, enhanced stability under physiological conditions, and long blood circulation time in mice [241].

Kimura and coworkers have also developed hybrid-type polymersomes composed of polysarcosine-*b*-PLA ("lactosomes") [242–243]. The lactosomes are fully biodegradable due to the equipped metabolic pathway for sarcosine and lactic acid. Hence, the lactosome is preferred for in vivo applications rather than for in vitro studies. Indeed, they have demonstrated a potential utility of lactosomes as a contrast agent for in vivo liver tumor imaging [243]. Lactosomes labeled with indocyanine green showed high escape ability from RES, were found to be stable in blood circulation, and gradually accumulated specifically at a liver tumor. The high tumor/liver imaging ratio was due to the enhanced permeability and retention (EPR) effect [244, 245] of lactosomes.

6 Nanogels

Nanogels are nanometer-sized hydrogel nanoparticles (less than about 100 nm) with three-dimensional networks of physically crosslinked polymer chains. They have attracted growing interest over the last decade because of their potential for applications in biomedical fields, such as DDS and bioimaging [246–249].

Physically crosslinked nanogels can be prepared using noncovalent interactions between polymer chains, such as hydrogen bonds, van der Waals forces, and electrostatic and hydrophobic interactions. It is generally well known that the preparation of stable physically crosslinked nanogels with controlled sizes using such associating polymers is difficult because of their relatively weak noncovalent interactions. Akiyoshi and coworkers proposed a self-assembling strategy for the preparation of physically crosslinked nanogels using the controlled association of hydrophobically modified polymers in dilute aqueous solution [250]. In fact, cholesterol-modified polysaccharides such as cholesterol–pullulan (CHP) formed stable monodispersive nanogels with a diameter of about 30 nm in water. This result suggests that the association of cholesteryl groups provides crosslinking points through hydrophobic interactions.

Physically crosslinked nanogels have been reported in various combinations such as cholesterol-bearing poly(amino acids) [251], cholesterol-bearing mannan [252], deoxycholic acid-bearing chitosan [253], bile acid-bearing Dex [254], synthetic polyelectrolytes with hydrophobic groups [255], and alkyl group-modified poly(*N*-isopropylacrylamide)–cholesterol-modified polysaccharide mixtures [256]. Physically crosslinked nanogels have advantages with respect to their biomedical applications because there is no need to use crosslinkers and/or catalysts, which may be toxic, and there are no byproducts in the preparation process.

Recently, Akiyoshi and coworker reported that lipase from *Pseudomonas cepacia* and the CHP nanogel were spontaneously complexed by simply mixing, leading to increased enzymatic activity after complexation. The complexation also led to a substantial increase in the thermal stability of the lipase (Fig. 13). This is a new type of nano-encapsulation of enzyme inside a hydrogel matrix. This simple and effective method is useful in enzyme engineering and bioengineering [257]. Akiyoshi et al. also proposed cholesterol group-modified enzymatically synthesized glycogen (CHESG) nanogels with approximately 35 nm in diameter, which were formed via a number of hydrophobic domains as physical crosslinking segments (Fig. 14). It was possible to provide strong complexation with a number of proteins, and high chaperone-like activity was seen for thermal stabilization of enzymes by CHESG-CD supramolecular systems. These functions are valuable for protein delivery systems [258].

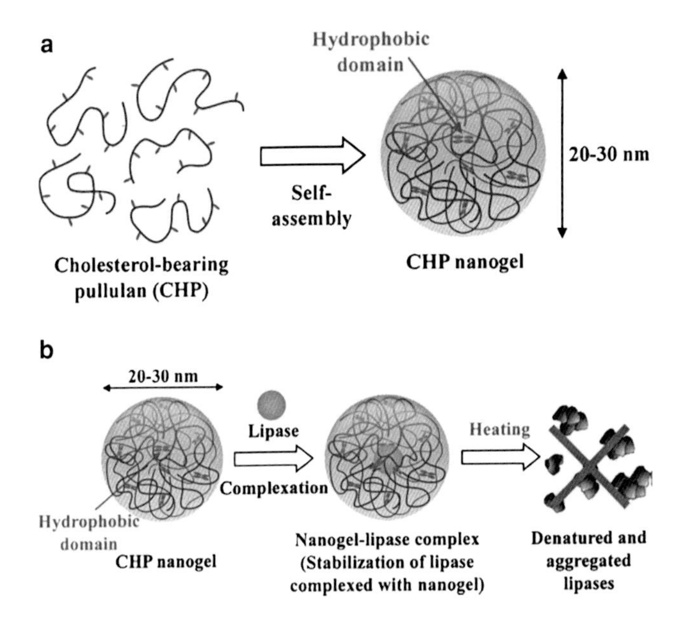

Fig. 13 (**a**) CHP nanogel and (**b**) the interaction between CHP nanogel and lipase. Reprinted from [257] with permission

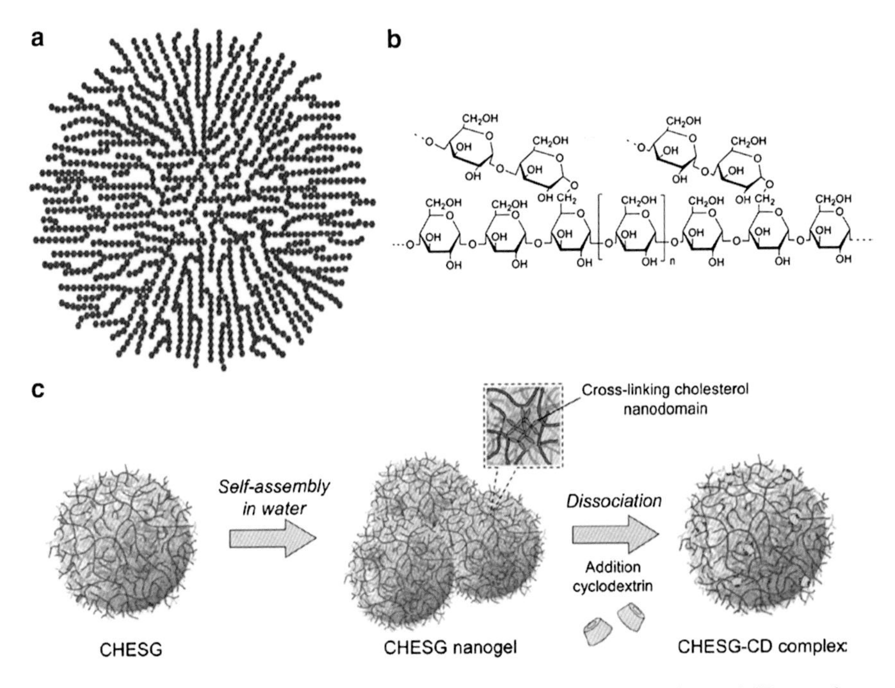

Fig. 14 (**a**) ESG, (**b**) chemical structure of ESG, and (**c**) CHESG association and CD complexation. Reprinted from [258] with permission

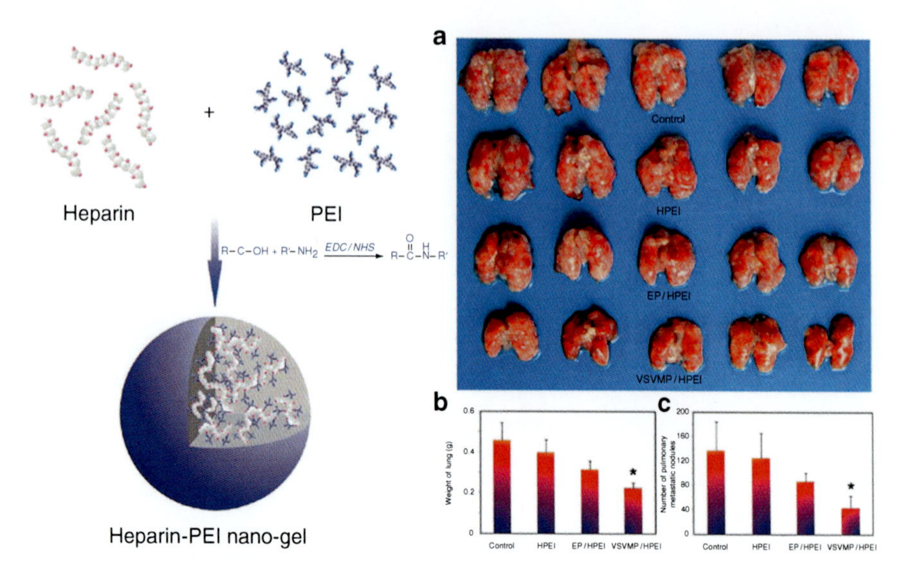

Fig. 15 *Left*: Heparin–PEI nanogel. *Right*: Effect of treatment with heparin–PEI nanogel (*HPEI*), pEP-loaded nanogel (*EP/HPEI*), and pVSVMP-loaded nanogel (*VSVMP/HPEI*). (**a**) Photograph of lungs bearing metastases of C-26 colon carcinomas in each treatment group. (**b**) Weight of lungs bearing pulmonary metastases of C-26 colon carcinoma in each treatment group. (**c**) Number of pulmonary metastatic nodules in each treatment group. Reprinted from [259] with permission

Wei et al. prepared heparin–polyethyleneimine (HPEI) nanogel, which was designed for pVSVMP (plasmid expressing vesicular stomatitis virus matrix protein) delivery to treat C-26 colon carcinoma in vitro and in vivo (Fig. 15). The HPEI nanogel is a novel nonviral gene vector, and able to efficiently transfect pVSVMP into C-26 colon carcinoma cells in vitro, inhibiting cell proliferation through apoptosis induction. Application of pVSVMP/HPEI complexes can efficiently inhibit the growth of colon carcinoma in vivo. A new and interesting genetic cancer therapy protocol for treatment of colon cancer using the HPEI nanogel was proposed [259]. Na et al. reported a HA-bearing photosensitizer moiety as a new type of photodynamic therapy agent [260]. After incubation of the HA/photosensitizer conjugate nanogels with cells, the degradation of the HA by enzymes in endosomes and lysosomes was monitored as a dissipation of the autoquenching behavior (Fig. 16). The nanogels also exhibited HA-induced tumor-homing properties, resulting in a rapid internalization rate. Moreover, the nanogels indicated high cytotoxicity against cancer cells under light emission, which was comparable with the free photosensitizer, but very low cytotoxicity without light.

We have reported biodegradable polysaccharide-*g*-PLA nanogels. PLA was chosen as physically crosslinking segments based on its hydrophobicity and biodegradability. Relatively short PLLA (oligolactide, OLLA) chains were introduced to pullulan or Dex by a coupling reaction [162–164]. We prepared protein-loaded Dex-*g*-OLLA nanogels using lysozyme as model protein and found that the

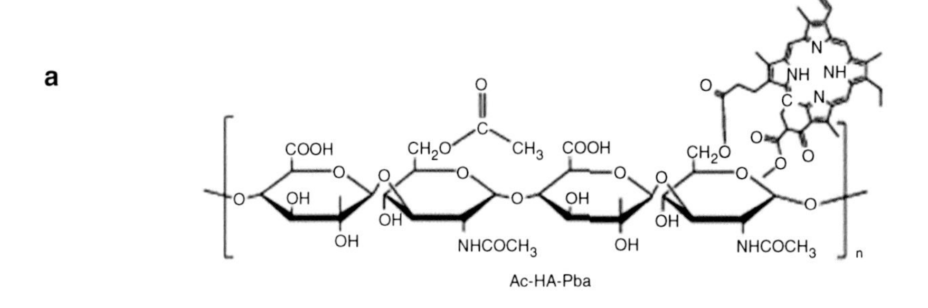

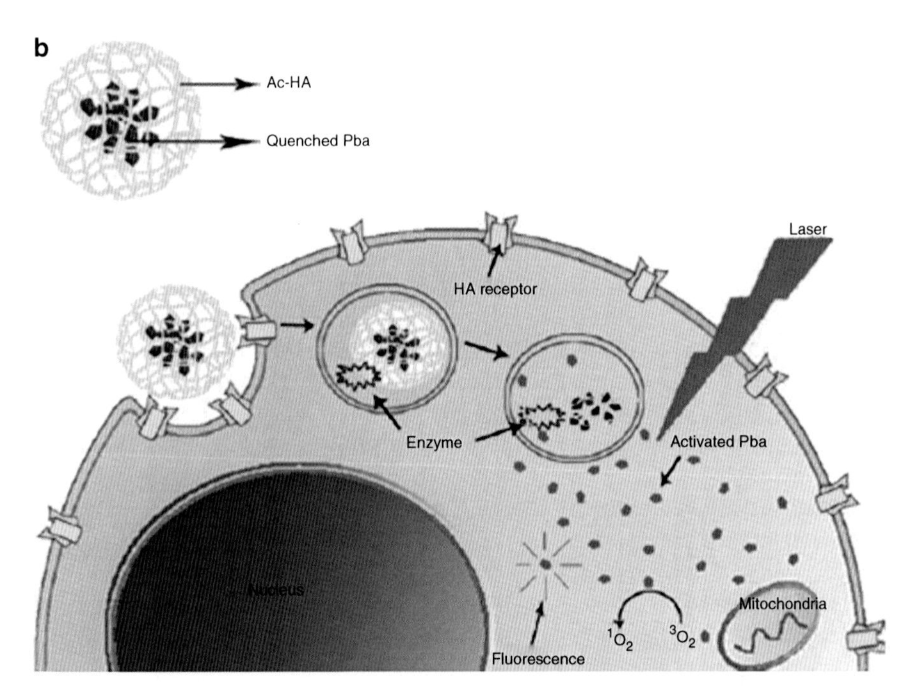

Fig. 16 (**a**) Chemical structure of HA/photosensitizer conjugate conjugates and (**b**) strategy for photodynamic therapy. Reprinted from [260] with permission

lysozyme-loaded nanogels showed a sustained release of lysozyme for 1 week without denaturation in PBS at 37 °C (Fig. 17) [164].

Usual PLAs have two enatiomeric forms, PLLA and poly(D-lactide) (PDLA), which are synthesized from LLA and the D,D-isomer of lactide (DLA), respectively. It is well known that 1:1 mixture of PLLA and PDLA can form a stable stereocomplex (SC), which has a higher T_m and higher mechanical strength than either of homopolymers [261]. We reported on the utility of SC formation to enhance the stability of nanogels. Monodisperse SC nanogels were obtained through the self-assembly of an equimolar mixture of Dex-*g*-OLLA and Dex-*g*-ODLA in a dilute aqueous solution.

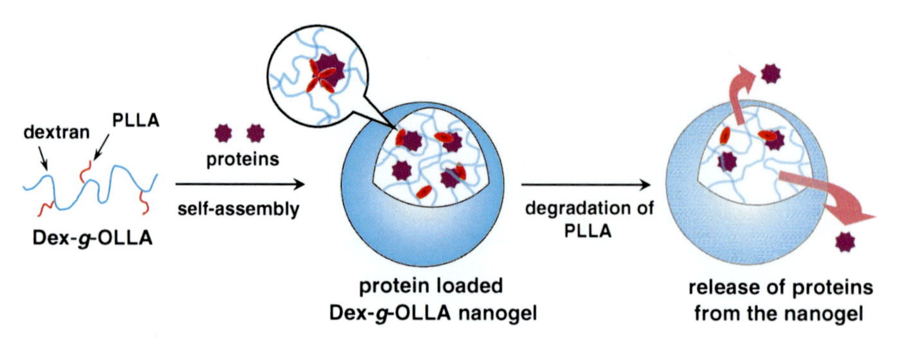

Fig. 17 Protein loaded Dex-*g*-PLLA nanogel. Reprinted from [164] with permission

The SC nanogels had 70 nm mean diameter with narrow size distribution, significantly lower critical aggregation concentration (CAC), and stronger thermodynamic stability compared with those of the corresponding L- or D-isomer nanogels [163].

As described above, PEG has been frequently used in biomedical applications. However, PEG is not biodegradable. Low molecular weight PEG (below ca. 30,000 Da) can be excreted from kidneys, but such excretion is difficult for higher molecular weight PEG. Therefore, it is valuable to provide biodegradable polymers having PEG-like high biocompatibility, as well as biodegradability. Moreover, PEG possesses reactive functional groups only at its termini, which results in a lack of functionality in the main chain. PEG-based nanogels prepared by physical crosslinking have not been reported. We reported the synthesis of PEG-like biodegradable polymers by polycondensation of dihydroxy bifunctional low molecular weight PEG and an amino-protected Asp derivative, poly(Asp-*alt*-PEG). Poly(Asp-*alt*-PEG)–capryl conjugates were synthesized as novel hydrophobically modified biodegradable PEG copolymers. The poly(Asp-*alt*-PEG)–capryl conjugates formed nanogels of approximately 15 nm in size by self-assembly at 20 °C in aqueous media, and the nanogel solutions displayed temperature-responsive phase transition. The reversible transition of the nanogel solution was tunable in the range 19–55 °C by changing the introduced amounts of capryl units and the solution concentrations. The nanogels gradually degraded within days in PBS at 37 °C [262].

7 Supramolecular Biodegradable Systems

Supermolecular interlocked macromolecules have been paid much attention as candidates of smart materials. Polyrotaxane (PRX) is a typical example. PEG/cyclodextrin (CD)-based polyrotaxane was firstly reported by Harada and coworkers by attachment of stoppers to pseudopolyrotaxane (pPRX) consisting of a PEG and CDs [263]. Subsequently, many CD-based PRXs have been designed and prepared as smart materials such as biomaterials, light-harvesting antennae, insulating polymers, stimuli-responsive molecular shuttles etc. [264–268].

In particular, a lot of research on PRX-based materials as biodegradable systems is being strenuously carried out. In PRX systems, there are two strategies for building supramolecular biodegradable systems: (1) using biodegradable bonds between stoppers and the ends of the main chain in the pPRX, and (2) using a biodegradable polymer as main chain.

The first strategy is the method most frequently used for preparation of biodegradable PRX materials. In the past, three types of degradation were reported for end-capping groups of PRXs for specific degradation at the target: (a) enzymatic cleavage, (b) hydrolytic cleavage (basic or acidic condition), and (c) cleavage under reductive conditions (Fig. 18).

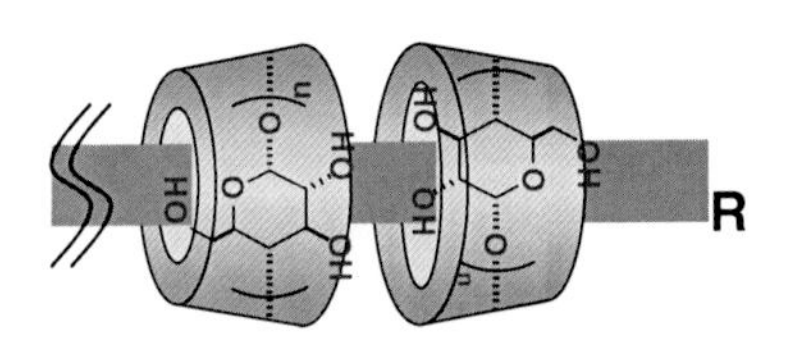

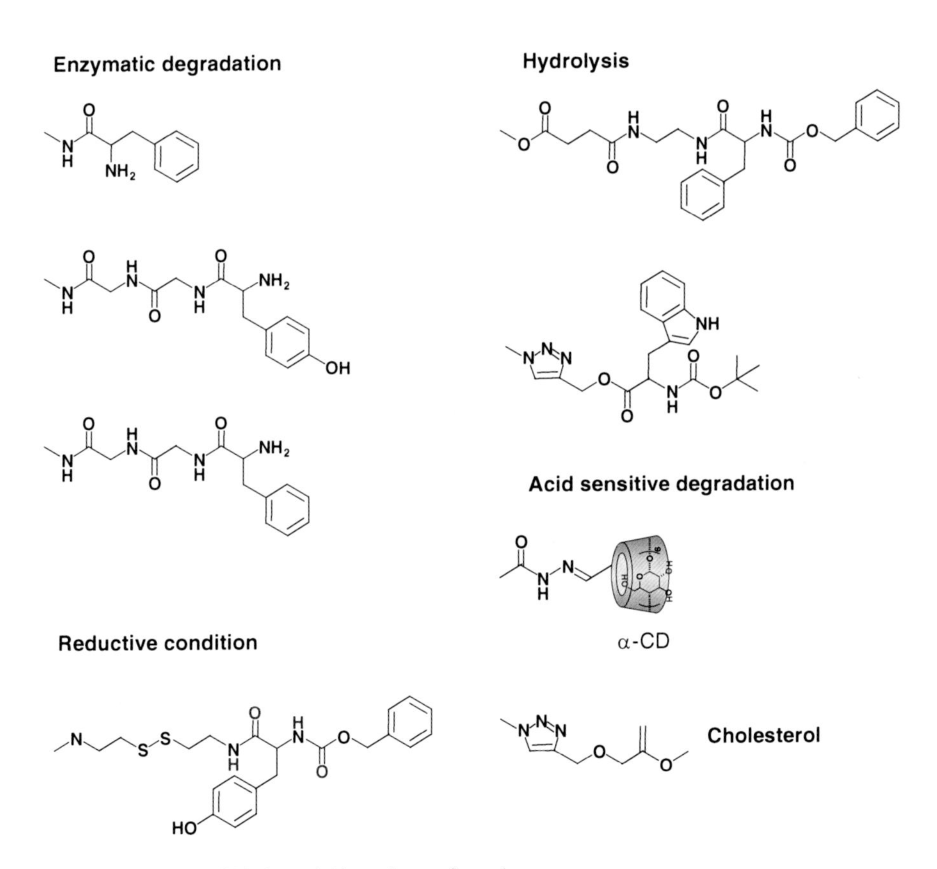

Fig. 18 Structure of biodegradable end caps for polyrotaxanes

Enzymatic degradation systems have been prepared by Yui and coworkers by end-capping pPRXs with several different bulky peptides as stoppers. In the early stages of research, PEG/CD-based PRXs were end-capped with L-Phe to give PRX exhibiting enzymatic degradability by papain [269]. Later, two kinds of enzymatically degradable tripeptides were employed as end-capping groups to enhance specific degradability of the PRXs. The PRXs end-capped with Tyr–Gly–Gly or Phe–Gly–Gly tripeptide with controlled threading percentages of α-CD were prepared by varying the reaction conditions. In some cases, the PRXs were converted to be water-soluble by modification of the threaded CDs with hydroxypropyl groups [270, 271]. These tripeptide end-capped PRXs showed significantly improved enzymatic degradation by aminopeptidase M compared with linear polymers having the same tripeptides. These results suggested that a more straight and rigid supramolecular structure improved the accessibility of enzyme to the terminal peptides.

Hydrolyzable PRXs were firstly reported by Yui and coworkers by introduction of ester linkages between the end groups of pPRX and stoppers. In this system, the relationship of hydrolysis rate of the end-capping groups and degree of acetylation of CDs was investigated. The degradation period could be prolonged from 80 to 1,000 h, when the degree of acetylation of CDs was ~30% [272, 273]. The polyrotaxane was used for regeneration of bone and cartilage by molding in porous hydrogel forms [267, 274]. Thompson and coworkers also synthesized PRXs end-capped with esters and examined the degradation kinetics [275]. There were two types of PRXs designed for acid-triggered cleavages. Water-soluble PRXs end-capped with hydrazone bond were synthesized by introduction of carboxyl groups of the CDs using succinic anhydride, and exhibited degradation in acidic solution [276]. Vinyl ethers are well-known acid-cleavable linkages that have been applied to DDS. PRXs end-capped with vinyl ether linkages were found to be stable under neutral pH, but degraded within about 45 min under acidic pH (4.0) [275].

Usually there are many kinds of thiol (–SH) compounds in cells, and intracellular (cytosol) conditions are reductive. PRXs that are cleavable under the reductive conditions of the cytosol were also prepared by introduction of disulfide bonds to the end-capping groups. Yui et al. reported the application of PRXs cleavable under reductive conditions for gene delivery systems [277]. Dimethylaminoethyl (DMAE)-modified PRX having disulfide end-capping groups formed a polyplex of 178–189 nm diameter with DNA. The polyplexes were completely dissociated upon the addition of polyanionic dextran sulfate in the presence of 10 mM dithiothreitol (DTT). Transfection experiments using PRX/rhodamine-labeled DNA polyplex were carried out. Effective escape of the DNA from the endosome/lysosme compartments and significant expression was confirmed.

Research on the second strategy has been reported for several types of PRX materials. Most studies concern pPRXs formation between α-CD and polyesters (Table 1) [278–290]. A pPRX of α-CD/PLLA was firstly demonstrated by Tonelli and coworkers [282]. Subsequently, we reported the pPRX formation of α-CDs and PLLA-*b*-PEG-*b*-PLLA triblock copolymer [288]. In this report, the formation of an inclusion complex and the stoichiometry of the amphiphilic biodegradable triblock

Table 1 Complex formation between cyclodextrins and polyesters

Polymer structure	CDs	Reference
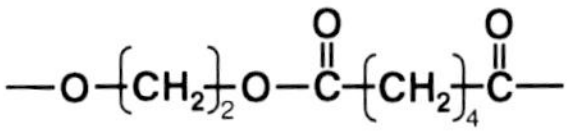 $-O-(CH_2)_2-O-\overset{O}{\overset{\parallel}{C}}-(CH_2)_4-\overset{O}{\overset{\parallel}{C}}-$ Poly(ethylene adipate) (PEA)	α-CD, γ-CD	[278]
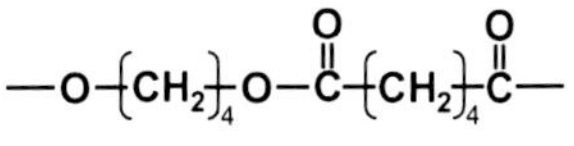 $-O-(CH_2)_4-O-\overset{O}{\overset{\parallel}{C}}-(CH_2)_4-\overset{O}{\overset{\parallel}{C}}-$ Poly(trimethylene adipate) (PTA)	α-CD	[279]
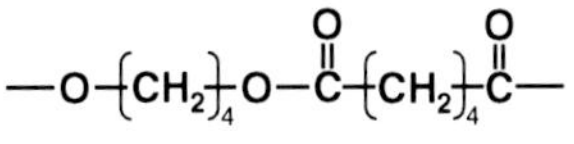 $-O-(CH_2)_4-O-\overset{O}{\overset{\parallel}{C}}-(CH_2)_4-\overset{O}{\overset{\parallel}{C}}-$ Poly(1,4-butylene adipate) (PBA)	α-CD	[280]
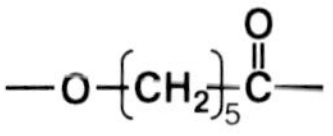 $-O-(CH_2)_5-\overset{O}{\overset{\parallel}{C}}-$ Poly(ε-caprolactone) [P(ε-CL)]	α-CD, γ-CD	[281]
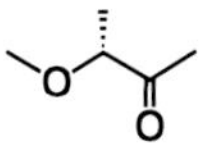 Poly(L-lactide) (PLLA)	α-CD	[282]
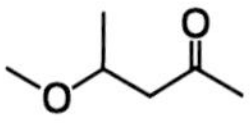 Atactic poly[(*R*, *S*)-3-hydroxy butyrate] (a-PHB)	α-CD	[283]
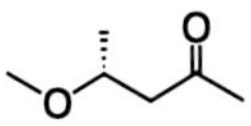 Poly[(*R*)-3-hydroxybutyrate] (i-PHB)	α-CD	[284]
P(ε-CL)–PEO–P(ε-CL)	α-CD, γ-CD	[285]
P(ε-CL)–*b*–PLLA	α-CD	[286]
P(ε-CL)–PPG–P(ε-CL)	α-CD, γ-CD	[287]
PLLA–PEG–PLLA	α-CD	[288]
PEG–PCL–PEI	α-CD	[289]
PHB–PEG–PHB	α-CD	[290]

copolymers with α-CDs were investigated by X-ray diffraction and solid-state ^{13}C CP/MAS (cross-polarization/magic angle spinning) NMR spectroscopic methods. The results suggested a channel-type crystalline structure with CDs due to the long chain nature of the triblock copolymers. We confirmed enantiospecific recognition of chiral PLLA by α-CD as a low molecular weight chiral host on pPRX formation [291]. PLLA could effectively form pPRX with α-CD, but PDLA did not.

We further synthesized biodegradable PRX composed of PLLA and α-CD (LA-PRX) by capping reaction of an amino group introduced at the termini of main chain PLLA [292]. The end-capping groups were attached through enzymatically degradable peptide linkages. Then, we investigated the enzymatic degradation behavior of the LA-PRX synthesized by each of the two strategies (using biodegradable links between main chain and end-capping groups or using biodegradable main chain). The PLLA chain in the obtained LA-PRX showed slower degradation behavior in the absence of papain, but rapidly degraded in the presence of papain compared with naked PLLA. These results mean that CDs can act as hydrolysis inhibitors in PRX form. However, cleavage of end-capping groups by papain and subsequent release of CDs led to rapid degradation of PLLA main chain (Fig. 19). Therefore, the LA-PRX system is expected to be applied to the development of biodegradable medical devices exhibiting specific stimuli-responsive degradation, drug release, or time-controlled excretion.

Zhuo and coworkers recently reported supramolecular hydrogels based on PRX [293]. They encapsulated several kinds of cells in the hydrogel. The in vitro cytotoxicity and histological studies demonstrated good biocompatibility and suggested that the hydrogels were good candidates for injectable scaffolds for tissue engineering and drug delivery devices.

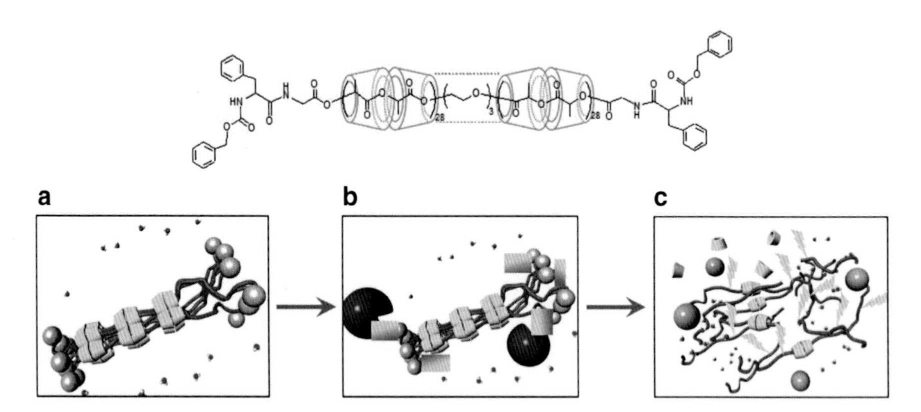

Fig. 19 Structure of LA-PRX (*above*) and degradation of LA-PRX (*below*). (**a**) Threaded α-CDs prevent hydrolysis of PLLA in LA-PRX. (**b**) LA-PRX converts into LA-pPRX by peptide linkage cleavage at bulky end-capping groups through action of papain. (**c**) Ester bond hydrolysis in the PLLA chain begins by an exposure of PLLA to water by release of α-CDs from LA-pPRX. Reprinted from [292] with permission

8 Biodegradable Stimuli-Responsive Systems

8.1 Biodegradable Temperature Responsive Systems

Stimuli-responsive polymers, especially temperature-responsive polymers, have attracted a great deal of attention in the last two decades as biomedical materials for drug delivery, separation of bioactive molecules, and tissue engineering [294, 295]. PNIPAAm is one of the most typical temperature-responsive polymers. PNIPAAm exhibits a rapid and reversible hydration–dehydration in response to small changes in the solution temperature around its lower critical solution temperature (LCST) of 32 °C [296]. PNIPAAm has been studied for development of smart materials such as stimuli-responsive particles, surfaces, and hydrogels [297–300]. However, PNIPAAm and its copolymers are nonbiodegradable, presenting an obstacle to their application as implantable biomaterials. Hence, several types of temperature-responsive biodegradable polymers have been recently developed [40, 301–304].

Some amino acid-based polymers possess unique properties and functions based on the polar side chains and amide bonds, which can form hydrogen bonds. Tachibana and coworkers have synthesized poly(amino acid)-based temperature-responsive polymers, poly(*N*-substituted α/β-asparagine), through coupling reaction of poly(succinimide) with a mixture of 5-aminopentanol and 6-aminohexanol [301]. The polymer showed LCST-type temperature responsiveness, and the LCST could be tuned in the range from 23 to 44 °C by varying the mixing ratio of 5-aminopentanol and 6-aminohexanol in the coupling reaction. Shimokuri and coworkers have also synthesized poly(γ-glutamic acid) derivatives having propyl amide groups as side chains, i.e., poly(α-propyl γ-glutamate) [302]. The polymer also showed LCST-type phase transition, and the LCST of the polymer was at around 30 °C. Moreover, chemical crosslinking of poly(α-propyl γ-glutamate) with hexamethylene diisocyanate (HMDI) produced biodegradable hydrogels having a temperature-responsive shrinking property [303].

As described above, we have synthesized PDP, alternating copolymers of an α-aspartic acid and a glycolic acid [poly(Glc–Asp)], having pendant carboxylic groups [100]. Generally, the temperature-dependent soluble–insoluble transition property of a polymer depends strongly on the hydrophobic/hydrophilic balance of the polymer. We have synthesized poly(Glc–Asp) substituted with moderately hydrophobic groups (isopropyl amide groups like NIPAAm). The resulting poly[Glc–Asn(*N*-isopropyl)] showed LCST-type temperature-responsive phase transition in water at 29 °C, as shown in Fig. 20 [304]. Degradation of poly [Glc–Asn(*N*-isopropyl)] to the monomer level occurred via cleavage of the ester bonds in the main chain, and the resulting degradation products were found to be nontoxic for cultured cells. Poly[Glc–Asn(*N*-isopropyl)] and related polymers exhibiting LCST between room temperature and body temperature, no toxicity, hydrolytic degradation, and chemical reactivity are expected to be applied in biomedical field.

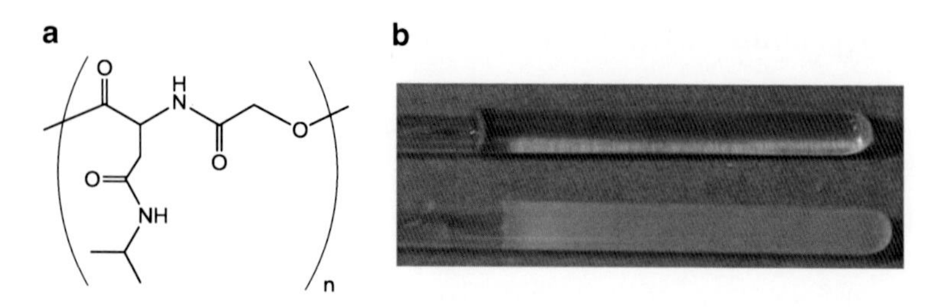

Fig. 20 (**a**) Chemical structures of poly[Glc–Asn(*N*-isopropyl)]. (**b**) Photographs showing temperature-responsiveness of poly[Glc–Asn(*N*-isopropyl)] 5% aqueous solution at below (*upper*) and above (*lower*) the LCST. [304]

Iwasaki and coworkers have synthesized temperature-responsive polyphosphoesters by ROP of two cyclic phosphoester monomers, ethyl ethylene phosphate (EP) and isopropyl ethylene phosphate (IPP) [40]. The obtained copolymers, poly(IPP-*co*-EP)s, showed reversible LCST-type temperature-responsiveness and the LCST linearly increased with an increase in the composition of IPP, indicating that the LCST of poly (IPP-*co*-EP)s can be controlled to physiological temperatures. Polyphosphoesters have been known to be degraded through enzymatic digestion of phosphate linkages under physiological conditions [305]. Thus, these properties indicate that poly(IPP-*co*-EP)s have potential as biodegradable smart biomedical materials.

8.2 Biodegradable Injectable Polymers

The temperature-responsive biodegradable polymers mentioned above are homopolymers or random copolymers exhibiting LCST-type phase transitions. Other types of temperature-responsive biodegradable polymers having amphiphilic block- or graft-type structures have been also reported [111, 127, 137, 306–321]. Some of these are reported to show temperature-responsive sol–gel type transitions, not soluble–insoluble-type transitions, and are called thermo-gelling polymers. They form a physically crosslinked hydrogel through noncovalent interactions, such as hydrophobic interactions, in aqueous solution triggered by a temperature change. Biodegradable thermo-gelling polymers with a sol–gel transition point between room temperature and body temperature are useful for injectable polymer (IP) systems in biomedical applications [127] because the polymer solution is in a sol state in a syringe at room temperature but then becomes a hydrogel in situ after injection into the body. Furthermore, they can be degraded into metabolizable monomers and/or low molecular weight water-soluble polymers, which can be excreted through the kidney.

IPs are very useful for DDS because IP systems allow easy entrapment of pharmaceuticals and bioactive agents and can be used as a depot for their sustained release after a simple injection with a syringe at the target site in a human body. Such a system can minimize the requirement for surgical operations for implantation of drug delivery depots, which should result in a better quality of life for patients, compared with the current regime of repeated administration. IP systems also should be useful for tissue engineering, whereby a polymer solution containing living cells and/or growth factors in suspension can be injected at room temperature at a target site to form a biodegradable hydrogel scaffold for cell growth and tissue repair. Biodegradable IP systems have also been reviewed recently [306, 307].

In 1997, Kim and coworkers first developed biodegradable IP systems using a triblock copolymer of PEG and PLLA, PEG-*b*-PLLA-*b*-PEG, and demonstrated sustained release of drugs from the hydrogel [127]. After this achievement, many kinds of biodegradable amphiphilic block copolymers (including multiblock copolymers) exhibiting temperature-responsive sol–gel transition have been reported [137, 308–318]. In this review, only several recent results are introduced.

Lee and coworkers have demonstrated bone regeneration using P(CL-LA)-*b*-PEG-*b*-P(CL-LA) hydrogels [310]. An aqueous solution of the P(CL-LA)-*b*-PEG-*b*-P(CL-LA) containing human mesenchymal stem cells and recombinant human bone morphogenetic protein-2 was injected into the backs of mice. After 7 weeks, mineralized tissue with high levels of alkaline phosphatase activity was found. Mitra and coworkers have developed a composite system DDS consisting of biodegradable MSs and IPs [311]. Ganciclovir-loaded PLGA MSs were dispersed in a PLGA-*b*-PEG-*b*-PLGA solution. The resulting composite hydrogel showed a slower drug release rate compared with that from PLGA MSs alone. Huang et al. reported sustained release of plasmid DNA from a PEG-*b*-PLGA-*b*-PEG hydrogel [312]. After a PEG-*b*-PLGA-*b*-PEG solution containing luciferase gene was applied to the skin of mice, the expression of luciferase was increased and reached a maximum at 24 h.

As mentioned above, PLLA and PDLA can form stable SC crystals [261]. The SC formation can be used as a driving force for gel formation and to stabilize the hydrogel. Kimura et al. reported temperature-responsive formation of a hydrogel from an enantiomeric mixture of the ABA-type triblock copolymers, PLLA–*b*-PEG–*b*-PLLA and PDLA–*b*-PEG–*b*-PDLA [313]. An equivalent volume of the aqueous solution of PLLA–*b*-PEG–*b*-PLLA and PDLA–*b*-PEG–*b*-PDLA were mixed and heated to induce a spontaneous gelation. Each individual solution of PLLA–*b*-PEG–*b*-PLLA or PDLA–*b*-PEG–*b*-PDLA did not show such temperature-responsive gelation. These results suggested that an increase in temperature triggered perturbation of the polymer micelle through dehydration of PEG (which formed the shell layer of the micelle) and led to intermicelle aggregation by SC formation of PLLA and PDLA segments. They also reported reverse-type temperature-responsive gelation phenomenon using the BAB-type triblock copolymers PEG–*b*-PLLA–*b*-PEG and PEG–*b*-PDLA–*b*-PEG [314].

Jeong and coworkers have reported peptide-based thermo-gelling systems using PEG-*b*-polyAla as an injectable cellular scaffold [315]. The polymer aqueous solution undergoes sol–gel transition as temperature increases. The fraction of the β-sheet structure of the polyAla dictated the population and thickness of fibrous nanostructure in the hydrogel, which affected the proliferation and protein expression of the encapsulated chondrocytes. We also reported that a block copolymer system of PDP and PEG exhibited temperature-responsive sol–gel transition. One of the merits of PDPs for design of temperature-responsive polymers compared with simple aliphatic polyesters is that the hydrophobicity can be controlled by choosing the amino acids [316]. Song and coworkers have shown a potential utility of polyphosphazene-based IP hydrogel system as a vehicle for cell delivery in cell-based therapy. They prepared a hydrogel entrapping pancreatic islets using temperature-responsive polyphosphazene. Rat islets in the hydrogel showed higher cell viability and insulin production over 28 days as compared to those for free rat islets [317]. In a subsequent study, polyphosphazene hydrogels were used to encapsulate hepatocytes as spheroids or single cells. The spheroid hepatocytes maintained a higher cell viability and produced albumin and urea over 28 days [318].

Some graft-type biodegradable copolymers were reported as thermo-gelling systems. We synthesized amphiphilic PLA-*g*-PEG by a coupling reaction of random copolymer of LA and DP having carboxylic acid side chains [poly(Glc–Asp)-*co*-LA] [111]. The obtained amphiphilic graft copolymers showed temperature-responsive sol–gel transition behavior and higher mechanical strength compared with usual linear block copolymer systems. Jeong and coworkers have demonstrated sustained release of insulin from a hydrogel of a PLGA-*g*-PEG/PEG-*g*-PLGA mixture [319]. After a single injection of insulin-containing PLGA-*g*-PEG aqueous solution, blood glucose levels could be adjusted for 5–16 days in diabetic rats. They also reported that chondrocyte-loaded PLGA-*g*-PEG hydrogels were useful to repair an articular cartilage defect [319]. The cartilage defect was completely repaired by PLGA-*g*-PEG hydrogel. The superior efficacy for cartilage defect repair was attributed to the favorable degradation profile of PLGA-*g*-PEG.

Usual linear triblock copolymer thermo-gelling systems have problems of low mechanical strength in the gel state. The storage moduli of linear IP systems in the gel state are usually less than 100 Pa. Multiblock copolymers and graft copolymers have somewhat improved mechanical strength. Recently, we have developed an IP system based on star-shaped branched block copolymers to improve mechanical strength. The branched structure should lead to efficient physical crosslinking during the gel formation. We synthesized eight-arm PEG-*b*-PLLA having a star-shaped branched structure using octa-functional PEG (eight-arm PEG) as macro-initiator [171–173], then hydrophobic cholesterol groups were attached to some of the ends of the star-shaped block copolymer to give eight-arm PEG-*b*-PLLA–cholesterol [320]. An aqueous solution of eight-arm PEG-*b*-PLLA–cholesterol (above 3 wt% in polymer concentration) exhibited temperature-responsive instantaneous gelation at 36 °C upon heating (Fig. 21), but the virgin eight-arm PEG-*b*-PLLA did not gel at any concentrations. The eight-arm PEG-*b*-PLLA–cholesterol showed significantly higher mechanical strength (storage modulus 5,000 Pa) compared with previous biodegradable thermo-gelling polymers. We then investigated the potential of the IP system as an injectable scaffold for tissue engineering. L929 cells encapsulated into the hydrogel were viable and proliferated three-dimensionally in the hydrogel, suggesting that the extracellular matrix (ECM)-like network structure of the hydrogel guided L929 cells into three-dimensional proliferation. The 10 wt% hydrogel eroded gradually in PBS at 37 °C

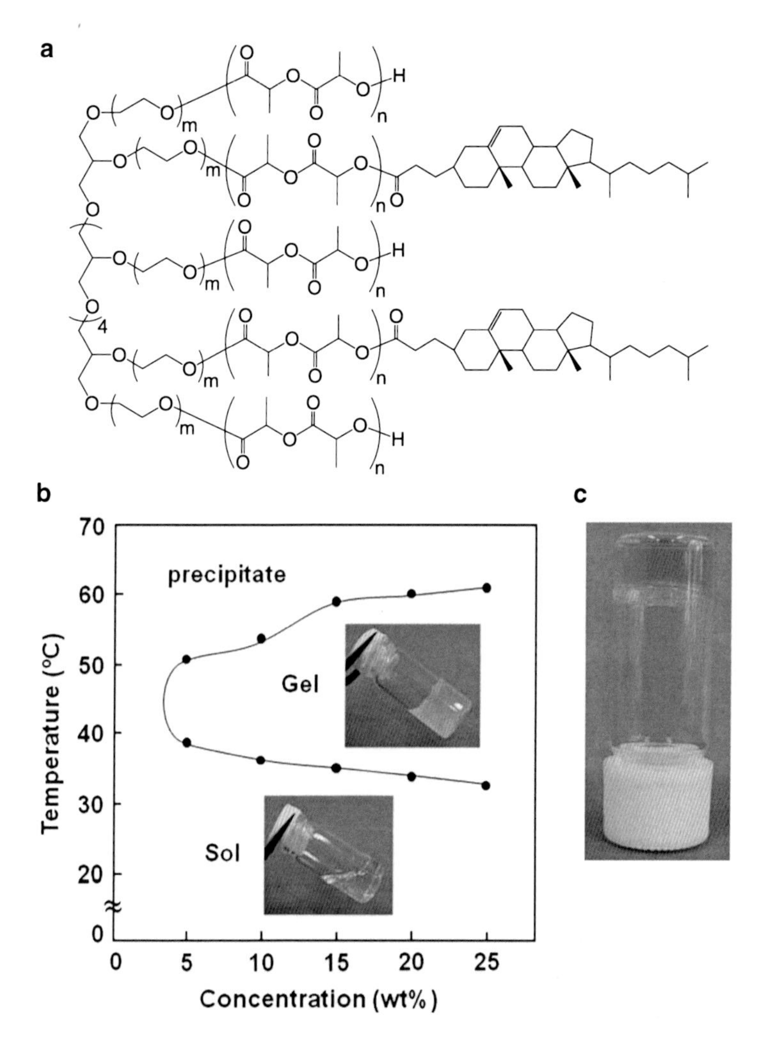

Fig. 21 (**a**) Chemical structure of eight-arm PEG-*b*-PLLA–cholesterol. (**b**) Phase diagram of eight-arm PEG-*b*-PLLA–choesterol aqueous solution. The photographs indicate the sol (flow) and gel (no-flow) states of 10 wt% polymer aqueous solution. (**c**) Eight-arm PEG-*b*-PLLA–cholesterol hydrogel in tissue culture medium containing live L929 cells. Reprinted from [320] with permission

over a month, after which time the gel was completely dissociated. These results indicate that the eight-arm PEG-*b*-PLLA–cholesterol can be a candidate as an injectable cellular scaffold for tissue regeneration.

To obtain biodegradable thermo-gelling system exhibiting even higher mechanical strength, we introduced SC in a hydrogel formation process. As mentioned above, Kimura et al. already reported a thermo-gelling system using SC formation on the linear triblock copolymer systems [313, 314]. We synthesized star-shaped triblock copolymers consisting of eight-arm PEG and PLLA or PDLA, i.e., eight-arm

PEG-*b*-PLLA-*b*-PEG or eight-arm PEG-*b*-PDLA-*b*-PEG [321]. An aqueous solution of a 1:1 mixture of these copolymers was in sol state at room temperature, but instantaneously formed a hydrogel in response to increasing temperature. The resulting hydrogel exhibited a significantly higher storage modulus (ca. 10 kPa) at 37 °C, twice that of the eight-arm PEG-*b*-PLLA–cholesterol system. Interestingly, once formed at the transition temperature, the hydrogel was stable even after cooling below the transition temperature. The hydrogel formation process was irreversible because of the formation of stable SC. In aqueous solution, gradual hydrolytic degradation was observed [321]. The rapid temperature-triggered irreversible hydrogel formation, high-mechanical strength, and degradation behavior render this polymer mixture system suitable for use in injectable biomedical materials such as a drug delivery depot or a biodegradable scaffold for tissue engineering.

8.3 Biodegradable Shape-Memory Polymers

Shape-memory polymers (SMPs) are a class of smart materials with the ability to change shape on demand in response to an environmental stimuli [322–325]. So far, the most commonly investigated SMPs are temperature-induced SMPs, whose shape-recovery behavior is triggered by thermal stimuli. Such SMPs have one shape at certain temperature and are converted to another shape at a different temperature (Fig. 22). Temperature-responsive SMPs usually require the combination

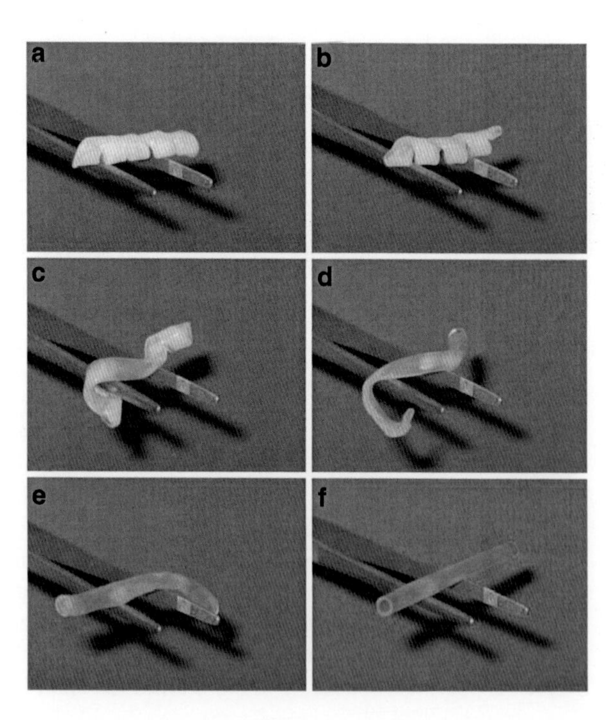

Fig. 22 (**a–f**) Time series photographs showing recovery of shape-memory tube from start to finish of the process; total time 10 s, at 50 °C. The tube was made of PCL-dimethacrylate polymer network that had been programmed to form a flat helix. Reprinted from [323] with permission

of a suitable polymer network structure composed of at least two separated phases: a crosslinking phase that determines the permanent shape, and a thermally reversible phase that fixes the temporary shape below the switching temperature (T_{trans}). Based on the temperature-induced shape-memory effects, such SMPs are expected to be a technological platform for development of multifunctional smart materials. SMPs with T_{trans} ranging between room temperature and body temperature are of special interest for biomedical applications. In addition to the appropriate T_{trans}, biodegradability is often required for SMPs designed as implantable materials to be used in the body. PCL and PLA have frequently been used for biodegradable SMPs as a components of thermally reversible phases, because the T_m of PCL is in the range of 46–64 °C depending on the molecular weight, and the T_g of PLA is in the range of 35–60 °C depending on the molecular weight and chirality [326–330]. Although the T_m and T_g of PCL and PLA phases might be high as T_{trans} for SMPs, it is possible to reduce these temperatures by copolymerization with other biodegradable polymers [328] or by introduction of branched architectures [168, 329].

One example of the application of SMPs is as an implant material for minimally invasive surgery. Current approaches for implanting biomedical materials often require complex surgery followed by material implantation. However, with the recent development of minimally invasive surgery and biodegradable SMPs, it is possible to place functional bulk materials inside of the body using a laparoscope. Bulk materials composed of biodegradable SMPs can be placed at the desired sites in the body in a compressed (temporary) shape through a small incision, and then they recover their application (permanent) shape when the temperature reaches above the T_{trans}. These types of surgical techniques may enable a bulky material to be implanted into the body in a convenient and minimally invasive way, producing innovative medical procedures.

Lendlein and coworkers have presented the first proposal of biodegradable SMPs for applications in biomedical materials [324]. SMPs based on PCL dimethacrylates and *n*-butyl acrylates induced angiogenesis and strong tissue integration in male mice 1 week after subcutaneous implantation [326]. Moreover, the SMPs proved their capability for autoinduced regeneration of a radical stomach wall defect in rats [327]. No gas leakage after gas insufflations could be detected, and fast and unfavorable degradation of the polymer did not occur. A tight connection between the SMP materials and the adjacent stomach was found, resulting in adequate mechanical stability under the extreme pathophysical conditions of the stomach milieu. In addition, the hydrolytic degradation rate of the SMP in PBS at 37 °C could be controlled by varying the monomer content in the copolymer [328].

Neuss and coworkers have reported the possibility of SMPs using PCL dimethacrylate copolymers as cellular scaffold for tissue engineering. Behaviors of different cells from three different species (human mesenchymal stem cells, human mesothelial cells, and rat mesothelial cells) on the matrices were investigated, and the differentiation capacity of mesenchymal stem cells on the matrices was also analyzed [329]. The SMPs proved biocompatibility for all tested cell types, supporting viability and proliferation. The SMPs also supported the osteogenic and adipogenic differentiation of human mesenchymal stem cells 3 weeks after induction.

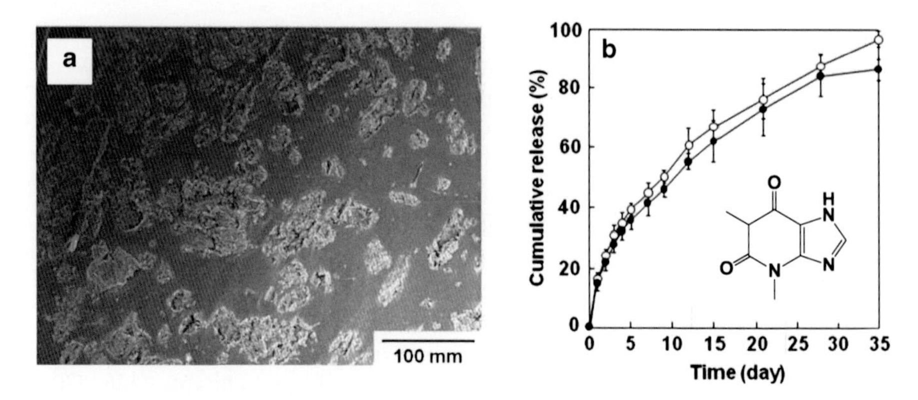

Fig. 23 (**a**) SEM image of cross-section of 10 wt% theophylline-loading XbOCL-10 film. (**b**) Release profile of theophylline from (*filled circles*) XbOCL-10 and (*open circles*) XbOCL-20 films in PBS at 37 °C. Reprinted from [328] with permission

We reported the possibility of biodegradable SMPs as DDS devices [330]. We prepared crosslinked branched oligocaprolactone (XbOCL) exhibiting remarkably rapid temperature-responsive shape recovery within a small temperature change: 90% of the permanent shape was recovered upon heating to within a 2 °C range (37–39 °C). The SMPs exhibited complete shape recovery from a temporary shape to the permanent shape within 10 s at 42 °C. We then chose theophylline as a model drug and prepared theophylline-loaded (10 and 20 wt%) SMPs (Fig. 23). The theophylline-loaded SMPs showed high shape-memory effect, as did the SMP without drug. A sustained release of the theophylline from the matrices was achieved over 1 month without an initial burst-release in PBS at 37 °C. A new type of smart biomedical material could be demonstrated by combining shape-memory effect with controlled drug release in a biodegradable SMP system.

9 Conclusion

In this chapter, the synthesis, preparation, and application of biodegradable polymeric self-assembly systems, including MSs, NSs, polymer micelles, nanogels, polymersomes, supramolecular (interlocked) systems, and stimuli-responsive systems, have been reviewed. There is no doubt that these polymeric assembly systems should play very important roles in the next generation of nanomedicine. Although they exist in a wide range of sizes from nano- to macroscales, it is extremely important to control the intermolecular interactions for design of their assembled structures and final functions. Biological systems use weak intermolecular interaction to maintain life. So, more precise control of intermolecular interactions should lead to higher levels of function, for example, by mimicking biological systems. Fine synthetic and assembly methods for polymers should be the powerful and promising tools for achieving precise design of polymers and their associated states. Although controlled radical polymerization gave a new chemistry for designing polymers consisting of C–C bonds, the usual biodegradable polymers cannot be synthesized in such a controlled manner. New synthetic strategies are needed to overcome the obstacles to design of biomedical materials with higher functions.

References

1. Auras R, Lim LK, Selke EM, Tsuji H (2010) Poly(lactic acid): synthesis, structures, properties, and applications. Wiley, Hoboken, NJ
2. Jeong B, Kim SW, Bae YH (2002) Adv Drug Deliv Rev 54:37
3. Kissel T, Li Y, Unger F (2002) Adv Drug Deliv Rev 54:99
4. Luten J, van Nostrum CF, De Smedt SC, Hennink WE (2008) J Control Release 126:97
5. Ulery BD, Nair LS, Laurencin CT (2001) J Polym Sci B Polym Phys 49:832
6. Nair LS, Laurencin CT (2006) Adv Biochem Eng Biotechnol 102:47
7. Nair LS, Laurencin CT (2007) Prog Polym Sci 32:762
8. Uhrich KE, Cannizzaro SM, Langer RS, Shakesheff KM (1999) Chem Rev 99:3181
9. Drumright RE, Gruber PR, Henton DE (2000) Adv Mater 12:1841
10. Dechy-Cabaret O, Martin-Vaca B, Bourissou D (2004) Chem Rev 104:6147
11. Jacobson GB, Shinde R, Contag CH, Zare RN (2008) Angew Chem Int Ed 47:7880
12. Wei G, Ma PX (2008) Adv Funct Mater 18:3568
13. Albertsson AC, Varma IK, Lochab B, Finne-Wistrand A, Kumar K (2010) In: Auras R, Lim LK, Selke EM, Tsuji H (eds) Poly(lactic acid): synthesis, structures, properties, and applications. Wiley, Hoboken, NJ, p 43
14. Ouchi T, Ohya Y (2004) J Polym Sci A Polym Chem 42:453
15. Sun H, Meng F, Dias AA, Hendriks M, Feijen J, Zhong Z (2011) Biomacromolecules 12:1937
16. Oh JK (2011) Soft Matter 7:5096
17. Feng Y, Lu J, Behl M, Lendlein A (2010) Macromol Biosci 10:1008
18. Pounder RJ, Dove AP (2010) Polym Chem 1:260
19. Gaucher G, Marchessault RH, Leroux J-C (2010) J Control Release 143:2
20. Chvapil M (1977) J Biomed Mater Res 11:721
21. Tabata Y, Ikada Y (1988) Adv Drug Deliv Rev 31:287
22. Rousou JA, Engelman RM, Breyer RH (1984) Ann Thorac Surg 38:409
23. Gopferich A (1997) Macromolecules 30:2598
24. Coulembier O, Degee P, Hedrick JL, Dubois P (2006) Prog Polym Sci 31:723
25. Hill JW (1932) J Am Chem Soc 54:4105
26. Rosen HB, Chang J, Wnek GE, Linhardt RJ, Langer R (1983) Biomaterials 4:131
27. Domb AJ, Gallardo CF, Langer R (1989) Macromolecules 22:3200
28. Pospiech D, Jomber H, Jehnichen D, Haussler L, Eckstein K, Scheibner H, Janke A, Kricheldorf HR, Petermann O (2005) Biomacromolecules 6:439
29. Qi M, Li X, Yang Y, Zhou S (2008) Eur J Pharm Biopharm 70:445
30. Heller J (1985) J Cotrol Release 2:167
31. Wang C, Ge Q, Ting D, Nguyen D, Shen H-R, Chen J, Eisen HN, Heller J, Langer R, Putnam D (2004) Nat Mater 3:190
32. Nguyen DN, Raghavan SS, Tashima LM, Lin EC, Fredette SJ, Langer RS, Wang C (2008) Biomaterials 29:2783
33. Heffernan MJ, Murthy N (2005) Bioconjug Chem 16:1340
34. Lee S, Yang SC, Heffernan MJ, Taylor WR, Murthy N (2007) Bioconjug Chem 18:4
35. Allcock HR, Kugel RL (1965) J Am Chem Soc 87:4216
36. Laurencin CT, Koh HJ, Neenan TX, Allcock HR, Langer RJ (1987) Biomed Mater Res 21:1231
37. Lapienis G, Penczek S (1974) Macromolecules 7:166

38. Kaluzynski K, Libisowski J, Penczek S (1976) Macromolecules 9:365
39. Wang YC, Tang LY, Sun TM, Li C-H, Xiong M-H, Wang J (2008) Biomacromolecules 9:388
40. Iwasaki Y, Wachiralarpphaithoon C, Akiyoshi K (2007) Macromolecules 40:8136
41. Chen R, Curran SJ, Curran JM, Hunt JA (2006) Biomaterials 27:4453
42. Athanasiou KA, Agrawal CM, Barber FA, Burkhart SS (1998) Arthroscopy 14:726
43. Wang YC, Lin MC, Wang DM, Hsieh HJ (2003) Biomaterials 24:1047
44. Porter JR, Henson A, Popat KC (2009) Biomaterials 30:780
45. Mo XM, Xu CY, Kotaki M, Ramakrishna S (2004) Biomaterials 25:1883
46. Katz AR, Turner RJ (1970) Surg Gynecol Obstet 131:701
47. Echeverria EA, Jimenez J (1970) Surgery 131:1
48. Knight S, Erggelet C, Endres M, Sittinger M, Kaps C, Stussi E (2007) J Biomed Mater Res B Appl Biomater 83:50
49. Wang L, Dormer NH, Bonewald LF, Detamore MS (2010) Tissue Eng A 16:1937
50. Dunne N, Jack V, O'Hara R, Farrar D, Buchanan FJ (2010) Mater Sci Mater Med 21:2263
51. Pihlajamaki HK, Salminen ST, Tynninen O, Bostman OM, Laitinen O (2010) Calcif Tissue Int 87:90
52. Erggelet C, Neumann K, Endres M, Haberstroh K, Sittinger M, Kaps C (2007) Biomaterials 28:5570
53. Frisbie DD, Lu Y, Kawcak CE, DiCarlo EF, Binette F, McIlwraith CW (2009) Am J Sports Med 37:71S
54. Mahmoudifar N, Doran PM (2010) Biomaterials 31:3858
55. Pihlajamaki H, Tynninen O, Karjalainen P, Rokkanen PJ (2007) Biomed Mater Res A 81:987
56. Xu L, Cao D, Liu W, Zhou G, Zhang WJ, Cao Y (2010) Biomaterials 31:3894
57. Ohara T, Itaya T, Usami K, Ando Y, Sakurai H, Honda MJ, Ueda M, Kagami HJ (2010) Biomed Mater Res A 94:800
58. Darney PD, Monroe SE, Klaisle CM, Alvarado A (1989) Am J Obstet Gynecol 160:1292
59. Gunatillake P, Mayadunne R, Adhikari R (2006) Biotechnol Annu Rev 12:301
60. Pankajakshan D, Philipose LP, Palakkal M, Krishnan K, Krishnan LK (2008) J Biomed Mater Res B Appl Biomater 87:570
61. Plikk P, Malberg S, Albertsson A-C (2009) Biomacromolecules 10:1259
62. Zuo Y, Yang F, Wolke JGC, Li Y, Jansen JA (2010) Acta Biomater 6:1238
63. Li J, Li L, Yu H, Cao H, Gao C, Gong Y (2006) ASAIO J 52:321
64. Garkhal K, Verma S, Tikoo K, Kumar NJ (2007) Biomed Mater Res A 82:747
65. Guarino V, Ambrosio L (2008) Acta Biomater 4:1778
66. Gunatillake P, Mayadunne R, Adhikari R (2006) Biotechnol Annu Rev 12:301
67. Miller RA, Brady JM, Cutright DE (1977) J Biomed Mater Res 11:711
68. Middleton JC, Tipton A (1998) J Med Plast Biomater 31
69. Lu XL, Sun ZJ, Cai W, Gao ZY (2008) J Mater Sci Mater Med 19:395
70. Ito Y, Ochii Y, Fukushima K, Sugioka N, Takada K (2010) Int J Pharm 384:53
71. Lee WL, Foo WL, Widjaja E, Loo SC (2010) J Acta Biomater 6:1342
72. Rieger J, Freichels H, Imberty A, Putaux J-L, Delair T, Jerome C, Auzely-Velty R (2009) Biomacromolecules 10:651
73. Richter A, Olbrich C, Krause M, Kissel T (2010) Int J Pharm 389:244
74. Mathisen T, Masus K, Albertsson AC (1989) Macromolecules 22:3842
75. Ryner M, Albertsson AC (2002) Biomacromolecules 3:601
76. Finne A, Andronova A, Albertsson AC (2003) Biomcromolecules 4:1451
77. Miura Y, Aoyagi S, Kusada Y, Miyamoto K (1980) J Biomed Mater Res 14:619
78. Ouchi T, Fujino A (1989) Macromol Chem 190:1523
79. Ohya Y, Kobayashi H, Ouchi T (1991) React Polym 15:153
80. Ohya Y, Hirai K, Ouchi T (1992) Makromol Chem 193:1881
81. Braud C, Vert M (1985) Polym Bull 13:293
82. Fournie P, Domurado D, Guerin P, Braud C, Vert M, Madelmont JC (1990) J Bioact Compat Polym 5:381

83. Kimura Y, Shirotani K, Yamane H, Kitao T (1989) Kobunshi Ronbunshu 46:281
84. Kimura Y, Shirotani K, Yamane H, Kitao T (1993) Polymer 34:1741
85. Yamaoka T, Hotta Y, Kobayashi K, Kimura Y (1999) Int J Biol Macromol 25:265
86. Mahmud A, Xiong XB, Lavasanifar A (2006) Macromolecules 39:9419
87. Xiong XB, Mahmud A, Uludag H, Lavasanifar A (2008) Pharm Res 25:2555
88. Xiong XB, Ma Z, Lavasanifar A (2010) Biomaterials 31:757
89. Nottelet B, Tommaso CD, Mondon K, Gurny R, Moller M (2010) J Polym Sci A Polym Chem 48:3244
90. Pratt RC, Nederberg F, Waymouth RM, Hedrick JL (2008) Chem Commun 114
91. Fukushima K, Pratt RC, Nederberg F, Tan JPK, Yang YY, Waymouth RM, Hedrick JL (2008) Biomacromolecules 9:3051
92. Suriano F, Pratt R, Tan JPK, Wiradharma N, Nelson A, Yang YY, Dubois P, Hedrick JL (2010) Biomaterials 31:2637
93. Stewart FHC (1969) Aust J Chem 22:1291
94. Helder J, Kohn FE, Sato S, van den Berg JW, Feijen J (1985) Makromol Chem Rapid Commun 6:9
95. Helder J, Feijen J, Lee SJ, Kim SW (1986) Makromol Chem Rapid Commun 7:193
96. In't Veld PJA, Dijkstra PJ, von Lochem JH, Feijen J (1990) Makromol Chem 191:1813
97. In't Veld PJA, Wei-Ping Y, Klap R, Dijkstra PJ, Feijen J (1992) Makromol Chem 193:1927
98. EP 322154, Phizer Inc. invs: Fung FN, Glowaky RC (1989) Chemcal Abstr 113:103447
99. Samyn C, van Baylen M (1988) Makromol Chem Macromol Symp 19:255
100. Ouchi T, Shiratani M, Jinno M, Hirao M, Ohya Y (1993) Makromol Chem Rapid Commun 14:825
101. In't Veld PJA, Dijkstra PJ, Feijen J (1993) Makromol Chem 193:2713
102. Ouchi T, Nozaki T, Okamoto Y, Shiratani M, Ohya Y (1996) Macromol Chem Phys 197:1823
103. Barrera DA, Zylstra E, Lansbury PT Jr, Langer R (1993) J Am Chem Soc 115:11010
104. Barrera DA, Zylstra E, Lansbury PT, Langer R (1995) Macromolecules 28:425
105. Cook AD, Hrkach JS, Gao NN, Johnson IM, Pajvani UB, Cannizzaro SM, Langer R (1997) J Biomed Mater Res 35:513
106. Ouchi T, Seike H, Nozaki T, Ohya Y (1998) J Polym Sci A Polym Chem 36:1283
107. Tasaka F, Miyazaki H, Ohya Y, Ouchi T (1999) Macromolecules 32:6386
108. Ohya Y, Matsunami H, Yamabe E, Ouchi T (2003) J Biomed Mater Res 65A:79
109. Ohya Y, Matsunami H, Ouchi T (2004) J Biomat Sci Polym Edn 15:111
110. Nagahama K, Ueda Y, Ouchi T, Ohya Y (2007) Biomacromolecules 8:3938
111. Nagahama K, Imai Y, Nakayama T, Ohmura J, Ouchi T, Ohya Y (2009) Polymer 50:3547
112. Gref R, Miralles G, Dellacherie E (1999) Polym Int 48:251
113. Zhu KJ, Xiangzhou L, Shilin YJ (1990) Appl Polym Sci 39:1
114. Penning JP, Dijkstra H, Pennings AJ (1993) Polymer 34:942
115. Zhu KJ, Bihai S, Shilin Y (1989) J Polym Sci A Polym Chem 27:2151
116. Piskin E, Kaitian X, Denkbas EB, Kucukyavuz ZJ (1995) Biomater Sci Polym Ed 7:359
117. Zhang X, Jackson JK, Burt HM (1996) Int J Pharm 132:195
118. Kimura Y, Matsuzaki Y, Yamane H, Kitao Y (1989) Polymer 30:1342
119. Vittaz M, Bazile D, Spenlehauer G, Verrecchia T, Veillard M, Puisieux F, Labarre D (1996) Biomaterials 17:1575
120. Rashkov I, Manolova N, Li SM, Espartero JL, Vert M (1996) Macromolecules 29:50
121. Park TG, Cohen S, Langer R (1992) Macromolecules 25:116
122. Chen X, McCarthy SP, Gross RA (1997) Macromolecules 30:4295
123. Li SM, Rashkov I, Espartero JL, Manolova N, Vert M (1996) Macromolecules 29:57
124. Hu DSG, Liu HJ (1994) Macromol Chem Phys 195:1213
125. Du YJ, Lemstra PJ, Nijenhui AJ, Aert HAMV, Bastiaansen C (1995) Macromolecules 28:2194
126. Hagan SA, Davis SS, Illum L, Davies MC, Garnett MC, Taylor DC, Irving MP, Tadros TF (1995) Langmuir 11:1482
127. Jeong B, Bae YH, Lee DS, Kim SW (1997) Nature 388:860

128. Lee DS, Shim MS, Kim SW, Lee H, Park I, Chang T (2001) Macromol Rapid Commun 22:587
129. Kissel T, Li YX, Volland C, Goerich S, Koneberg R (1996) J Control Release 39:315
130. Li Y, Kissel T (1998) Polymer 39:4421
131. Hiemstra C, Zhong Z, Dijkstra PJ, Feijen J (2005) Macromol Symp 224:119
132. Cai C, Wang L, Dong CM et al (2006) J Polym Sci A Polym Chem 44:2034
133. Chen W, Luo W, Wang S, Bei J (2003) Polym Adv Tech 14:245
134. Ohya Y, Nakai T, Nagahama K, Ouchi S, Tanaka S, Kato K (2006) J Bioact Compat Polym 21:557
135. Yamaoka T, Takahashi Y, Ohta T, Miyamoto M, Murakami A, Kimura Y (1999) J Polym Sci A Polym Chem 37:1513
136. Yamaoka T, Takahashi Y, Fujisato T, Lee CW, Tsuji T, Ohta T, Murakami A, Kimura Y (2001) J Biomed Mater Res 54:470
137. Lee J, Bae YH, Sohn YS, Jeong B (2006) Biomacromolecules 7:1729
138. Ouchi T, Miyazaki H, Arimura H, Tasaka F, Hamada A, Ohya Y (2002) J Polym Sci A Polym Chem 40:1218
139. Ouchi T, Miyazaki H, Arimura H, Tasaka F, Hamada A, Ohya Y (2002) J Polym Sci A Polym Chem 40:1426
140. Ouchi T, Seike H, Miyazaki H, Tasaka F, Ohya Y (2000) Design Monom Polym 3:279
141. Ouchi T, Uchida T, Arimura H, Ohya Y (2003) Biomacromolecules 4:477
142. Arimura H, Ohya Y, Ouchi T (2004) Macromol Rapid Commun 25:743
143. Arimura H, Ohya Y, Ouchi T (2005) Biomacromolecules 6:720
144. Schatz C, Lecommandoux S (2010) Macromol Rapid Commun 31:1664
145. Li BG, Zhang LM (2008) Carbohydr Polym 74:390
146. Liu JY, Zhang LM (2007) Carbohydr Polym 69:196
147. Sun H, Guo B, Li X, Cheng R, Meng F, Liu H, Zhong Z (2010) Biomacromolecules 11:848
148. Wang CH, Wang CH, Hsiue GH (2005) J Control Release 108:140
149. Kohori F, Sakai K, Aoyagi T, Yokoyama M, Sakurai Y, Okano T (1998) J Control Release 55:87
150. Nakayama M, Okano T, Miyazaki T, Kohori F, Sakai K, Yokoyama M (2006) J Control Release 115:46
151. You Y, Hong C, Wang W, Lu W, Pan C (2004) Macromolecules 37:9761
152. Hales M, Barner CK, Davis TP, Stenzel MM (2004) Langmuir 20:10809
153. Hsiue GH, Lo CL, Cheng CH, Lin CP, Huang CK, Chen HH (2007) J Polym Sci A Polym Chem 45:688
154. Huang CK, Lo CL, Chen HH, Hsiue GH (2007) Adv Funct Mater 17:2291
155. Watanabe J, Eriguchi T, Ishihara K (2002) Biomacromolecules 3:1375
156. Ohya Y, Maruhashi S, Ouchi T (1998) Macromolecules 31:4662
157. Ohya Y, Maruhashi S, Ouchi T (1998) Macromol Chem Phys 199:2017
158. Ouchi T, Kontani T, Ohya Y (2003) Polymer 44:3927
159. Ouchi T, Kontani T, Ohya Y (2003) J Polym Sci A Polym Chem 41:2462
160. Ouchi T, Kontani T, Saito T, Ohya Y (2005) J Biomater Sci Polym Ed 16:1035
161. Ouchi T, Kontani T, Aoki R, Saito T, Ohya Y (2006) J Polym Sci A Polym Chem 44:6402
162. Ouchi T, Minari T, Ohya Y (2004) J Polym Sci A Polym Chem 42:5482
163. Nagahama K, Mori Y, Ohya Y, Ouchi T (2007) Biomacromolecules 8:2135
164. Nagahama K, Ouchi T, Ohya Y (2008) Macromol Biosci 8:1044
165. de Jong SJ, De Smedt SC, Wahls MWC, Demeester J, Kettenes-van den Bosch JJ, Hennink WE (2000) Macromolecules 33:3680
166. de Jong SJ, van Eerdenbrugh B, van Nostrum CF, Kettenes-van den Bosch JJ, Hennink WE (2001) J Control Release 71:261
167. Hennink WE, De Jong SJ, Bos GW, Veldhuis TFJ, van Nostrum CF (2004) Int J Pharm 277:99
168. Ouchi T, Ichimura S, Ohya Y (2006) Polymer 47:429
169. Tasaka F, Ohya Y, Ouchi T (2001) Macromolecules 34:5494

170. Stolt M, Viljanmaa A, Sodergard A, Tormala P (2004) J Appl Polym Sci 91:196
171. Nagahama K, Ohya Y, Ouchi T (2006) Macromol Biosci 6:412
172. Nagahama K, Ohya Y, Ouchi T (2006) Polym J 38:852
173. Nagahama K, Nishimura Y, Ohya Y, Ouchi T (2007) Polymer 48:2649
174. Nagahama K, Shimizu K, Ouchi T, Ohya Y (2009) React Funct Polym 69:891
175. Tasaka F, Ohya Y, Ouchi T (2001) Macromol Rapid Commun 22:820
176. Ogawa Y, Okada H, Heya T (1989) J Pharm Pharmacol 41:439
177. Ogawa Y, Okada H, Yamamoto M, Shimamoto T (1988) Chem Pharm Bull 36:2576
178. Okada H (1997) Adv Drug Deliv Rev 28:43
179. Okada H, Heya T, Ogawa Y, Shimamoto T (1988) J Pharm Exp Ther 244:744
180. Ouchi T, Hamada A, Ohya Y (1999) Macromol Chem Phys 200:436
181. Ouchi T, Toyohara M, Arimura H, Ohya Y (2002) Biomacromolecules 3:885
182. Ouchi T, Sasakawa M, Arimura H, Ohya Y (2004) Polymer 45:1583
183. Ouchi T, Saito T, Kontani T, Ohya Y (2004) Macromol Biosci 4:458
184. Zhoua S, Liaob X, Lia X, Denga X, Lib H (2003) J Control Release 86:195
185. Liang H-F, Chen CT, Chen SC, Kulkarni AR, Chiu YL, Chen MC, Sung HW (2006) Biomaterials 27:2051
186. Saeed AO, Magnusson JP, Moradi E, Soliman M, Wang W, Stolnik S, Thurecht KJ, Howdle SM, Alexander C (2011) Bioconjug Chem 22:156
187. Liu SQ, Yang YY, Liu XM, Tong YW (2003) Biomacromolecules 4:1784
188. Yokoyama M, Inoue S, Kataoka K, Yui N, Okano T, Sakurai Y (1987) Makromol Chem Rapid Commun 8:431
189. Yokoyama M, Miyauchi M, Yamada N, Okano T, Sakurai Y, Kataoka K, Inoue S (1990) J Control Release 11:269
190. Yokoyama M, Okano T, Sakurai Y, Ekimoto H, Shibazaki C, Kataoka K (1991) Cancer Res 51:3229
191. Yokoyama M, Okano T, Sakurai Y, Fukushima S, Okamoto K, Kataoka K (1999) J Drug Target 7:171
192. Kataoka K, Harada A, Nagasaki Y (2001) Adv Drug Deliv Rev 47:113
193. Torchilin VP (2002) Adv Drug Deliv Rev 54:235
194. Blanazs A, Armes SP, Ryan AJ (2009) Macromol Rapid Commun 30:267
195. Shin HC, Alani AWG, Rao DA, Rockich NC, Kwon GS (2009) J Control Release 140:294
196. Mikihail AS, Allen C (2011) Biomacromolecules 11:1273
197. Ding H, Sumer BD, Kessinger CW, Dong Y, Huang G, Boothman DA (2011) Gao J 151:271
198. Kim SH, Tan JPK, Nederberg F, Fukushima K, Colson J, Yang C, Nelson A, Yang YY, Hedrick JL (2010) Biomaterials 31:8063
199. Wang F, Wang YC, Yan LF, Wang J (2009) Polymer 50:5048
200. Liu J, Pang Y, Huang W, Huang X, Meng L, Zhu X, Zhou Y, Yan D (2011) Biomacromolecules 12:1567
201. Ohya Y, Takeda S, Shibata Y, Ouchi T, Maruyama A (2010) Macromol Chem Phys 211:1750
202. Ohya Y, Takeda S, Shibata Y, Ouchi T, Kano A, Iwata T, Mochizuki S, Taniwaki Y, Maruyama A (2011) J Control Release155:104
203. Discher DE, Eisenberg A (2002) Science 297:967
204. Meier W (2000) Chem Soc Rev 29:295
205. Torchilin VP (2005) Nat Rev Drug Discov 4:145
206. Onaca O, Enea R, Hughes DW, Meier W (2009) Macromol Biosci 9:129
207. Christian DA, Cai S, Bowen DM, Kim Y, Pajerowski D, Discher DE (2009) Eur J Pharm Biopharm 71:463
208. De Cock LJ, De Koker S, De Geest BG, Grooten J, Vervaet C, Remon JP, Sukhorukov GB, Antipina MN (2010) Angew Chem Int Ed 49:6954
209. Chang TMS (2005) Nat Rev Drug Discov 4:221
210. Kishimura A, Koide A, Osada K, Yamasaki Y, Kataoka K (2007) Angew Chem Int Ed 46:6085
211. Price AD, Zelikin AN, Wang Y, Caruso F (2009) Angew Chem Int Ed 48:329

212. Johnston APR, Zelikin AN, Caruso F (2007) Adv Mater 19:3727
213. Bally M, Bailey K, Sugahara K, Grieshaber D, Vörös J, Stäler B (2010) Small 6:2481
214. Holowka E, Sun VZ, Kamei DT, Deming TJ (2007) Nat Mater 6:52
215. Bangham AD (1993) Chem Phys Lipids 64:275
216. Ringsdorf H, Schlarb B, Venzmer J (1988) Angew Chem Int Ed 27:113
217. Yoshina-Ishii C, Miller GP, Kraft ML, Kool ET, Boxer SG (2005) J Am Chem Soc 127:1356
218. Discher BM, Won YY, Ege DS, Lee JCM, Bates FS, Discher DE, Hammer DA (1999) Science 284:1143
219. Lee JC, Bermudez H, Discher BM, Sheehan MA, Won YY, Bates FS, Discher DE (2001) Biotechnol Bioeng 73:135
220. Lee Y, Chang JB, Kim HK, Park TG (2006) Macromol Res 14:359
221. Pata V, Dan N (2003) Biophys J 85:2111
222. Wu J, Eisenberg A (2006) J Am Chem Soc 128:2880
223. Borchert U, Lipprandt U, Bilang M, Kimpfler A, Rank A, Peschka-Suss R, Schubert R, Lindner P, Forster S (2006) Langmuir 22:5843
224. Checot F, Lecommandoux S, Gnanou Y, Klok HA (2002) Angew Chem Int Ed 41:1339
225. Qin SH, Geng Y, Discher DE, Yang S (2006) Adv Mater 18:2905
226. Meng FH, Hiemstra C, Engbers GHM, Feijen J (2003) Macromolecules 36:3004
227. Meng FH, Engbers GHM, Feijen J (2005) J Control Release 101:187
228. Ghoroghchian PP, Li GZ, Levine DH, Davis KP, Bates FS, Hammer DA, Therien MJ (2006) Macromolecules 39:1673
229. Ahmed F, Discher DE (2004) J Control Release 96:37
230. Ahmed F, Pakunlu RI, Srinivas G, Brannan A, Bates F, Klein ML, Minko T, Discher DE (2006) Mol Pharm 3:340
231. Rameez S, Alosta H, Palmer AF (2008) Bioconjug Chem 19:1025
232. Kim Y, Tewari M, Pajerowski JD, Cai SS, Sen S, Williams J, Sirsi S, Lutz G, Discher DE (2009) J Control Release 134:132
233. Nagahama K, Saito T, Ouchi T, Ohya Y (2011) J Biomed Sci Polym Ed 22:407
234. Nagahama K, Ohmura J, Sakaue H, Ouchi T, Ohya Y, Yui N (2010) Chem Lett 39:250
235. Ooya T, Eguchi M, Yui N (2003) J Am Chem Soc 125:13016
236. Yui N (2009) Macromol Symp 279:158
237. Bellomo EG, Wyrsta MD, Pakstis L, Pochan DJ, Deming T (2004) Nat Mater 3:244
238. Makino A, Yamahara R, Ozeki E, Kimura S (2007) Chem Lett 36:1220
239. Tanisaka H, Kizaka-Kondoh S, Makino A, Tanaka S, Hiraoka M, Kimura S (2008) Bioconjug Chem 19:109
240. Koide A, Kishimura A, Osada K, Jang WD, Yamasaki Y, Kataoka K (2006) J Am Chem Soc 128:5988
241. Anraku Y, Kishimura A, Oba M, Yamasaki Y, Kataoka K (2010) J Am Chem Soc 131:1631
242. Makino A, Kimura S (2011) React Funct Polym 71:272
243. Makino A, Kizaka-Kondoh S, Yamahara R, Hara I, Kanzaki T, Ozeki E, Hiraoka M, Kimura S (2009) Biomaterials 30:5156
244. Matsumura Y, Maeda H (1986) Cancer Res 46:6387
245. Maeda H, Matsumura Y (1989) Crit Rev Ther Drug Carrier Syst 6:193
246. Oha JK, Lee DI, Park JM (2009) Prog Polym Sci 34:1261
247. Sasaki Y, Akiyoshi K (2010) Chem Rec 10:366
248. van de Manakker F, Vermonden T, van Nostrum CF, Hennink WE (2009) Biomacromolecules 10:3157
249. Kabanov AV, Vinogradov SV (2009) Angew Chem Int Ed 48:5418
250. Akiyoshi K, Deguchi N, Moriguchi N, Yamaguchi S, Sunamoto J (1993) Macromolecules 26:3062
251. Akiyoshi K, Ueminami A, Kurumada S, Nomura Y (2000) Macromolecules 33:6752
252. Akiyama E, Morimoto N, Kujawa P, Ozawa Y, Winnik FM, Akiyoshi K (2007) Biomacromolecules 8:2366

253. Lee KY, Jo WH, Kwon IC, Kim Y, Jeong SY (1998) Macromolecules 31:378
254. Nichifor M, Lopes A, Carpov A, Melo E (1999) Macromolecules 32:7078
255. Yusa S, Kamachi M, Morishima Y (1998) Langmuir 14:6059
256. Morimoto N, Qiu XP, Winnik FM, Akiyoshi K (2008) Macromolecules 41:5985
257. Sawada S, Akiyoshi K (2010) Macromol Biosci 10:353
258. Takahashi H, Sawada S, Akiyoshi K (2011) ACS Nano 5:337
259. Gou ML, Men K, Zhang J, Li YH, Song J, Luo S, Shi HS, Wen YJ, Guo G, Huang MJ, Zhao X, Qian ZY, Wei YQ (2010) ACS Nano 4:5573
260. Li F, Bae B, Na K (2010) Bioconjug Chem 21:1312
261. Tsuji H, Ikada Y (1993) Macromolecules 26:6918
262. Nagahama K, Hashizume M, Yamamoto H, Ouchi T, Ohya Y (2009) Langmuir 25:9734
263. Harada A, Li J, Kamachi M (1992) Nature 356:325
264. Nepogodiev SA, Stoddart JF (1998) Chem Rev 98:1959
265. Wenz G, Han BH, Müller A (2006) Chem Rev 106:782
266. Harada A, Hashidzume A, Yamaguchi H, Takashima Y (2009) Chem Rev 109:5974
267. Loethen S, Kim JM, Thompson DH (2007) Polym Rev 47:383
268. Li J, Loh XJ (2008) Adv Drug Deliv Rev 60:1000
269. Ooya T, Yui N (1998) Macromol Chem Phys 199:2311
270. Ooya T, Arizono K, Yui N (2000) Polym Adv Technol 11:642
271. Ooya T, Eguchi M, Yui N (2001) Biomacromolecules 2:200
272. Watanabe J, Ooya T, Yui N (1998) Chem Lett 1031
273. Watanabe J, Ooya T, Yui N (1999) J Biomater Sci Polym Ed 10:1275
274. Tachaboonyakiat W, Furubayashi T, Katoh M, Ooya T, Yui N (2004) J Biomater Sci Polym Ed 15:1389
275. Loethen S, Ooya T, Choi HS, Yui N, Thompson DH (2006) Biomacromolecules 7:2501
276. Ooya T, Ito A, Yui N (2005) Macromol Biosci 5:379
277. Ooya T, Choi HS, Yamashita A, Yui N, Sugaya Y, Kano A, Maruyama A, Akita H, Ito R, Kogure K, Harashima H (2006) J Am Chem Soc 128:3852
278. Harada A, Nishiyama T, Kawaguchi Y, Okada M, Kamachi M (1997) Macromolecules 30:7115
279. Harada A, Kawaguchi Y, Nishiyama T, Kamachi M (1997) Macromol Rapid Commun 18:535
280. Huang L, Allen E, Tonelli AE (1998) Polymer 39:4857
281. Kawaguchi Y, Nishiyama T, Okada M, Kamachi M, Harada A (2000) Macromolecules 33:4472
282. Rusa CC, Tonelli AE (2000) Macromolecules 33:5321
283. Shuai X, Porbeni FE, Wei M, Bullions T, Tonelli AE (2002) Macromolecules 35:3126
284. Shuai X, Porbeni FE, Wei M, Bullions T, Tonelli AE (2002) Macromolecules 35:3778
285. Lu J, Shin ID, Nojima S, Tonelli AE (2000) Polymer 41:5871
286. Shuai X, Porbeni FE, Wei M, Shin ID, Tonelli AE (2001) Macromolecules 34:7355
287. Shuai X, Porbeni FE, Wei M, Bullions T, Tonelli AE (2002) Macromolecules 35:2401
288. Choi HS, Ooya T, Sasaki S, Yui N, Ohya Y, Nakai T, Ouchi T (2003) Macromolecules 36:9313
289. Shuai X, Merdan T, Unger F, Kissel T (2005) Bioconjug Chem 16:322
290. Liu KL, Goh SH, Li J (2008) Macromolecules 41:6027
291. Ohya Y, Takamido S, Nagahama K, Ouchi T, Ooya T, Katoono R, Yui N (2007) Macromolecules 40:6441
292. Ohya Y, Takamido S, Nagahama K, Ouchi T, Katoono R, Yui N (2009) Biomacromolecules 10:2261
293. Wu DQ, Wang T, Lu B, Xu XD, Cheng SX, Jiang XJ, Zhang XZ, Zhuo RX (2008) Langmuir 24:10306
294. Hoffman AS (2002) Adv Drug Deliv Rev 54:3
295. Jeong B, Kim SW, Bae YH (2002) Adv Drug Deliv Rev 54:37

296. Heskins M, Guillent JE, James E (1968) J Macromol Sci Chem A2:1441
297. Chen GH, Hoffman AS (1995) Nature 373:49
298. Yamato M, Akiyama Y, Kobayashi J, Yang J, Kikuchi A, Okano T (2007) Prog Polym Sci 32:1123
299. Yoshida R, Uchida K, Kaneko Y, Sakai K, Kikuchi A, Sakurai Y, Okano T (1995) Nature 374:240
300. Heras CDL, Pennadam S, Alexander C (2005) Chem Soc Rev 34:276
301. Tachibana Y, Kurisawa M, Uyama H, Kakuchi T, Kobayashi S (2003) Chem Commun 106
302. Shimokuri T, Kaneko T, Serizawa T, Akashi M (2004) Macromol Biosci 4:407
303. Shimokuri T, Kaneko T, Akashi M (2004) J Polym Sci A Polym Chem 42:4492
304. Ohya Y, Toyohara M, Sasakawa M, Arimura H, Ouchi T (2005) Macromol Biosci 5:273
305. Renier ML, Kohn DH (1997) J Biomed Mater Res 34:95
306. Nguyen MK, Lee DS (2010) Macromol Biosci 10:563
307. Joo MK, Park MH, Choi BG, Jeong B (2009) J Mater Chem 19:5891
308. Jeong B, Bae YH, Kim SW (2000) J Control Release 63:155
309. Kim YJ, Choi S, Koh JJ, Lee M, Ko KS, Kim SW (2001) Pharm Res 18:548
310. Kim HK, Shim WS, Kim SE, Lee K, Kang E, Kim J, Kim K, Kwon IC, Lee DS (2009) Tissue Eng 15:923
311. Duvvuri S, Janoria KG, Mitra AK (2005) J Control Release 108:282
312. Li Z, Ning W, Wang J, Choi A, Lee P, Tyagi P, Huang L (2003) Pharm Res 20:884
313. Fujiwara T, Mukose T, Yamaoka T, Yamane H, Sakurai S, Kimura Y (2001) Macromol Biosci 1:204
314. Mukose T, Fujiwara T, Nakano J, Taniguchi I, Miyamoto M, Kimura Y, Teraoka I, Lee CW (2004) Macromol Biosci 4:361
315. Choi BG, Park MH, Cho SH, Joo MK, Oh HJ, Kim EH, Park K, Han DK, Jeong B (2010) Biomaterials 31:9266
316. Ohya Y, Yamamoto H, Nagahama K, Ouchi T (2009) J Polym Sci A Polym Chem 47:3892
317. Park K, Song S (2005) J Biomater Sci Polym Ed 16:1421
318. Park K, Song S (2006) J Biosci Bioeng 101:238
319. Jeong B, Lee KM, Gutowska A, An YH (2002) Biomacromolecules 3:865
320. Nagahama K, Ouchi T, Ohya Y (2008) Adv Funct Mater 18:1220
321. Nagahama K, Fujiura K, Enami S, Ouchi T, Ohya Y (2008) J Polym Sci A Polym Chem 46:6317
322. Osada Y, Matsuda A (1995) Nature 376:219
323. Langer R, Tirrell DA (2004) Nature 428:487
324. Lendlein A, Langer R (2002) Science 296:1673
325. Lendlein A, Kelch S (2002) Angew Chem Int Ed 41:2035
326. Lendlein A, Schmidt AM, Langer R (2001) Proc Natl Acad Sci USA 98:842
327. Rickert D, Lendlein A, Kelch S, Moses MA, Franke RP (2005) Clin Hemorheol Microcirc 32:117
328. Kelch S, Steuer S, Schmidt AM, Lendlein A (2007) Biomacromolecules 8:1018
329. Neuss S, Blomenkamp I, Stainforth R, Boltersdorf D, Jansen M, Butz N, Perez-Bouza A, Knüchel R (2009) Biomaterials 30:1697
330. Nagahama K, Ueda Y, Ouchi T, Ohya Y (2009) Biomacromolecules 10:1789

Adv Polym Sci (2012) 247: 115–140
DOI: 10.1007/12_2011_154

Published online: 25 October 2011

PEGylation Technology in Nanomedicine

Yutaka Ikeda and Yukio Nagasaki

Abstract PEGylation refers to the covalent attachment of polyethylene glycol to proteins to reduce immunogenicity and extend their time in blood circulation. PEGylation is recognized as a promising method for increasing the therapeutic efficacy of medicines in clinical settings. The main advantages of PEGylation are (1) an increase in the size of drug molecule, resulting in reduced filtration by kidneys, (2) an increase in solubility, and (3) protection from enzymatic digestion and recognition by antibodies. A variety of molecules, such as small molecules, peptides, proteins, enzymes, antibodies and their fragments, and nanoparticles have been modified with PEG. Several PEGylated drugs have been approved by the US Food and Drug Administration (FDA) and several more are being tested in clinical settings. This review summarizes the methodologies and effects of PEGylation on drug delivery and highlights recent developments in PEGylated drugs.

Keywords Drug delivery · Nanomedicine · Nanoparticle · PEGylated drug · PEGylation

---

Y. Ikeda
Department of Materials Science, Graduate School of Pure and Applied Sciences, University of Tsukuba, Ten-noudai 1-1-1, Tsukuba, Ibaraki 305-8573, Japan

Y. Nagasaki (✉)
Department of Materials Science, Graduate School of Pure and Applied Sciences, University of Tsukuba, Ten-noudai 1-1-1, Tsukuba, Ibaraki 305-8573, Japan

Master's School of Medical Sciences, Graduate School of Comprehensive Human Sciences, University of Tsukuba, Tennoudai 1-1-1, Tsukuba, Ibaraki 305-8573, Japan

Satellite Laboratory, International Center for Materials Nanoarchitectonics (MANA), National Institute of Materials Science (NIMS), Tennoudai 1-1-1, Tsukuba, Ibaraki 305-8573, Japan
e-mail: nagasaki@nagalabo.jp

Contents

1 Introduction

In 1977, Abuchowski et al. reported that covalent attachment of polyethylene glycol (PEG) to albumin reduced the immunogenicity of albumin [1]. Subsequently, this group also found that PEGylated biomolecules had a longer blood circulation time than the corresponding normal biomolecules [2]. On the basis of this discovery, PEGylation has been widely recognized as one of the more promising methods for exploration of therapeutic drugs. This exploration includes developments in the methodologies of PEGylation [3]. In the first generation of PEGylated molecules, the target molecule was nonspecifically and irreversibly PEGylated with linear PEG chains (Fig. 1). In the second generation, the molecule was PEGylated with branched PEG chains at specific positions and covalently bound, so that PEG could be released by stimuli from the outside environment. A variety of molecules, including small molecules, peptides, proteins, enzymes, antibodies, antibody fragments, and nanoparticles have been modified with PEG. At present, 11 PEGylated drugs have been approved for clinical use by the US Food and Drug Administration (FDA) and several more trials in clinical settings are ongoing.

2 PEGylation Chemistry

At present, the most frequently used methods for PEGylation are chemical conjugation between reactive groups in the drug, such as the primary amine of lysine in protein, and end-reactive PEG derivatives, such as the *N*-hydroxysuccinimide-

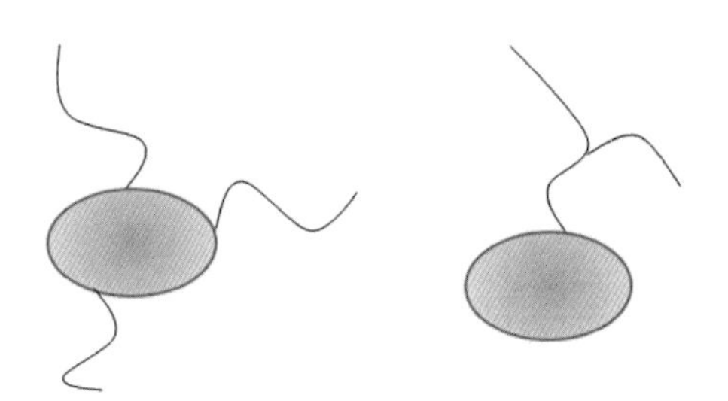

Fig. 1 PEGylation technologies. Nonspecific and irreversible PEGylation is associated with several limitations, such as altered drug properties. To overcome this problem, a new generation of PEGylation technologies that enable highly specific and reversible PEGylation have been developed

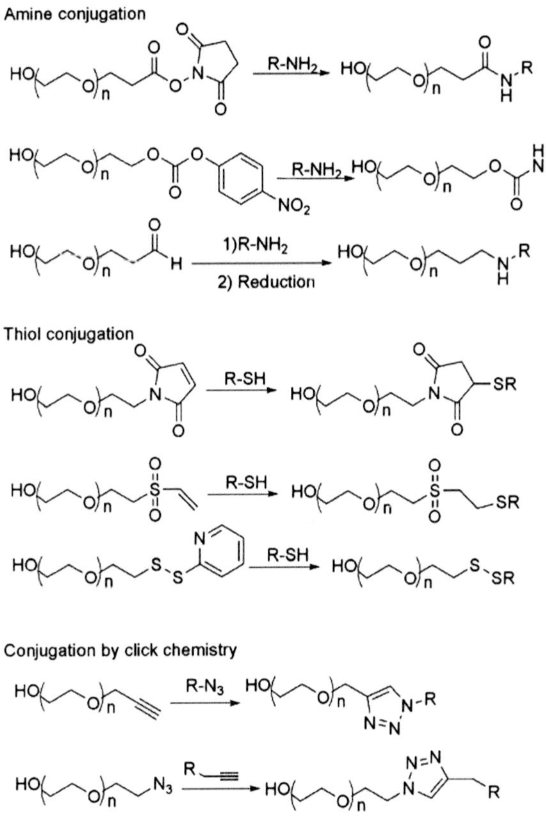

Amine conjugation

Thiol conjugation

Conjugation by click chemistry

Fig. 2 Examples of activated PEG derivatives commonly used for PEGylation

terminated PEG derivative. These conventional methods are summarized in Fig. 2. Amine-specific and thiol-specific reagents are efficient and afford a good yield of PEGylated products. These simple methods can be applied to PEGylation of various molecules. Indeed, several PEGylated drugs have been approved for clinical use by the FDA. However, nonselective conjugations of the amino and/or thiol groups in protein molecules results in product heterogeneity, which often causes significant deactivation of the product. For example, 20–70% of native interferon-beta-1b (IFN-β-1b) antiviral activity was retained in mono-PEGylated IFN-β-1b, but the activity was greatly reduced or disappeared almost completely in multi-PEGylated IFN-β-1b [4]. Therefore, the development of site-specific PEGylation technology is quite important for developing more active and safer drugs.

2.1 Site-Specific PEGylation

PEGylation of drugs with an amine- or thiol-specific reagent is effective and thus the most popular method in current use. The thiol group of cysteine is often used for PEGylation of protein because PEGylation at cysteine is more specific than that of the amino group of lysine. Generally, cysteine forms dithiol linkages and the free thiol group is not available. To utilize the free thiol group for conjugation, it is necessary to engineer a new and free cysteine into the protein via recombination techniques. Although this approach works well, the genetic engineering involved in the process requires high skill, and protein misfolding and aggregation often occur during the purification process. To overcome this problem, Brocchini et al. developed site-specific PEGylation technology [5, 6]. An outline of their technology is shown in Fig. 3. Briefly, this technique involves the synthesis of a novel bis-thiol-specific PEG reagent (PEG monosulfone) containing a thiol-specific bis-alkylating group, which comprises an α, β-unsaturated carbonyl group possessing a sulfomethyl group at the α-position of the unsaturated double bond. After the reduction of the disulfide bond in the protein, both free reactive thiol groups react with the PEG reagent to produce disulfide-bridging PEGylation with a three-carbon bridge.

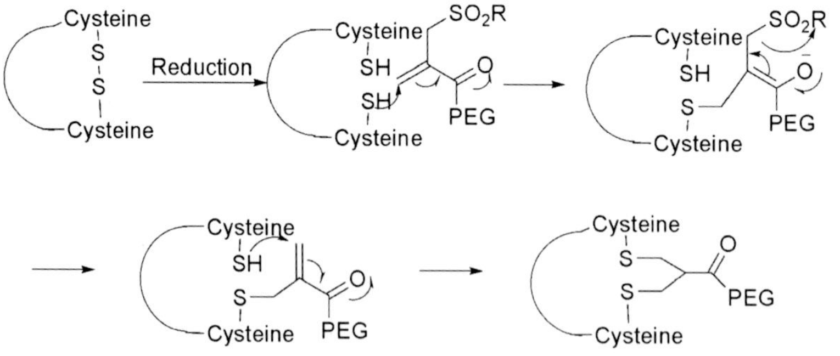

Fig. 3 Chemistry of site-specific PEGylation developed by Brocchini et al. [5, 6]. After cleavage of the native disulfide bond between two cysteine thiols by reduction, the free cysteines are reacted with an α, β-unsaturated PEG derivative to produce a PEG conjugate via a three-carbon bridge

The N-terminal methionine residue of protein can also be employed for selective PEGylation using aldehyde-terminated PEG via a reductive amination reaction, because the N-terminal primary amine has a lower p*K*a of 7.8 than other amines such as lysines, whose p*K*a is 10.1 [7]. After reaction with aldehyde-terminated PEG at low pH, the resultant imine is reduced with sodium cyanoborohydrate to provide PEGylated protein (Fig. 4) [8, 9]. This technique was used for the production of Neulasta, which was approved for use by the FDA in 2002 [10].

2.2 *Enzymatic PEGylation*

Novel methods have been proposed by various researchers to achieve site-specific PEGylation using enzymatic PEGylation reactions. Sato et al. utilized transglutaminase (TGase; protein-glutamine γ-glutamyltransferase), an emerging enzyme [11]. This enzyme catalyzes an acyl transfer reaction between the γ-carboxyamide group of the glutamine residue (acyl donor) in a protein and a variety of primary amines (acyl acceptor) (Fig. 5) [12]. TGase is believed to require special sequential structures of the acyl donor for efficient modification; however, it is interesting to note that a variety of primary amines are accepted as acyl donors. Thus, TGase is a very useful reagent for protein modification [13]. An amino derivative of PEG, such as PEG-NH_2, can be used as an acyl acceptor for PEGylation of proteins. Several

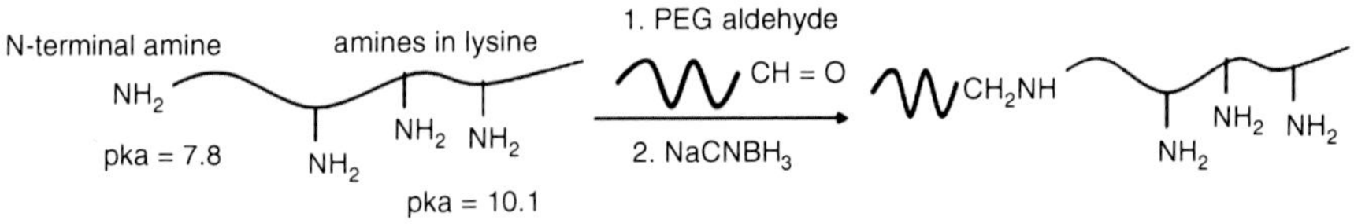

Fig. 4 PEGylation at the N-terminal methionine residue. The difference in p*K*a between the N-terminal amine and other amines in the protein enables site-specific PEGylation. After reaction with aldehyde-terminated PEG at low pH, reduction of the resultant imine produces PEGylated protein

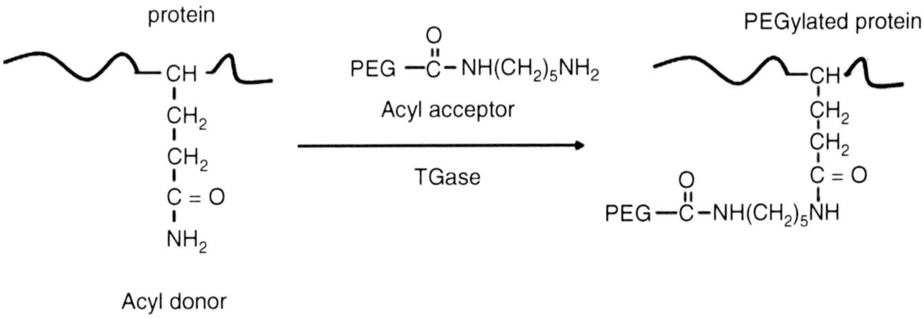

Fig. 5 Enzymatic site-specific PEGylation by transglutaminase (*TGase*). The alkylamine derivative of PEG can be introduced into proteins in a site-specific manner

clinically important proteins, including human growth hormone and interleukin-2, have been PEGylated using TGase [14].

2.3 Heterobifunctional PEG

Heterobifunctional PEG, which possesses different functional groups at the α- and ω-chain ends, is very useful in the field of drug delivery [15]. For example, heterobifunctional PEG can conjugate drug-containing nanoparticles with a targeting ligand (Fig. 6). In one study, a method for the synthesis of heterobifunctional PEG by direct ring-opening polymerization of ethylene oxide (EO) using a metal alkoxide initiator with a protected functional group was developed (Fig. 7) [16, 17]. This useful method was further developed by Akiyama et al., [18–21].

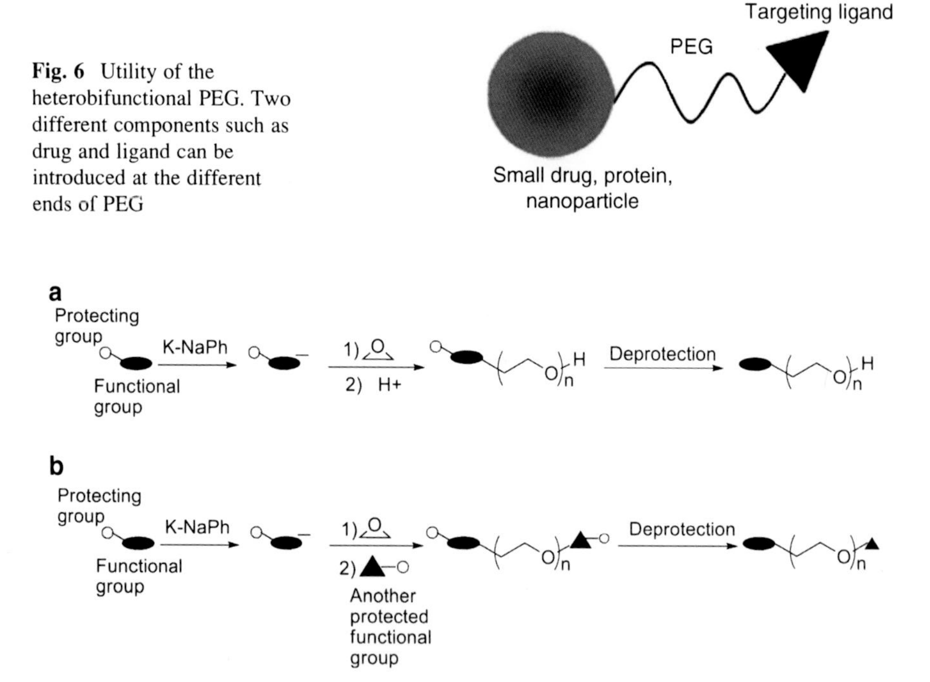

Fig. 6 Utility of the heterobifunctional PEG. Two different components such as drug and ligand can be introduced at the different ends of PEG

Fig. 7 Synthesis of heterobifunctional PEG. (**a**) Nagasaki et al. developed a method for the polymerization of EO using an initiator containing defined functionalities [16, 17]. (**b**) Akiyama et al. further developed a synthetic route to prepare a series of heterobifunctional PEGs [18–21]. After the ring-opening polymerization of ethylene oxide, a second functional group was introduced at the ω-end of PEG

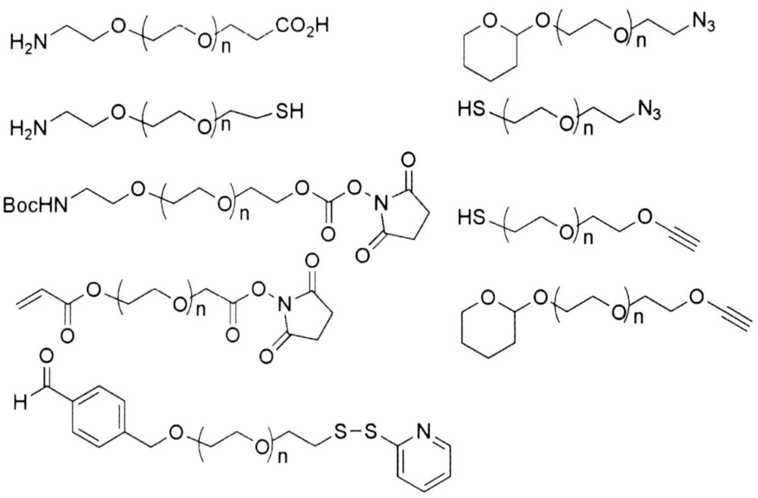

Fig. 8 Examples of heterobifunctional PEGs popularly used for PEGylation

To date, a variety of heterobifunctional PEGs have been reported. Some popular derivatives are shown in Fig. 8. Several PEG derivatives are now commercially available from NOF, Japan.

2.4 Linear and Branched PEGs

Early PEGylation technology utilized linear PEG chains for conjugation. As stated above, multiple and nonspecific conjugations often change the activity of native proteins significantly, and it has been reported that the large molecular weight of PEG causes a tendency to accumulate in the liver [22]. Branched PEG derivatives are effective candidates for solving these issues. The same PEGylation effect on a pharmaceutical can be obtained by introducing a smaller branched PEG with fewer conjugation points. The second generation of PEGylation technology often utilized branched PEGs because branched PEGylated products circulate longer in the blood than linear PEGylated products [23]. This effect is thought to be because of the steric hindrance of branched PEG [24]. PEGylation with branched-chain PEG has been adopted in the development of FDA-approved drugs, including PEGASYS [25], Macugen [26], and Cimzia [27].

Size-exclusion chromatography (SEC) showed no significant difference in size between branched and linear PEGylated proteins [28]. Therefore, the longer in vivo half-life of branched PEGylated drugs was not due to the size of the conjugate in solution, but probably to the more effective masking of the protein surface by branched PEGs.

Although the detailed mechanism of the longer circulation time of branched PEGylated protein is unclear, the architecture of PEG affects the release profile, the pharmacokinetics of the drug [29], and the behavior of the protein at the interface (e.g., protein absorption on hydrophobic surfaces [30]).

2.5 Releasable PEGylation

Although covalent attachments of PEG to drugs prolongs the lifetime of the drug in vivo, they often have the opposite effect on biological and pharmacological properties because the active site of drugs is inactivated due to shielding by massive PEG chains [31, 32]. Even optimized site-specific PEGylation often results in insufficient pharmacological properties. It also prevents internalization of the drug into the cell [33]. New emerging technologies, such as de-PEGylation from complex drugs, have been developed. Specific biodegradable linkages between the drug and PEG chains are introduced to allow de-PEGylation. After the release of PEG chains, drugs such as small molecules and proteins recover their original structures, activities, and cellular-uptake capacities.

Roberts et al. reported the synthesis of PEG–drug conjugates via an ester bond between the drug and PEG chain [34]. Despite its simplicity and efficacy, it is difficult to regulate the release specificity because numerous esterases exist in the cellular environment. In addition, many biologically active compounds often lack a hydroxyl or carboxyl group, which is required for ester formation. In this system, residues of linking reagent that connect PEG to the drug were left on the parent molecules, even after the cleavage of PEG chains [34]. These residues may affect the biological activities of the drug and might be a potential source of immunogenicity.

Shulman et al. synthesized bifunctional linking reagents containing 2-sulfo-9-fluorenyl-methoxycarbonyl (FMS) to produce a PEG conjugate that can be cleaved by spontaneous hydrolysis under physiological conditions, on the basis of the FMS cleavable system (Scheme 1) [35]. Amino groups in drugs can be utilized for conjugation to PEG in this case.

Zalipsky et al. reported a drug–PEG conjugate via a benzyl carbamate linkage (Scheme 2) [36]. This linkage is cleaved by a benzyl elimination reaction initiated by the thiolytic cleavage of disulfide in the *para* or *ortho* position. Filpula et al. developed a series of releasable PEG linkers that enable the controlled release of drugs [37]. In their system, the cleavage reaction is initiated by a trigger reaction, such as ester bond cleavage by esterase (Scheme 3). By controlling the rate-determining step with an optimized linker structure, the release rate of PEG chains can be controlled. For example, the introduction of steric hindrance, which slows the triggered hydrolysis reaction of esters, results in a diminishing release rate [37].

ProLynx LLC has developed another type of releasable PEGylation technology based on the β-elimination reaction shown in Scheme 4. In this system, the release rate of the native drug is determined by the acidity of the proton adjacent to the

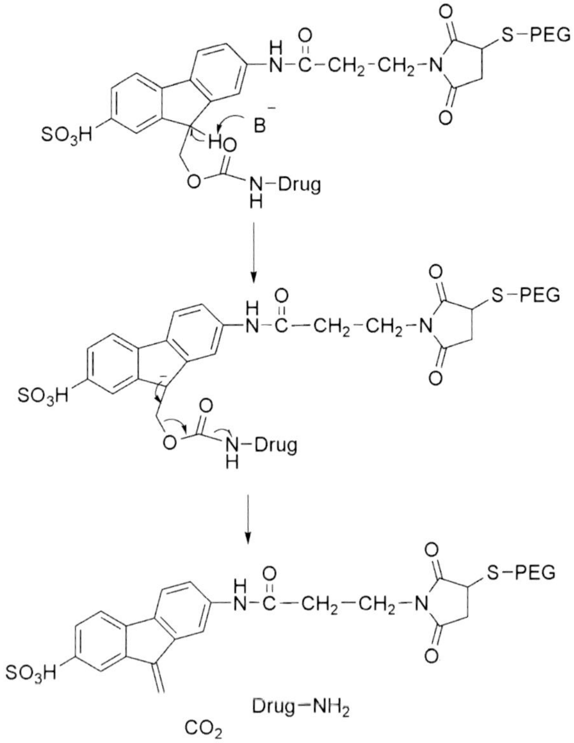

Scheme1 Releasable PEGylation based on the FMS principle

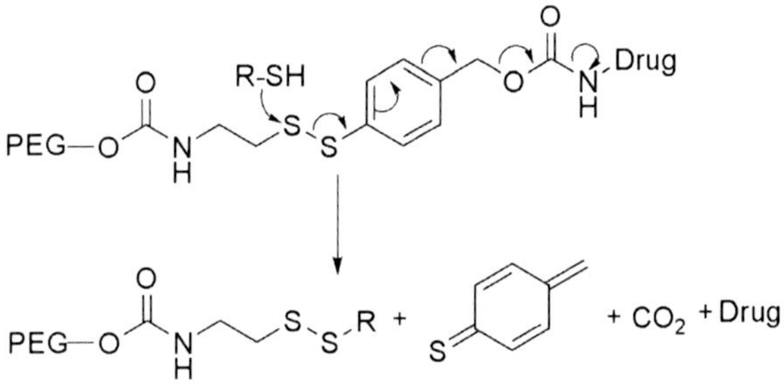

Scheme 2 Drug–PEG conjugate via a benzyl carbamate linkage and its thiolytic cleavage

modulator. Half-lives of molecules prepared using this approach range from several hours to several weeks.

The concept of de-PEGylation can be applied to the development of nanoparticle-based drug delivery systems. PEG is used for the modification of liposomes to increase their blood circulation time [38]. However, it also prevents cellular uptake, resulting in a decrease in therapeutic efficiency; thus, modifications of the liposome surface with PEG interfere with membrane fusion to the cell membrane and liposome decomposition [39]. One of the possible strategies to solve this problem is to cleave the PEG chains after the nanoparticle reaches the target site (Fig. 9). This system of de-PEGylation of liposomes is also useful in avoiding the immune

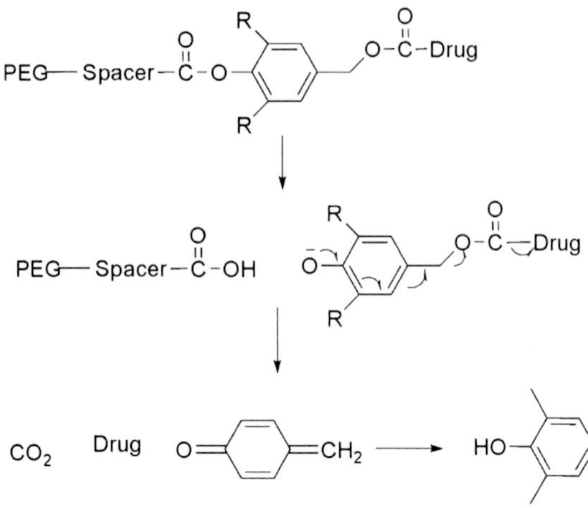

Scheme 3 Releasable PEGylation based on 1,6-benzyl elimination prodrug strategy

Scheme 4 Releasable PEGylation based on the β-elimination reaction

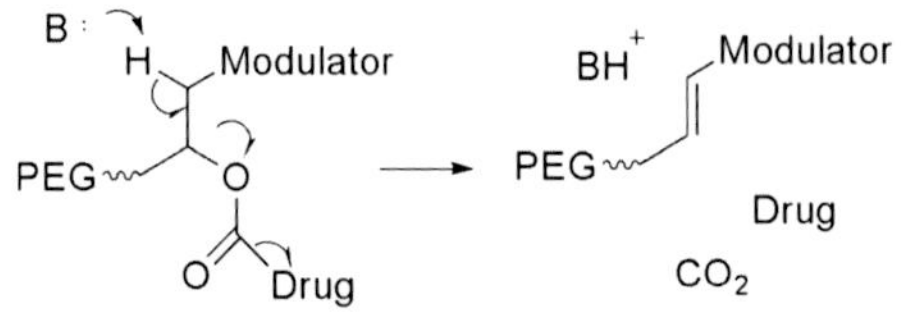

response called the accelerated blood clearance phenomenon (ABC phenomenon), in which the circulation time of a second dose of injected PEGylated liposome is substantially reduced [40, 41]. Harashima et al. developed a multifunctional envelope-type nanodevice (MEND) for a nonviral gene delivery system [39, 42]. In this system, multiple device functions are assembled into a single system. In particular, PEGs on the surface of liposomes were designed to be cleaved by enzymes such as matrix metalloproteinases (MMPs) that are specifically expressed at tumor sites. Removing PEGs resulted in an improvement of cellular uptake and subsequent endosomal escape of the liposome.

3 Effect of PEGylation on Pharmaceuticals

3.1 Reduction of Renal Clearance

One of the remarkable properties of PEGylation is to increase the hydrodynamic radius in order to decrease renal clearance. For example, Kubetzko et al. have reported that antibody fragments (theoretically 29 kDa) showed retention volumes corresponding to a size range of 200–300 kDa by SEC upon mono-PEGylation with

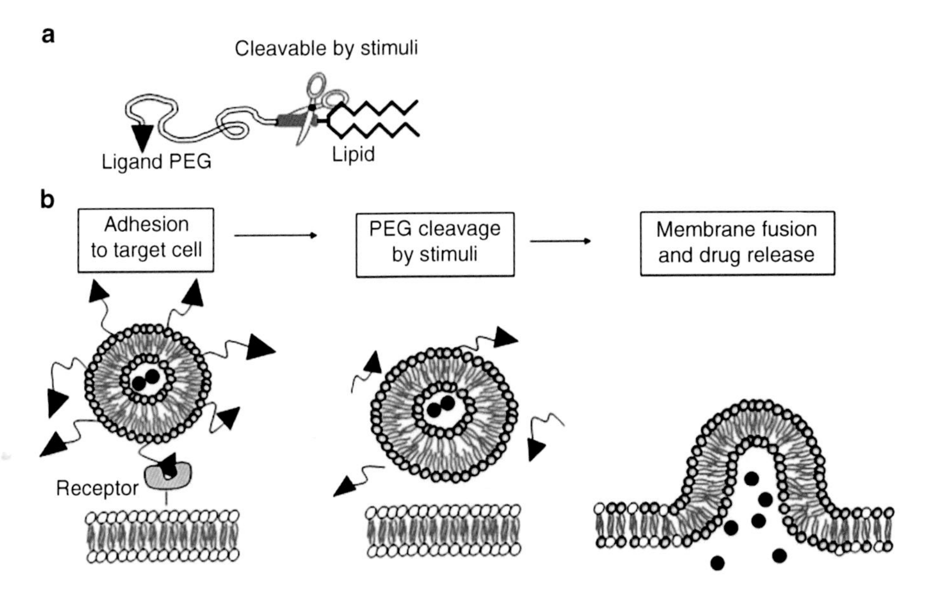

Fig. 9 Utility of de-PEGylation technology in liposomes. (**a**) PEG derivative possessing a lipid moiety. The covalent bond between PEG and the lipid moiety can be cleaved by stimuli such as those within the acid environment of cancer and inflammation. (**b**) After binding the target cell via specific recognition of the receptor by the ligand, PEG molecules on the surface of the liposome are cleaved. The release of PEG facilitates membrane fusion of the liposome and liposome decomposition, resulting in efficient drug delivery

PEG (20 kDa) [43]. This demonstrates the strong hydrodynamic properties of PEGylated molecules. The increase in hydrodynamic radius significantly decreases renal clearance. Although the threshold of the molecular weight cut-off of renal filtration of protein is about 65 kDa, the 30-kDa PEG demonstrates minimal renal permeability [44].

3.2 *Molecular Recognition of PEGylated Molecules*

Although PEGylated molecules in the blood stream have longer half-lives than the parent molecules, studies have reported conflicting conclusions about changes in the binding affinities of PEGylated molecules. Chapman et al. demonstrated that site-specific modification of antibody fragments at the termini of PEG diminishes the loss of activity of the antibody fragment [45]. In contrast, Kubetzko et al. reported a fivefold decrease in apparent affinity upon attachment of the 20-kDa PEG molecule at the C-terminus of the antibody fragment [43]. By analyzing the binding kinetics, they found that the reduction in affinity was mainly due to a slower association rate constant, whereas the dissociation rate constant was nearly unchanged. Using a mathematical model, intramolecular and/or intermolecular blocking by tethered PEG were proposed as the main factors behind the decrease in the observed association rate constant. The model suggested that more than 90% of PEGylated ligands are not capable of binding the target, indicating that accessibility to PEGylated molecules is significantly restricted [43]. This model can partially explain the lower immunogenicity and higher enzymatic resistance of PEGylated molecules.

Although the affinity was decreased fivefold upon PEGylation, PEGylated antibody fragments showed an 8.5-fold higher accumulation in tumors than unmodified antibody fragments, because of a longer serum half-life [46].

3.3 *PEGylation on the Surface*

PEGylation technology is also relevant for solid surfaces. Immobilization of proteins, antibody fragments, and whole antibodies has been widely used in biosensing and bioseparation systems [47]. There are several factors that affect the ability of these systems. These factors include the quantity, density, conformation, and orientation of the immobilized molecules. A common method of immobilization of a protein is based on the reaction between reactive residues in the protein, such as lysine, and the reactive surface. Yoshimoto et al. immobilized antibody fragments (Fab′) on a gold surface through S–Au linkage [48]. However, after the initial absorption of Fab′ onto the gold surface, reactions between the interactive residues of Fab′ and the gold surface changed the conformation and orientation of Fab′, resulting in the inactivation of the antigen-binding function of

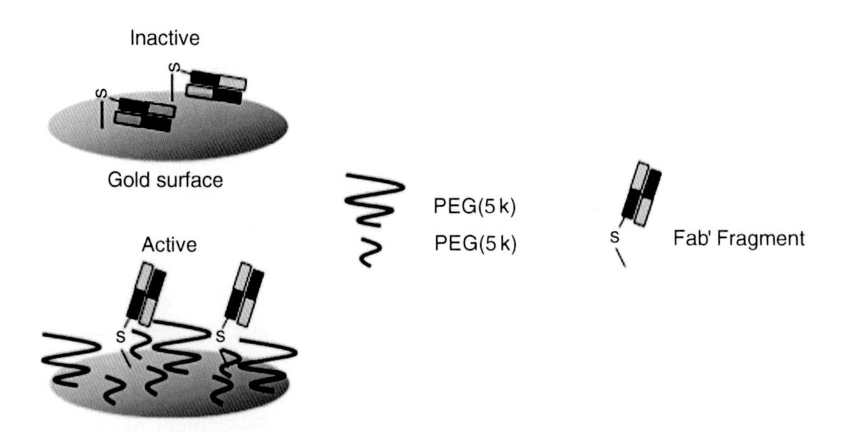

Fig. 10 PEGylation on the surface. A highly dense PEG layer composed of mixed-PEG prevents nonspecific protein interactions and inactivation of Fab′

Fab′. To overcome this problem, they used a mixed-PEG layer formation in which different molecular weights of PEG (2 and 5 kDa) were used for the formation of a densely packed PEG layer. Formation of a mixed-PEG layer was originally developed by our group [49]. A highly dense mixed-PEG layer almost completely prevented nonspecific protein absorption and facilitated biospecific interactions (Fig. 10) [48]. This methodology can be applied to the construction of targeted drug delivery systems, in which a ligand is needed on the surface of nanoparticles to bind the target molecule with high specificity and efficiency.

4 PEGylated Drugs

PEGylation technology has been applied to the development of various kinds of drugs, including small molecules, peptides, proteins, antibody fragments, whole antibodies, oligonucleotides, and macromolecules such as polymer micelles and liposomes. Currently, there are ten PEGylated drugs, which utilize proteins, enzymes, antibody fragments, and oligonucleotides, and a PEGylated nanoparticle named Doxil available on the market (Table 1). PEGylated small drugs are currently under investigation in clinical tests; however, there are no approved drugs available on the market. Sections 4.1–4.4 describe various PEGylated drugs that have been approved or are in clinical trials.

4.1 Small Molecules

PEGylation of small drugs has several advantages. First, PEGylation improves pharmacokinetic properties due to the increased blood circulation time. Second, the immunogenicity of immunogenic small drugs is reduced. Third, PEGylation increases drug accumulation in tumors via the enhanced permeability and retention (EPR) effect. Finally, the toxicity of the drug is reduced by the massive PEG molecule.

Enzon pharmaceuticals have developed a PEGylated drug called SN-38 (7-ethyl-10-camptothecin) [50]. SN-38 is an active metabolite of irinotecan and is produced by hydrolysis of CPT-11 [51]. Several problems arise in the development of drugs using SN-38 or CPT-11. First, carboxylesterase-2 is thought to be the main esterase that hydrolyzes CPT-11; however, the expression of this enzyme in the blood is low. Accordingly, only 1–9% of injected CPT-11 is converted to the active form SN-38. The second problem associated with CPT-11 is the opening reaction of the lactone-E ring, which results in a form that is inactive against the target protein. Finally, SN-38 cannot be used for systemic applications because of poor solubility. This problem was solved by PEGylation of SN-38. PEGylation of SN-38 by acylation of the 20-hydroxyl functional group improved water solubility and preserved the active lactone form in the circulation. To increase the loading of

Table 1 List of approved PEGylated drugs and their properties

PEG drug in formation	PEG	PEGylation	PEG size (kDa)	Year of approval
Adagen (adenosine deaminase)		Nonspecific, amine PEGylation	5	1990 (USA)
Oncaspar (L-asparaginase)		Nonspecific, amine PEGylation	5	1994 (USA, EU)
Doxil (PEGylated liposome)		PEG possessing lipid moiety is used as a component of liposome	2	1995, 1999 (USA), 1996 (EU)
PEG-INTRON (interferon-α-2b)		Histidine 34 (major), amine PEGylation	12	2000 (EU), 2001 (USA)

PEGASYS (interferon-α-2a)	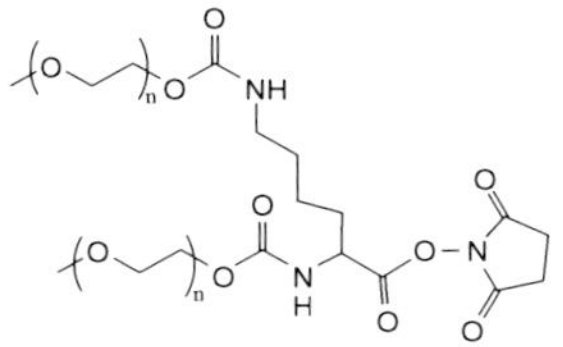	Lysines 31, 121, 131, or 134, amine PEGylation	40	2002 (USA, EU)
Neulasta (G-CSF)	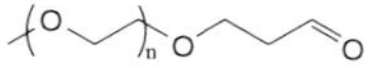	N-Terminal methionine, amine PEGylation	20	2002 (USA), 2003 (EU)
Somavert (hG11 antagonist B2036)	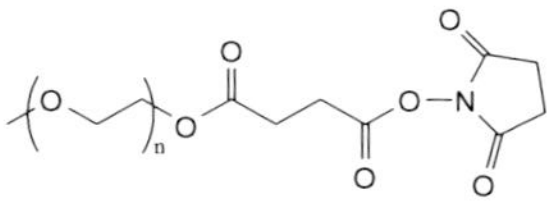	Nonspecific, amine PEGylation	5	2002 (EU), 2003 (USA)
Macugen (anti-VEGF, aptamer)	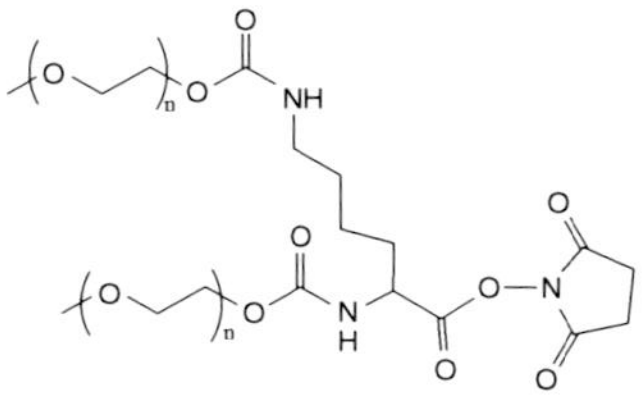	5′-Terminus of modified oligo RNA, amine PEGylation	40	2004 (USA), 2006 (EU)
Mircera (continuous erythropoietin receptor activator)	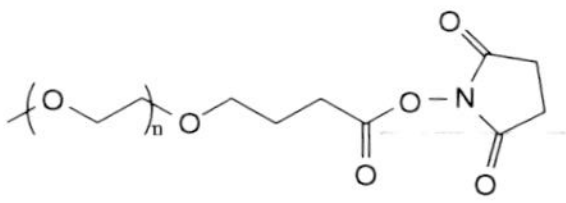	Lysines 52 or 46, amine PEGylation	30	2007 (USA, EU)

(continued)

Table 1 (continued)

PEG drug in formation	PEG	PEGylation	PEG size (kDa)	Year of approval
Cimzia (anti-TNFα, Fab′)	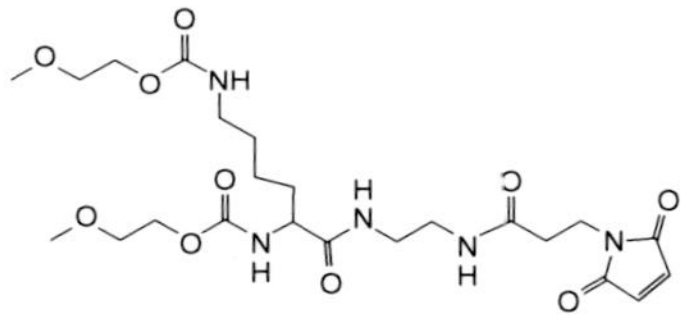	C-Terminal cysteine, thiol PEGylation	40	2008, 2009 (USA), 2009 (EU)
Krystexxa (mammalian urate oxidase)	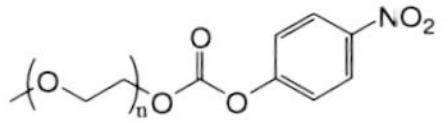	Nonspecific, amine PEGylation	10	2010 (USA)

the drug onto PEG, multi-arm PEG (a four-armed-PEG derivative) was used to enable the conjugation of four drugs to one molecule [52]. This conjugate will now be assessed in further preclinical development and clinical trials [53].

4.2 Peptides, Proteins, Antibodies, and Antibody Fragments

The majority of approved PEGylated drugs are categorized into this group. Eight PEGylated proteins, including antibody fragment conjugates, have been approved to date. Adagen is the first approved PEGylated product in which bovine adenosine deaminase is randomly conjugated with a 5-kDa mono-methoxy PEG [54]. This conjugate is synthesized using PEG succinimidyl succinate. This activated ester group can be reacted with nucleophilic amino acid units such as lysine. Since the approval of Adagen, seven other protein–PEG conjugates have been approved. Although nonspecific PEGylation has been reported to decrease the activity of protein in some cases, several approved drugs employ nonspecific PEGylation. Krystexxa, which was approved in 2010 for the management of treatment-resistant gout and hyperuricemia, was also prepared using nonspecific PEGylation. A conjugate containing six strands of 10-kDa PEG per subunit was found to have a significantly longer half-life in blood and dramatic urate-lowering potency [55].

In contrast to nonspecific PEGylation, several protein–PEG conjugates have adopted site-specific PEGylation. Cimzia is a PEGylated anti-tumor necrosis factor (TNF)-α antibody fragment used for the treatment of Crohn's disease and rheumatoid arthritis [27]. Recent progress in biotechnology has enabled low-cost production of the recombinant antibody fragment by *Escherichia coli* expression [56]. However, such proteins, which are obtained and purified from *E. coli,* often possess immunogenicity. PEGylation on the protein reduces the immunogenicity of the recombinant non-human protein. To prepare Cimzia, the C-terminal cysteine is reacted with maleimide, which is introduced at the end of the 40-kDa branched PEG chain [27].

Another type of PEG–antibody fragment is PEGylated di-Fab, in which two antibody fragments are attached to PEG [57]. CDP791 is prepared using a bis-maleimide PEG and a humanized antibody fragment, resulting in a divalent PEGylated Fab fragment. This unique architecture enables high affinity for vascular endothelial growth factor receptor 2 (VEGFR-2), resulting in reduction of solid tumors [58].

4.3 Oligonucleotide–PEG Conjugates

Oligonucleotide-based drugs such as antisense drugs, aptamers, and small interfering RNA (siRNA) have attracted considerable attention as promising therapeutic agents for the treatment of various human diseases [59]. To develop therapeutic oligonucleotides, several issues must be addressed [60]. These issues include the instability of native oligonucleotides and their rapid clearance from the blood. To overcome these problems, PEGylation of the oligonucleotide is very useful. For example, Macugen, which is used for the treatment of the wet form of age-related macular degeneration (ARMD), is an aptamer drug modified with branched 40-kDa PEG at the 5′-terminus [26]. A number of PEGylated oligonucleotides are now at various stages of clinical trials.

4.4 *PEGylated Nanoparticles*

Precise design of the surface of nanoparticles is very important for the efficient and specific delivery of the drug by the nanocarrier. Surface modification of nanoparticles, such as micelles and liposomes, with PEG represents an essential strategy for reducing nonspecific interactions with serum proteins and endothelial cells in the blood stream, as well as avoiding recognition by immune system components such as the reticuloendothelial system (RES) [61]. Thus, stabilized nanocarriers tend to yield long blood circulation times and facilitate accumulation in the tumor tissue through the effects of EPR. Tamura et al. reported the influence of the surface PEG density on nanoparticles in the blood circulation [62] by using a nanogel composed of chemically crosslinked poly[2-*N*,*N*-(diethylamino)ethyl methacrylate] (PEAMA) gel cores surrounded by PEG palisade layers [63]. Because of their chemically crosslinked polyamine gel core, the PEGylated nanogels show higher stability against extremely dilute and high salt conditions than self-assembled nanocarriers such as liposomes and micelles. This stable nanoparticle is suitable for studying the influence of the physicochemical properties of nanoparticles on pharmacokinetics. The density of PEG on the surface of the nanogels was controlled by the post-PEGylation method. It was clearly demonstrated that the blood circulation time of the post-PEGylated nanogels was definitely prolonged as the PEG content was increased [62].

4.4.1 PEGylated Liposomes

Several liposome-based drugs have been approved for clinical application [64]. One of the clinically approved liposomes is Doxil, a PEGylated liposome containing doxorubicin (DOX), which is used for the treatment of a number of diseases [65]. As shown in this case, in the field of liposome drug development, PEG is widely used to protect the liposome from recognition by opsonins, thereby reducing liposome clearance.

A number of PEGylation reactions with liposomes have been developed [66]. One of the methods utilizes lipophilic compounds that possess reactive groups such as amino and carboxyl groups. By incorporating these components into the bilayer membrane, 500–2,000 functional groups can be introduced onto the liposome

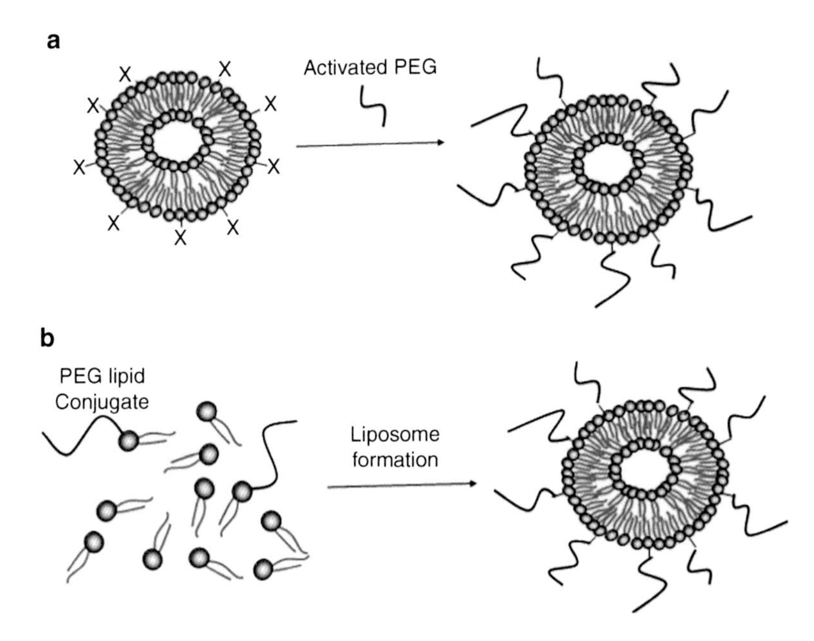

Fig. 11 Methods for the construction of PEGylated liposomes. (**a**) Liposomes possessing reactive groups, such as amino and carboxyl groups, can be prepared by incorporating lipophilic components containing these functional groups into a bilayer membrane. Functionalized liposomes can be PEGylated by reaction with activated PEG derivatives. (**b**) Preparation of PEGylated liposomes using PEG derivatives possessing lipid moieties

surface. This functionalized liposome can be used for the preparation of PEGylated liposomes (Fig. 11).

Another method for the preparation of PEGylated liposomes utilizes PEG, which possesses a lipid moiety at one end, in conjunction with low molecular weight lipid molecules during the preparation of the liposome (Fig. 11). This method was originally reported by Kilbanov et al. in 1990 [67]. They conjugated phosphatidylethanolamine with PEG possessing activated carboxyl groups. For liposome preparation, 7.4 mol% of the PEG lipid was incubated. At this ratio of components, no increase in the leakage of the entrapped compound was observed. This methodology was adopted for the preparation of Doxil.

4.4.2 Micelles

Polymeric micelles containing anticancer drugs were originally developed by Kataoka and Kabanov, independently [68, 69]. Anticancer drugs are incorporated into micelles via physical entrapment or chemical conjugation. A number of micelles are being assessed in clinical trials [70], and progress on the polymer micelle system is emerging. Nishiyama et al. reported the development of a

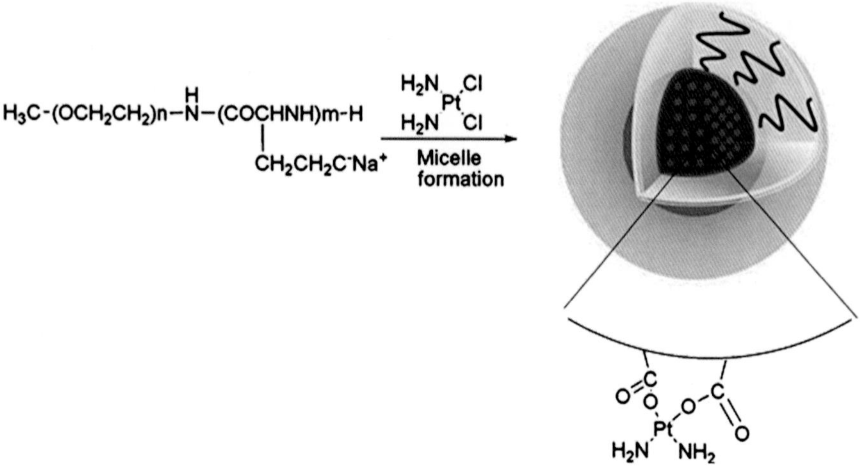

Fig. 12 Cisplatin-loaded micelle developed by Nishiyama et al. [71]. Cisplatin is bonded to block polymers via coordination by carboxylate groups in the core of the micelle

Fig. 13 The structure of DOX-conjugated PEG-*b*-poly (aspartic acid) diblock copolymer. DOX molecules are covalently bonded to a diblock copolymer via hydrazine linkage, which can be cleaved in acidic conditions, enabling the release of DOX in a site-specific manner

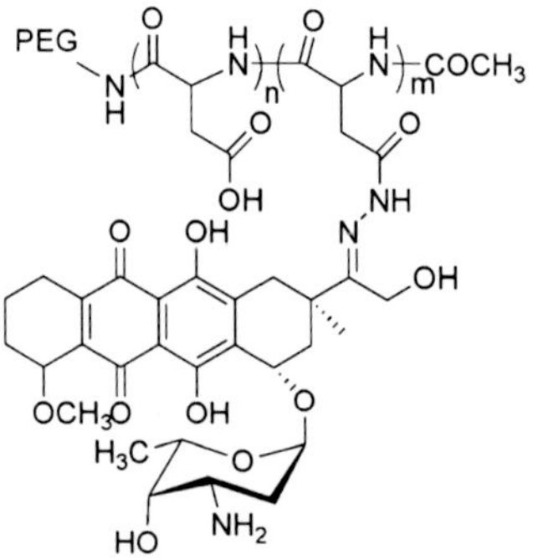

cisplatin-loaded micelle [71], in which platinum is coordinated by carboxylate groups in block copolymers consisting of PEG and polyaspartate (Fig. 12). DOX has been also loaded into micelles by chemical conjugation. Bae et al. developed a novel method for the conjugation of DOX with PEG-*b*-poly(aspartic acid) diblock copolymers [72], via a hydrazine linkage, which enables the release of DOX in acidic environments such as acidosis and endocytosis. This property facilitates the stimuli-responsive delivery of the drug (Fig. 13).

4.4.3 Inorganic Nanoparticles

Inorganic nanoparticles have also attracted interest in the field of drug delivery [73]. These inorganic nanoparticles include calcium phosphate, gold, silicon oxide, and iron oxide. They can be prepared easily with controllable size and can be readily functionalized. Inorganic nanoparticles, however, are generally unstable and may be toxic in biological systems. Accordingly, surface modification is needed to improve the biological stability and biocompatibility.

Surface modification of inorganic nanoparticles with PEG is a very useful way to overcome this problem. For instance, thiol groups are suitable anchors on gold nanoparticles. A variety of drugs, such as small compounds, oligonucleotides, and proteins, have been delivered by gold nanoparticles that carry drugs co-immobilized with thiol-PEG [74, 75]. For instance, recombinant human TNF-α was immobilized on PEGylated colloidal gold nanoparticles [76] to facilitate preferential accumulation of TNF-α in tumors and minimal uptake in healthy organs. This TNF-α-immobilized gold nanoparticle has been evaluated in a Phase I clinical trial [77]. Our group has investigated the stabilization of nanoparticles by block copolymers possessing PEG as one of the segments. For example, PEG-*b*-poly[2-(*N*,*N*-dimethylamino)ethyl methacrylate] significantly improves the stabilization of gold nanoparticles under physiological conditions [78, 79]. Multi-anchoring of amino groups in the poly[2-(*N*,*N*-dimethylamino)ethyl methacrylate] segment strongly improves adsorption efficiency. Luminescent nanoparticles have also been modified by several block copolymers [80, 81].

5 Conclusions and Future Prospects

Since PEGylation of proteins was first reported in the 1970s, extensive research on PEGylation technology and pharmaceutical development of PEGylated molecules has been conducted. A variety of molecules, including small organic molecules, proteins, antibody fragments, and nanoparticles have been modified with PEG. Currently, 11 PEGylated drugs have been marketed, and many other PEGylated drugs are in clinical trials. In recent years, the success rate for bringing new drugs to market has been decreasing [82]. One of the reasons for this is that the FDA is highly focused on the safety of new drugs. In this regard, PEGylation is very useful because PEG is categorized as "generally regarded as safe" (GRAS) by the FDA. Although there are potential concerns regarding non-degradability, product heterogeneity, and accumulation of large linear PEG chains in the liver [22], PEG provides substantial benefits, such as reduced immunogenicity and antigenicity of the drug. As in the cases of PEG-INTRON, Neulasta, and Doxil, emerging drugs can be developed by PEGylation of previously commercialized non-PEGylated drugs. By further development of cost-effective PEGylation technologies that enable more controlled release of PEG from the drug and site-specific modifications to deliver homogeneous products, the market of PEGylated drugs will continue to grow.

References

1. Abuchowski A, van Es T, Palczuk NC, Davis FF (1977) Alteration of immunological properties of bovine serum albumin by covalent attachment of polyethylene glycol. J Biol Chem 252:3578–3581
2. Abuchowski A, McCoy JR, Palczuk NC, van Es T, Davis FF (1977) Effect of covalent attachment of polyethylene glycol on immunogenicity and circulating life of bovine liver catalase. J Biol Chem 252:3582–3586
3. Roberts MJ, Bentley MD, Harris JM (2002) Chemistry for peptide and protein PEGylation. Adv Drug Deliv Rev 54:459–476
4. Basu A, Yang K, Wang M, Liu S, Chintala R, Palm T, Zhao H, Peng P, Wu D, Zhang Z, Hua J, Hsieh MC, Zhou J, Petti G, Li X, Janjua A, Mendez M, Liu J, Longley C, Zhang Z, Mehlig M, Borowski V, Viswanathan M, Filpula D (2006) Structure-function engineering of interferon-beta-1b for improving stability, solubility, potency, immunogenicity, and pharmacokinetic properties by site-selective mono-PEGylation. Bioconjug Chem 17:618–630
5. Balan S, Choi JW, Godwin A, Teo I, Laborde CM, Heidelberger S, Zloh M, Shaunak S, Brocchini S (2007) Site-specific PEGylation of protein disulfide bonds using a three-carbon bridge. Bioconjug Chem 18:61–76
6. Brocchini S, Godwin A, Balan S, Choi JW, Zloh M, Shaunak S (2008) Disulfide bridge based PEGylation of proteins. Adv Drug Deliv Rev 60:3–12
7. Wong SS (1991) Chemistry of Protein Conjugation and Cross- linking. CRC Press, Boston
8. Hu J, Sebald W (2011) N-terminal specificity of PEGylation of human bone morphogenetic protein-2 at acidic pH. Int J Pharm 413:140–146
9. Lee H, Jang IH, Ryu SH, Park TG (2003) N-terminal site-specific mono-PEGylation of epidermal growth factor. Pharm Res 20:818–825
10. Renwick W, Pettengell R, Green M (2009) Use of filgrastim and pegfilgrastim to support delivery of chemotherapy: twenty years of clinical experience. BioDrugs 23:175–186
11. Sato H (2002) Enzymatic procedure for site-specific pegylation of proteins. Adv Drug Deliv Rev 54:487–504
12. Lorand L, Parameswaran KN, Stenberg P, Tong YS, Velasco PT, Jönsson NA, Mikiver L, Moses P (1979) Specificity of guinea pig liver transglutaminase for amine substrates. Biochemistry 18:1756–1765
13. Griffin M, Casadio R, Bergamini CM (2002) Transglutaminases: nature's biological glues. Biochem J 368:377–396
14. Fontana A, Spolaore B, Mero A, Veronese FM (2008) Site-specific modification and PEGylation of pharmaceutical proteins mediated by transglutaminase. Adv Drug Deliv Rev 60:13–28
15. Otsuka H, Nagasaki Y, Kataoka K (2003) PEGylated nanoparticles for biological and pharmaceutical applications. Adv Drug Deliv Rev 55:403–419
16. Nagasaki Y, Iijima M, Kato M, Kataoka K (1995) Primary amino-terminal heterobifunctional poly(ethylene oxide). Facile synthesis of poly(ethylene oxide) with a primary amino group at one end and a hydroxyl group at the other end. Bioconjug Chem 6:702–704
17. Nagasaki Y, Kutsuna T, Iijima M, Kato M, Kataoka K, Kitano S, Kadoma Y (1995) Formyl-ended heterobifunctional poly(ethylene oxide): synthesis of poly(ethylene oxide) with a formyl group at one end and a hydroxyl group at the other end. Bioconjug Chem 6:231–233
18. Akiyama Y, Nagasaki Y, Kataoka K (2004) Synthesis of heterotelechelic poly(ethylene glycol) derivatives having alpha-benzaldehyde and omega-pyridyl disulfide groups by ring opening polymerization of ethylene oxide using 4-(diethoxymethyl)benzyl alkoxide as a novel initiator. Bioconjug Chem 15:424–427
19. Akiyama Y, Otsuka H, Nagasaki Y, Kato M, Kataoka K (2000) Selective synthesis of heterobifunctional poly(ethylene glycol) derivatives containing both mercapto and acetal terminals. Bioconjug Chem 11:947–950

20. Hiki S, Kataoka K (2007) A facile synthesis of azido-terminated heterobifunctional poly (ethylene glycol)s for "click" conjugation. Bioconjug Chem 18:2191–2196
21. Hiki S, Kataoka K (2010) Versatile and selective synthesis of "click chemistry" compatible heterobifunctional poly(ethylene glycol)s possessing azide and alkyne functionalities. Bioconjug Chem 21:248–254
22. Pasut G, Veronese FM (2007) Polymer-drug conjugation, recent achievements and general strategies. Progr Polym Sci 32:933–961
23. Zhao H, Yang K, Martinez A, Basu A, Chintala R, Liu HC, Janjua A, Wang M, Filpula D (2006) Linear and branched bicin linkers for releasable PEGylation of macromolecules: controlled release in vivo and in vitro from mono- and multi-PEGylated proteins. Bioconjug Chem 17:341–351
24. Veronese FM, Caliceti P, Schiavon O (1997) Branched and linear poly(ethylene glycol): Influence of the polymer structure on enzymological, pharmacokinetic and immunological properties of proteinconjugates. J Bioact Compatible Polym 12:196–207
25. Aghemo A, Rumi MG, Colombo M (2010) Pegylated interferons alpha2a and alpha2b in the treatment of chronic hepatitis C. Nat Rev Gastroenterol Hepatol 7:485–494
26. Campa C, Harding SP (2011) Anti-VEGF compounds in the treatment of neovascular age related macular degeneration. Curr Drug Targets 12:173–181
27. Patel AM, Moreland LW (2010) Certolizumab pegol: a new biologic targeting rheumatoid arthritis. Expert Rev Clin Immunol 6:855–866
28. Fee CJ (2007) Size comparison between proteins PEGylated with branched and linear poly (ethylene glycol) molecules. Biotechnol Bioeng 98:725–731
29. Veronese FM, Schiavon O, Pasut G, Mendichi R, Andersson L, Tsirk A, Ford J, Wu G, Kneller S, Davies J, Duncan R (2005) PEG-doxorubicin conjugates: influence of polymer structure on drug release, in vitro cytotoxicity, biodistribution, and antitumor activity. Bioconjug Chem 16:775–784
30. Pinholt C, Bukrinsky JT, Hostrup S, Frokjaer S, Norde W, Jorgensen L (2011) Influence of PEGylation with linear and branched PEG chains on the adsorption of glucagon to hydrophobic surfaces. Eur J Pharm Biopharm 77:139–147
31. Somack R, Saifer MG, Williams LD (1991) Preparation of long-acting superoxide dismutase using high molecular weight polyethylene glycol (41,000–72,000 daltons). Free Radic Res Commun 12–13:553–562
32. Knauf MJ, Bell DP, Hirtzer P, Luo ZP, Young JD, Katre NV (1988) Relationship of effective molecular size to systemic clearance in rats of recombinant interleukin-2 chemically modified with water-soluble polymers. J Biol Chem 263:15064–15070
33. Brandenberger C, Mühlfeld C, Ali Z, Lenz AG, Schmid O, Parak WJ, Gehr P, Rothen-Rutishauser B (2010) Quantitative evaluation of cellular uptake and trafficking of plain and polyethylene glycol-coated gold nanoparticles. Small 6:1669–1678
34. Robers MJ, Harris JM (1998) Attachment of degradable Poly(ethylene glycol) to proteins has the potential to increase therapeutic efficacy. J Pharm sci 11:1440–1445
35. Peleg-Shulman T, Tsubery H, Mironchik M, Fridkin M, Schreiber G, Shechter Y (2004) Reversible PEGylation: a novel technology to release native interferon alpha2 over a prolonged time period. J Med Chem 47:4897–4904
36. Zalipsky S, Qazen M, Walker JA 2nd, Mullah N, Quinn YP, Huang SK (1999) New detachable poly(ethylene glycol) conjugates: cysteine-cleavable lipopolymers regenerating natural phospholipid, diacyl phosphatidylethanolamine. Bioconjug Chem 10:703–707
37. Filpula D, Zhao H (2008) Releasable PEGylation of proteins with customized linkers. Adv Drug Deliv Rev 60:29–49
38. Yatuv R, Robinson M, Dayan I, Baru M (2010) Enhancement of the efficacy of therapeutic proteins by formulation with PEGylated liposomes; a case of FVIII, FVIIa and G-CSF. Expert Opin Drug Deliv 7:187–201
39. Hatakeyama H, Akita H, Harashima H (2011) A multifunctional envelope type nano device (MEND) for gene delivery to tumours based on the EPR effect: a strategy for overcoming the PEG dilemma. Adv Drug Deliv Rev 63:152–160

40. Ishida T, Kiwada H (2008) Accelerated blood clearance (ABC) phenomenon upon repeated injection of PEGylated liposomes. Int J Pharm 354:56–62
41. Xu H, Wang KQ, Deng YH, da Chen W (2010) Effects of cleavable PEG-cholesterol derivatives on the accelerated blood clearance of PEGylated liposomes. Biomaterials 31:4757–4763
42. Hatakeyama H, Akita H, Ito E, Hayashi Y, Oishi M, Nagasaki Y, Danev R, Nagayama K, Kaji N, Kikuchi H, Baba Y, Harashima H (2011) Systemic delivery of siRNA to tumors using a lipid nanoparticle containing a tumor-specific cleavable PEG-lipid. Biomaterials 32:4306–4016
43. Kubetzko S, Sarkar CA, Plückthun A (2005) Protein PEGylation decreases observed target association rates via a dual blocking mechanism. Mol Pharmacol 68:1439–1454
44. Yamaoka T, Tabata Y, Ikada Y (1994) Distribution and tissue uptake of poly(ethylene glycol) with different molecular weights after intravenous administration to mice. J Pharm Sci 83:601–606
45. Chapman AP, Antoniw P, Spitali M, West S, Stephens S, King DJ (1999) Therapeutic antibody fragments with prolonged in vivo half-lives. Nat Biotechnol 17:780–783
46. Kubetzko S, Balic E, Waibel R, Zangemeister-Wittke U, Plückthun A (2006) PEGylation and multimerization of the anti-p185HER-2 single chain Fv fragment 4D5: effects on tumor targeting. J Biol Chem 281:35186–35201
47. Andresen H, Bier FF (2009) Peptide microarrays for serum antibody diagnostics. Methods Mol Biol 509:123–134
48. Yoshimoto K, Nishio M, Sugasawa H, Nagasaki Y (2010) Direct observation of adsorption-induced inactivation of antibody fragments surrounded by mixed-PEG layer on a gold surface. J Am Chem Soc 132:7982–7989
49. Uchida K, Otsuka H, Kaneko M, Kataoka K, Nagasaki Y (2005) A reactive poly(ethylene glycol) layer to achieve specific surface plasmon resonance sensing with a high S/N ratio: the substantial role of a short underbrushed PEG layer in minimizing nonspecific adsorption. Anal Chem 77:1075–1080
50. Yu D, Peng P, Dharap SS, Wang Y, Mehlig M, Chandna P, Zhao H, Filpula D, Yang K, Borowski V, Borchard G, Zhang Z, Minko T (2005) Antitumor activity of poly(ethylene glycol)-camptothecin conjugate: the inhibition of tumor growth in vivo. J Control Release 110:90–102
51. Chabot GG (1997) Clinical pharmacokinetics of irinotecan. Clin Pharmacokinet 33:245–259
52. Zhao H, Rubio B, Sapra P, Wu D, Reddy P, Sai P, Martinez A, Gao Y, Lozanguiez Y, Longley C, Greenberger LM, Horak ID (2008) Novel prodrugs of SN38 using multiarm poly(ethylene glycol) linkers. Bioconjug Chem 19:849–859
53. Pastorino F, Loi M, Sapra P, Becherini P, Cilli M, Emionite L, Ribatti D, Greenberger LM, Horak ID, Ponzoni M (2010) Tumor regression and curability of preclinical neuroblastoma models by PEGylated SN38 (EZN-2208), a novel topoisomerase I inhibitor. Clin Cancer Res 16:4809–4821
54. Hershfield MS (1995) PEG-ADA: an alternative to haploidentical bone marrow transplantation and an adjunct to gene therapy for adenosine deaminase deficiency. Hum Mutat 5:107–112
55. Burns CM, Wortmann RL (2011) Gout therapeutics: new drugs for an old disease. Lancet 377:165–177
56. Laden JC, Philibert P, Torreilles F, Pugnière M, Martineau P (2002) Expression and folding of an antibody fragment selected in vivo for high expression levels in *Escherichia coli* cytoplasm. Res Microbiol 153:469–474
57. Jayson GC, Parker GJ, Mullamitha S, Valle JW, Saunders M, Broughton L, Lawrance J, Carrington B, Roberts C, Issa B, Buckley DL, Cheung S, Davies K, Watson Y, Zinkewich-Péotti K, Rolfe L, Jackson A (2005) Blockade of platelet-derived growth factor receptor-beta by CDP860, a humanized, PEGylated di-Fab', leads to fluid accumulation and is associated with increased tumor vascularized volume. J Clin Oncol 23:973–981

58. Ton NC, Parker GJ, Jackson A, Mullamitha S, Buonaccorsi GA, Roberts C, Watson Y, Davies K, Cheung S, Hope L, Power F, Lawrance J, Valle J, Saunders M, Felix R, Soranson JA, Rolfe L, Zinkewich-Peotti K, Jayson GC (2007) Phase I evaluation of CDP791, a PEGylated di-Fab' conjugate that binds vascular endothelial growth factor receptor 2. Clin Cancer Res 13:7113–7118
59. Davidson BL, McCray PB Jr (2011) Current prospects for RNA interference-based therapies. Nat Rev Genet 12:329–340
60. Lares MR, Rossi JJ, Ouellet DL (2010) RNAi and small interfering RNAs in human disease therapeutic applications. Trends Biotechnol 28:570–579
61. Joralemon MJ, McRae S, Emrick T (2010) PEGylated polymers for medicine: from conjugation to self-assembled systems. Chem Commun (Camb) 46:1377–1393
62. Tamura M, Ichinohe S, Tamura A, Ikeda Y, Nagasaki Y (2011) In vivo and in vitro characteristics of core-shell-type nanogel particles: optimization of core cross-linking density and surface PEG density in PEGylated nanogels. Acta Biomaterialia. doi:10.1016/j.actbio.2011.05.027
63. Oishi M, Nagasaki Y (2010) Stimuli-responsive smart nanogels for cancer diagnostics and therapy. Nanomedicine (Lond) 5:451–468
64. Elbayoumi TA, Torchilin VP (2008) Liposomes for targeted delivery of antithrombotic drugs. Expert Opin Drug Deliv 5:1185–1198
65. Jiang W, Lionberger R, Yu LX (2011) In vitro and in vivo characterizations of PEGylated liposomal doxorubicin. Bioanalysis 3:333–344
66. Torchilin VP (2005) Recent advances with liposomes as pharmaceutical carriers. Nat Rev Drug Discov 4:145–160
67. Klibanov AL, Maruyama K, Torchilin VP, Huang L (1990) Amphipathic polyethyleneglycols effectively prolong the circulation time of liposomes. FEBS Lett 268:235–237
68. Yokoyama M, Okano T, Sakurai Y, Ekimoto H, Shibazaki C, Kataoka K (1991) Toxicity and antitumor activity against solid tumors of micelle-forming polymeric anticancer drug and its extremely long circulation in blood. Cancer Res 51:3229–3236
69. Alakhov V, Klinski E, Lemieux P, Pietrzynski G, Kabanov A (2001) Block copolymeric biotransport carriers as versatile vehicles for drug delivery. Expert Opin Biol Ther 1:583–602
70. Matsumura Y, Kataoka K (2009) Preclinical and clinical studies of anticancer agent-incorporating polymer micelles. Cancer Sci 100:572–579
71. Nishiyama N, Okazaki S, Cabral H, Miyamoto M, Kato Y, Sugiyama Y, Nishio K, Matsumura Y, Kataoka K (2003) Novel cisplatin-incorporated polymeric micelles can eradicate solid tumors in mice. Cancer Res 63:8977–8983
72. Bae Y, Kataoka K (2009) Intelligent polymeric micelles from functional poly(ethylene glycol)-poly(amino acid) block copolymers. Adv Drug Deliv Rev 61:768–784
73. Karakoti AS, Das S, Thevuthasan S, Seal S (2011) PEGylated inorganic nanoparticles. Angew Chem Int Ed Engl 50:1980–1994
74. Brown SD, Nativo P, Smith JA, Stirling D, Edwards PR, Venugopal B, Flint DJ, Plumb JA, Graham D, Wheate NJ (2010) Gold nanoparticles for the improved anticancer drug delivery of the active component of oxaliplatin. J Am Chem Soc 132:4678–4684
75. Rosi NL, Giljohann DA, Thaxton CS, Lytton-Jean AK, Han MS, Mirkin CA (2006) Oligonucleotide-modified gold nanoparticles for intracellular gene regulation. Science 312:1027–1030
76. Visaria RK, Griffin RJ, Williams BW, Ebbini ES, Paciotti GF, Song CW, Bischof JC (2006) Enhancement of tumor thermal therapy using gold nanoparticle-assisted tumor necrosis factor-alpha delivery. Mol Cancer Ther 5:1014–1020
77. Libutti SK, Paciotti GF, Byrnes AA, Alexander HR Jr, Gannon WE, Walker M, Seidel GD, Yuldasheva N, Tamarkin L (2010) Phase I and pharmacokinetic studies of CYT-6091, a novel PEGylated colloidal gold-rhTNF nanomedicine. Clin Cancer Res 16:6139–149
78. Ishii T, Otsuka H, Kataoka K, Nagasaki Y (2004) Preparation of functionally Pegylated gold nanoparticles with narrow distribution through autoreduction of auric cation by alpha-biotinyl-PEG-block-[poly (2-(N, N-dimethylamino) ethyl methacrylate)]. Langmuir 20:561–564

79. Miyamoto D, Oishi M, Kojima K, Yoshimoto K, Nagasaki Y (2008) Completely dispersible PEGylated gold nanoparticles under physiological conditions: modification of gold nanoparticles with precisely controlled PEG-b-polyamine. Langmuir 24:5010–5017
80. Kamimura M, Miyamoto D, Saito Y, Soga K, Nagasaki Y (2008) Design of poly(ethylene glycol)/streptavidin co-immobilized upconversion nanophosphors and their application to fluorescence biolabeling. Langmuir 24:8864–8870
81. Kamimura M, Kanayama N, Tokuzen K, Soga K, Nagasaki Y (2011) Near-infrared (1550 nm) In vivo bioimaging based on rare-earth doped ceramic nanophosphors modified with PEG-b-poly (4-vinylbenzyl phosphonate). Nanoscale. doi:10.1039/C1NR10466G
82. Mullard A (2011) 2010 FDA drug approvals. Nat Rev Drug Discov 10:82–85

Adv Polym Sci (2012) 247: 141–166
DOI: 10.1007/12_2011_151

Published online: 22 September 2011

Cytocompatible Hydrogel Composed of Phospholipid Polymers for Regulation of Cell Functions

Kazuhiko Ishihara, Yan Xu, and Tomohiro Konno

Abstract We propose a cell encapsulation matrix for use with a cytocompatible phospholipid polymer hydrogel system for control of cell functions in three-dimensional (3D) cell engineering and for construction of well-organized tissue *in vivo*. In cell engineering fields, it is important to produce cells with highly cell-specific functions. To realize this, we consider that new devices are needed for cell culture. So, we have designed soft biodevices using a spontaneously forming and cytocompatible polymer hydrogel system. A water-soluble phospholipid polymer bearing a phenylboronic acid unit, poly(2-methacryloyloxyethyl phosphorylcholine-*co*-*n*-butyl methacrylate-*co*-*p*-vinylphenylboronic acid) (PMBV), was prepared. This polymer, in aqueous solution, spontaneously converted to a hydrogel by addition of poly(vinyl alcohol) (PVA) aqueous solution due to reversible bonding between the boronate groups in PMBV and the hydroxyl groups in PVA.

K. Ishihara (✉)
Department of Materials Engineering, The University of Tokyo, Bunkyo-ku, Tokyo, Japan

Department of Bioengineering, The University of Tokyo, Bunkyo-ku, Tokyo, Japan

Center for NanoBio Integration, The University of Tokyo, Bunkyo-ku, Tokyo, Japan

Core Research of Evolutional Science and Technology (CREST), Japan Science and Technology Agency (JST), Chiyoda-ku, Tokyo, Japan
e-mail: ishihara@mpc.t.u-tokyo.ac.jp

Y. Xu
Department of Applied Chemistry, School of Engineering, Bunkyo-ku, Tokyo, Japan

Center for NanoBio Integration, The University of Tokyo, Bunkyo-ku, Tokyo, Japan

T. Konno
Department of Bioengineering, The University of Tokyo, Bunkyo-ku, Tokyo, Japan

Center for NanoBio Integration, The University of Tokyo, Bunkyo-ku, Tokyo, Japan

Core Research of Evolutional Science and Technology (CREST), Japan Science and Technology Agency (JST), Chiyoda-ku, Tokyo, Japan

The PMBV/PVA hydrogel was dissociated by an exchange reaction with low molecular weight diol compounds such as D-fructose, which have high binding affinity to the phenylboronic acid unit. Cells were encapsulated easily in the PMBV/PVA hydrogel, and the cells in the hydrogel kept their original morphology and slightly proliferated during the preservation period. After dissociation of the hydrogel, the cells could be recovered as a cell suspension and cultured under conventional cell culture conditions as usual. Embryonic stem cells could be encapsulated without any adverse effects from the polymer hydrogel, i.e., the cells maintained their undifferentiated character during preservation in the PMBV/PVA hydrogel. Cell preservation and activity in the hydrogel were also investigated using microfluidic chips. The results clearly indicated that the PMBV/PVA hydrogels provide a useful platform for 3D encapsulation of cell culture systems without any reduction of their bioactivity.

Keywords Cell encapsulation · Cell engineering · Cytocompatibility · Hydrogel · Phospholipid polymers

Contents

1 Polymer Hydrogels in Cell Engineering and Tissue Engineering

Cell engineering and tissue engineering has made progress because of recent developments not only in biotechnology and molecular biology, but also in biomaterials engineering [1–5]. From these two fields will develop a new medical area, i.e., regenerative medicine. Synthetic and natural polymer materials can be applied as scaffolds and nanoparticles in a wide range of cell and tissue engineering applications [6, 7].

Recently, it has been reported that several functional cells, e.g., embryonic stem (ES) cells and induced pluripotent stem (iPS) cells, were established and produced in culture [8–10]. These cells are expected to be used as a tool in regenerative medicine and cell engineering. The ES cells are generally preserved by classic cryopreservation after *in vitro* cell culture. However, it is well known that the recovery rate after thawing is quite low, and that some organic solvents contained in the general cryopreservation medium cause serious damage and the cells lose their function. Thus, developing a cytocompatible hydrogel that can preserve the cell functions without any adverse effects on bioactivity is extremely important in the field of cell engineering and tissue engineering. Polymeric hydrogels can be used for biomedical applications because these soft biomaterials can provide a three-dimensional (3D) network that supports biological materials (Table 1). The water permeability and gas permeability of hydrogels are important to living cells in the network. Further, the cytocompatibility of hydrogels depends on the chemical structure of the polymer chains and network size of the gel structure.

In order to replace or restore physiological functions lost in diseased or damaged organs, tissue engineering typically involves fabrication of tissue structures using cells and polymer scaffolds. The polymer scaffolds are designed to provide mechanical support for the cells, which can then perform the appropriate tissue functions; however, in practice, the simple addition of cells to porous polymer scaffolds is often inadequate for reproducing sufficient tissue function. One approach for increasing the functionality of these tissue-engineered constructs relies on attempts to mimic both the architecture of tissues and the environment around cells within the living organism.

Tissues consist of smaller repeating units on the scale of hundreds of micrometers *in vivo*. The 3D architecture of these repeating tissue units underlies the coordination of multicellular processes, emergent mechanical properties, and integration with other organ systems via the microcirculation [11]. Furthermore, the local cellular environment presents biochemical, cellular, and physical stimuli that orchestrate cellular fate processes such as proliferation, differentiation, migration, and apoptosis. Thus, successful fabrication of a fully functional tissue must include both an appropriate environment for cell viability and function at the microscale

Table 1 Advantages and disadvantages of hydrogels as tissue engineering matrices

Advantages	Aqueous environment can protect cells and fragile drugs (peptides, proteins, oligonucleotides, DNA)
	Good transport of nutrient to cells and products from cells
	Can be easily modified with cell adhesion ligands
	Can be injected in vivo as a liquid that gets at body temperature
	Usually biocompatible
Disadvantages	Can be hard to handle
	Usually mechanically weak
	Can be difficult to load drugs and cells and then crosslink in vitro as a prefabricated matrix
	Can be difficult to sterilize

level, as well as macroscale level properties that allow sufficient transport of nutrients, provide adequate mechanical properties, and facilitate coordination of multicellular processes.

Fabrication approaches have been previously used in two-dimensional (2D) micropatterned model systems and have led to insights on the effect of cell–cell and cell–polymer scaffold interactions on hepatocyte and endothelial cell fate. Extending these studies, the application of 3D fabrication techniques may also prove useful for studying structure–function relationships in model tissues.

Many strategies currently proposed for tissue engineering depend on employing a polymer scaffold. These scaffolds serve as a synthetic extracellular matrix (ECM) to organize cells into a 3D architecture and to present stimuli, which direct the growth and formation of a desired tissue [12]. Depending on the tissue of interest and the specific application, the required polymer scaffold and its properties will be quite different. A commonly used polymer scaffold is poly(lactide-*co*-glycolide) (PLG); however, a variety of hydrogels are being employed as scaffold materials. They are composed of hydrophilic polymer chains, which are either synthetic or natural in origin. The structural integrity of hydrogels depends on crosslinks formed between polymer chains via various chemical bonds and physical interactions. Hydrogels used in these applications can be processed under relatively mild conditions, have mechanical and structural properties similar to many tissues, and the ECM, and can be delivered in a minimally invasive manner [13].

A wide variety range of polymer compositions have been used to fabricate hydrogels [14]. Table 2 summarizes the many varied compositions. The compositions can be classified into natural polymer hydrogels and synthetic polymer hydrogels. Natural polymer hydrogels have been formed from collagen, gelatin, agarose, alginic acid, chitosan, and hyaluronic acid. Synthetic polymer hydrogels include, among others, poly(ethylene glycol) (PEG), poly(vinyl alcohol) (PVA),

Table 2 Polymer component for constituting hydrogels as scaffolds

Origin	Constituent	Biological response
Natural	Collagen	++
	Gelatin	++
	Hyaluronic acid	+
	Alginic acid	+
	Agarose	+
	Chitosan	+
	Dextran sulfate	+
Synthetic	Poly(L-lysine)	++
	Poly(lactic acid)	+
	Poly(*N*-isopropylacrylamide)	+
	Poly(2-hydroxyethyl methacrylate)	+
	Poly(acrylic acid)	+
	Poly(vinyl alcohol)	+
	Poly(ethylene glycol)	–
	Poly(2-methacryloyloxyethyl phosphorylcholine)	–

+ Tissue response-inducible, ++ cell adhesion-inducible

poly(2-hydroxyethyl methacrylate) (PHEMA), and poly(*N*-isopropylacrylamide) (PNIPAAm), and poly(2-methacryloyloxyethyl phosphorylcholine) (PMPC) segments. These synthetic polymer networks can be synthesized using various polymerization techniques. The polymer chemist can design and synthesize polymer networks with molecular-scale control over structure such as crosslinking density, and with tailored properties such as biodegradation, mechanical strength, and chemical and biological response to stimuli. Cells can be encapsulated homogeneously within these synthetic polymer hydrogels.

From the viewpoint of the actions of the hydrogels toward cells, they can be classified into three groups: cell adhesion-inducible matrices, tissue response-inducible matrices, and bioinert matrices. Collagen, gelatin, and poly(L-lysine) contain cell-adhesive ligands in their structure and are classified as cell adhesion-inducible matrices. Hydrogels composed of PEG and the MPC polymer are classified as bioinert matrices. Other hydrogels such as polysaccharide derivatives are classified as tissue response-inducible matrices. PEG-based hydrogels are particularly intriguing because of their bioinert property, hydrophilicity, and ability to be customized by changing the chain length to tune transport properties or by incorporating biologically relevant molecules [15]. They have been used to immobilize various cell types including osteoblasts and fibroblasts that can attach, grow, and produce matrix. PEG-based hydrogels can be customized by incorporation of adhesion domains of ECM proteins to promote cell adhesion, growth factors to modulate cell function, and degradable linkages [16]. Hydrogels for tissue engineering is a rapidly growing field because of their chemical flexibility for customization and the resulting tissue-like physical properties.

Collagen is the most widely used tissue-derived natural polymer, and is a main component of the ECM of tissues. However, these gels lack physical strength, are potentially immunogenic, and can be expensive. Furthermore, there can be big variations between produced collagen batches. However, collagen meets many of the biological design parameters, as it is composed of specific combinations of amino acid sequences that are recognized by cells and degraded by enzymes (collagenase) secreted from the cells.

Agarose is another type of marine algal polysaccharide, but unlike alginate it forms thermally reversible gels. The gel structure is thought to be bundles of associated double helices, and the junction zones to consist of multiple chain aggregation. The physical structure of the gels can be mainly controlled by using a range of agarose concentrations, which results in various pore sizes. The large pores and low mechanical stiffness of the gels at low concentrations of agarose may enable the migration and proliferation of cells, and these factors have been found to affect neurite growth *in vitro*.

Chitosan has found many biomedical applications, including tissue engineering approaches. Enzymes such as chitosanase and lysozyme can degrade chitosan. However, chitosan is easily soluble in the presence of acid, and generally insoluble in neutral conditions as well as in most organic solvents due to the existence of amino groups and the high crystallinity. Therefore, many derivatives have been reported to enhance the solubility and processability of this polymer.

2 Spontaneously Forming Reversible Hydrogels

In situ gelling hydrogels can be subdivided into two main categories: systems that are created upon irradiation with light, and systems in which the polymer molecules self-assemble [17]. Photoinduced polymerization can form a hydrogel in situ; however, they are not self-gelling. Spontaneously forming hydrogels based on molecular interactions of polymer components are formed spontaneously or after certain triggers under biological conditions, such as a change in temperature, pH, ionic strength or molecular interactions. Also, reversibility of the gelation/dissociation process is important for polymer scaffolds applied in cell and tissue engineering. Table 3 summarizes the types of spontaneously forming reversible hydrogel systems [21–25, 27, 30, 31, 33–39, 42–51].

PEG-based hydrogels are the most widely used materials. PEG-based hydrogels are nontoxic and nonimmunogenic, and they can be covalently crosslinked using various methods to form hydrogels. It was reported that PEG-poly(propylene glycol) (PEG-PPG) triblock copolymers (Pluronics or Poloxamers) turn into hydrogels at physiological temperature by forming a liquid crystalline phase [42]. PEG–peptide bioconjugates are amenable to proteolytic degradation in response to secreted proteases, such as matrix metalloproteases (MMPs) from cells [33, 34]. Another important synthetic polymer is PVA. PVA hydrogels are stable and elastic gels that can be prepared by repeated freezing and thawing cycles [31]. The physically crosslinked PVA hydrogels are biodegradable, and thus can be used for various applications in cell engineering.

Table 3 Classification of spontaneously forming and reversible hydrogels

Hydrogel	Crosslinking mode	Dissociation signal	References
Collagen/gelatin	Entanglement	Temperature	[18, 19]
Poly(L-lysine)-based hydrogel	Ionic interaction	pH	[20]
Poly(lactic acid)-based hydrogel	Stereocomplex formation	pH	[21–25]
	Hydrophobic interaction	Hydrolysis	[26]
Alginic acid/divalent cation complex	Ionic interaction	pH	[27]
Agarose hydrogel	Entanglement	Hydrolysis	[28]
Chitosan hydrogel	Ionic interaction	pH	[29]
PNIPAAm-based hydrogel	Hydrophobic interaction	Temperature	[30]
PVA hydrogel	Hydrogen bonding	Temperature	[31]
	Boronate/diol reaction	Chemical stimulation (e.g., sugar)	[32]
PEG-based hydrogel	Biological interaction	Protease	[33, 34]
MPC polymer-based hydrogel PMA/PMB hydrogel	Hydrogen bonding	pH	[35–39]
PMBV/PVA hydrogel	Boronate/diol reaction	Chemical stimulation (e.g., sugar)	[40, 41]

PMA poly(MPC-*co*-methacrylic acid), *PMB* poly(MPC-*co*-BMA)

Stimuli-responsive hydrogels that undergo abrupt changes in volume in response to external stimuli such as pH, temperature, and solvent composition have potential applications in cell engineering. Temperature-responsive hydrogels are some of the most widely studied stimuli-responsive hydrogel systems. These systems, which are mostly based on PNIPAAm, undergo a reversible volume phase transition with a change in the temperature of the environmental conditions [30].

Alginate is a linear polysaccharide that is widely used for drug delivery and in cell engineering. Alginate forms a hydrogel under benign conditions, which makes it attractive for cell encapsulation. When divalent cations such as Ca^{2+}, Mg^{2+}, Ba^{2+}, and Sr^{2+} are added into alginate aqueous solution, the hydrogel is obtained [27]. The hydrogels have also been used for transplantation of cells such as chondrocytes, hepatocytes, and Langerhans islets to treat diabetes.

The molecular design of polymers for spontaneously forming hydrogels has been carried out using a MPC unit. Watanabe et al. reported a MPC polymer porous scaffold prepared through stereocomplex formation between poly(D-lactic acid) and PLA [21–25]. Kimura et al. and Nam et al. reported a spontaneously forming hydrogel by mixing two kinds of MPC polymer solutions [35–38]. Hydrophobic domains were formed by water-soluble MPC polymer containing a hydrophobic unit, poly(MPC-*co*-*n*-butyl methacrylate) (PMB), in aqueous solution. Another water-soluble MPC polymer containing methacrylic acid, poly(MPC-*co*-methacrylic acid) (PMA) was introduced into the hydrophobic domain. After being gently shaken, the gelation occurred due to the formation of dimers by hydrogen bonding, which acts as a physical crosslinking of the polymer chains. The formed hydrogel (PMA/PMB hydrogel) is expected to be useful as a drug reservoir for a pH-responsive drug delivery system. Also, a PMA/PMB hydrogel has been applied for preventing peritendinos adhesion after physical operation [39].

3 Cytocompatible Polymer Hydrogels Composed of Water-Soluble Phospholipid Polymers

3.1 Polymer Design and Fundamental Properties

Similarly to the phospholipid polymers, the MPC polymers show excellent biocompatibility and blood compatibility [43–48]. These properties are based on the bioinert character of the MPC polymers, i.e., inhibition of specific interaction with biomolecules [49, 50]. Recently, the MPC polymers have been applied to various medical and pharmaceutical applications [44–47, 51–55]. The crosslinked MPC polymers provide good hydrogels and they have been used in the manufacture of soft contact lenses. We have applied the MPC polymer hydrogel as a cell-encapsulation matrix due to its excellent cytocompatibility. At the same time, to prepare a spontaneously forming reversible hydrogel, we focused on the reversible covalent bonding formed between phenylboronic acid and polyol in an aqueous system.

It is known that boronic acids can bind with hydroxyl compounds, including polyols such as PVA, through the complex formation of a reversible covalent bonding [56, 57].

Using this property, a glucose-responsive hydrogel system to control release of insulin has been prepared. The MPC polymer bearing a phenylboronic acid moiety, poly[MPC-*co*-*n*-butyl methacrylate (BMA)-*co*-*p*-vinylphenylboronic acid (VPBA)] (PMBV), can spontaneously form a hydrogel with PVA, even when the polymers are dissolved in cell culture medium [40, 41]. The PMBV/PVA hydrogel can be dissociated by addition of low molecular weight sugar compounds based on the exchange reaction with PVA.

PMBV was synthesized by a conventional radical polymerization. The monomer unit compositions of the PMBV were 0.64, 0.25, and 0.11 unit mole fractions for MPC, BMA, and VPBA, respectively. The number-averaged molecular weight and weight-averaged molecular weight were 6.2×10^4 and 6.5×10^4, respectively. This PMBV was completely water-soluble due to hydrophilic MPC units in the polymer. Figure 1 shows the chemical structure of PMBV.

When the PMBV solution and PVA solution were mixed, a hydrogel was formed within a short term. Dissociation of the PMBV/PVA hydrogel occurred through addition of a sugar compound such as D-sorbitol. This spontaneous gelation and dissociation mechanism is useful for 3D cell immobilization, and the PMBV/PVA hydrogel is a promising platform as a soft biodevice for 3D cell engineering. Figure 2 shows the concept of 3D cell engineering based on a PMBV/PVA hydrogel system.

The gelation mechanism is shown in Fig. 3. The spontaneous gelation was visually confirmed when the polymer concentration was 5 wt% PMBV solution and 2.5 wt% PVA solution.

Figure 4 indicates the dynamic viscoelasticity of the PMBV/PVA hydrogel. The dynamic viscoelasticity was measured immediately after mixing of the two polymer solutions. It was confirmed that the cross-point between the storage modulus (G′) and loss modulus (G″) is at 37 s. This result indicated that the mixture formed a crosslinking network, and that the mixture finally produced a hydrogel structure. Furthermore, gelation even occurred in the cell culture medium. Also, the hydrogel was reversibly dissociated by the addition of sugar molecules. The hydrogel had good network structure and the pore size was a few micrometers.

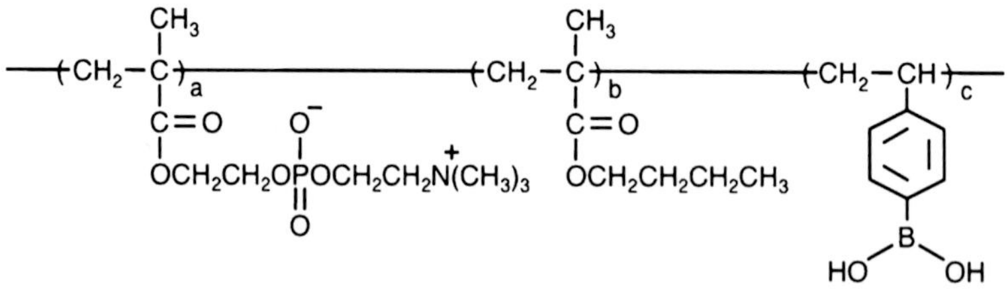

Fig. 1 Chemical structure of water-soluble poly(2-methacryloyloxyethyl phosphorylcholine-*co*-*n*-butyl methacrylate-*co*-*p*-vinylphenylboronic acid (PMBV)

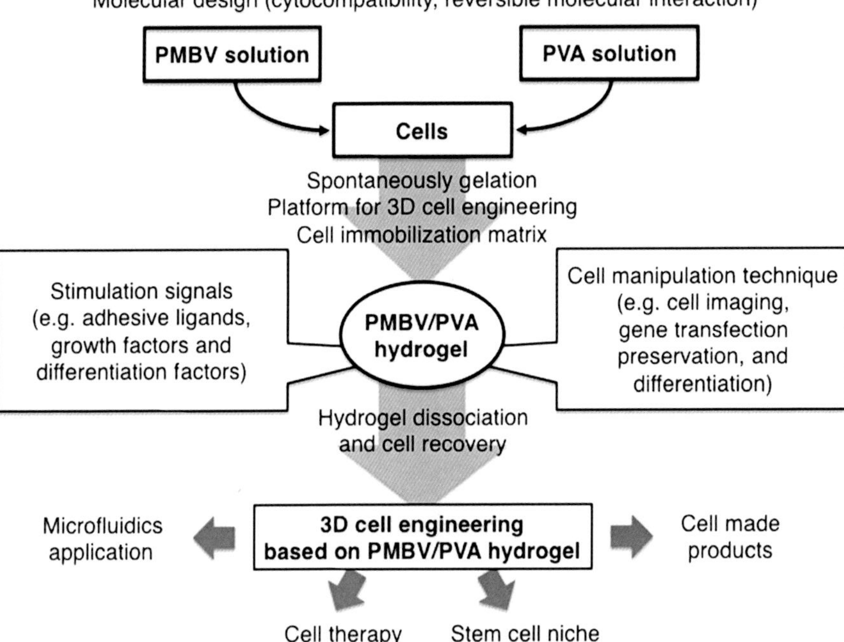

Fig. 2 Concept of 3D cell engineering based on a spontaneously forming and reversible PMBV/PVA hydrogel with high cytocompatibility as soft biodevice

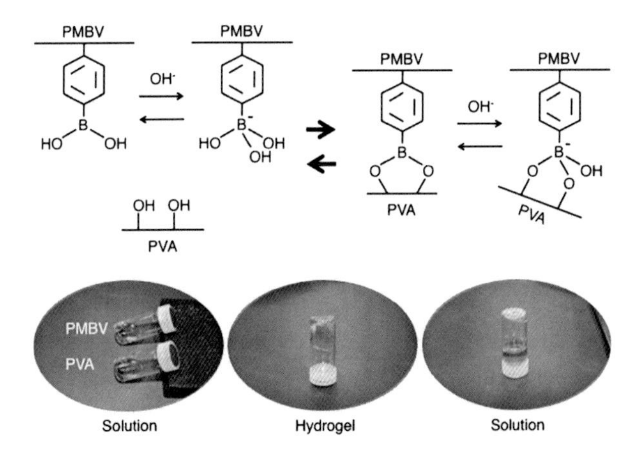

Fig. 3 *Above*: Spontaneous gelation mechanism between the phenylboronic acid moiety (boronate ion) in water-soluble PMBV and the hydroxyl groups (diol units) in PVA. *Below*: Photoimages of spontaneously forming PMBV/PVA hydrogel, before gelation (*left*), after gelation (*center*), and after dissociation (*right*)

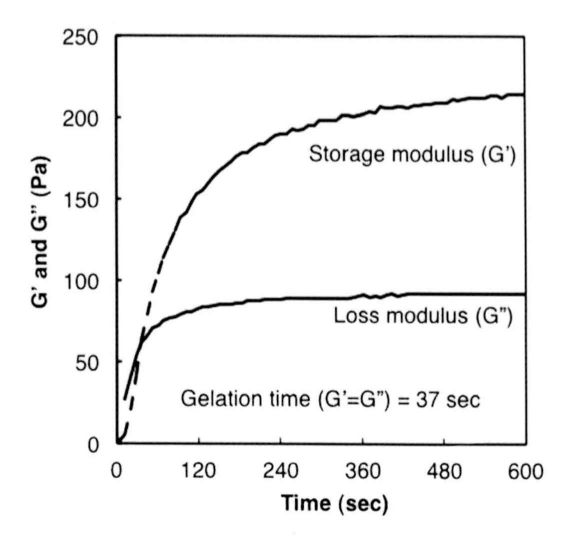

Fig. 4 Dynamic viscoelasticity measurements after equal mixing of 5 wt% PMBV and 2.5 wt% PVA

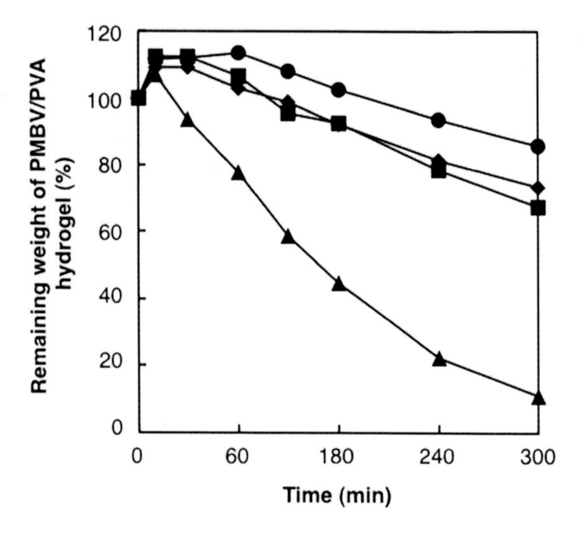

Fig. 5 Dissociation process of PMBV/PVA hydrogel after addition of 0.2 M of various sugar solutions: *circles* phosphate buffered saline, *squares* D-glucose, *diamonds* D-galactose, *triangles* D-sorbitol

The dissociation property of the PMBV/PVA hydrogel was evaluated by weight measurements of the hydrogel in phosphate-buffered saline (PBS) containing the sugar molecules. The weight change of the PMBV/PVA hydrogel in the presence of various sugar molecules against the incubation time is shown in Fig. 5. The hydrogel was initially swollen under all conditions. After the initial swelling, the weight change of the hydrogel strongly depended on the sugar molecules. The weight of the hydrogel immediately decreased, especially in the case of the solution containing D-fructose. The order of dissociation of hydrogel was D-fructose > D-galactose > D-glucose. This order corresponded to the complex formation constants of phenylboronic acid, which were reported by Lorand, et al. [58]. The order of formation constants of polyol complexes were reported as follows:

D-fructose (4370 M^{-1}) > D-galactose (276 M^{-1}) > D-glucose (110 M^{-1}). From these results, it was confirmed that the PMBV/PVA hydrogel was formed under biological conditions, and that the reversible properties of the hydrogel corresponded to the formation of a complex between the phenylboronic acid group in PMBV and the diol moiety in PVA.

4 Cells in PMBV/PVA Hydrogels

4.1 Encapsulation Technique

PMBV and PVA can spontaneously form a hydrogel without using any crosslinkers. Even in cell culture conditions, the gelation can be confirmed. Therefore, it was possible to encapsulate cells in the PMBV/PVA hydrogel. The encapsulation method is illustrated in Fig. 6.

The encapsulation technique was briefly as follows: L929 fibroblast cells were suspended in 5 wt% of PMBV solution dissolved in the culture medium. The PMBV solution containing the cells and 2.5 wt% of PVA solution were equally mixed, and the PMBV/PVA hydrogel containing L929 cells was formed. The PMBV/PVA hydrogel immobilizing the cells was stored in an incubator. After 72 h, 0.2 M of D-fructose solution was added to the PMBV/PVA hydrogel to

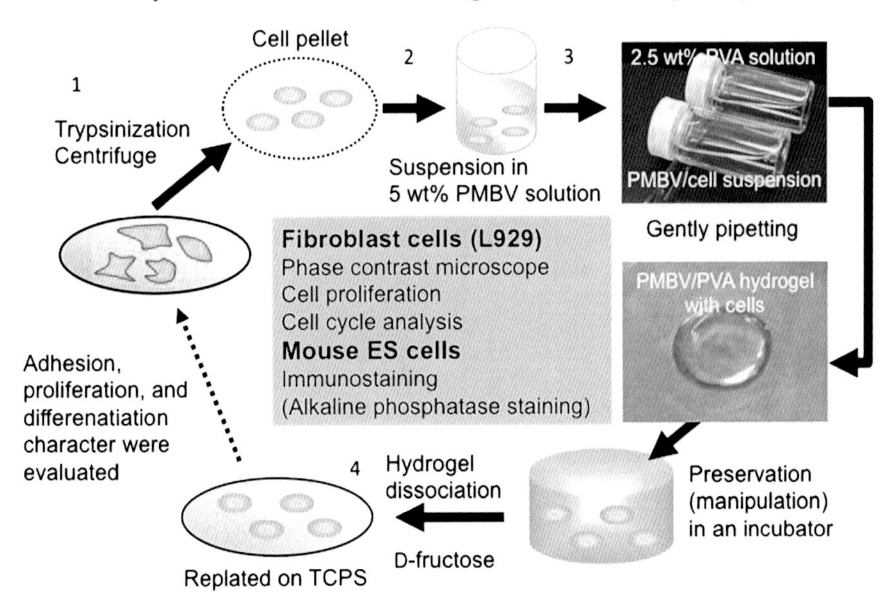

Fig. 6 Cell encapsulation using PMBV/PVA hydrogel

dissociate the hydrogel and collect the immobilized cells. After dissociation of the PMBV/PVA hydrogel, the recovered cells were collected by centrifugation, and the survival ratio was evaluated.

To evaluate cell proliferation in the hydrogel, the L929 cells were immobilized at a density of 1.0×10^5 cells/mL. As a control sample, L929 cells were seeded onto a conventional cell culture plate at a density of 0.5×10^4 cells/mL.

4.2 *Morphology of Cells in the Hydrogels*

The shape and morphology of the L929 cells were observed using a phase contrast microscope. The L929 cells proliferated and formed cell clusters derived by expansion of a single cell without forming aggregations in the PMBV/PVA hydrogel.

After 7 days encapsulated without changing the medium for fresh medium, the PMBV/PVA hydrogel was dissociated by the addition of D-sorbitol solution, and the L929 cells recovered from the PMBV/PVA hydrogel and seeded on a conventional cell culture plate of tissue-culture polystyrene (TCPS). Figure 7 shows phase contrast microscopic images of the L929 cells on TCPS and in the PMBV/PVA hydrogel. The L929 cells adhered, spread, and proliferated on the TCPS. On the other hand, morphology of the L929 cells kept a round shape during the culture period. Figure 8 shows a phase contrast image of the L929 cells recovered from the PMBV/PVA hydrogel. It was possible to recover the cell clusters derived by expansion of single cells.

4.3 *Proliferation of Encapsulated Cells*

Figure 9 shows the proliferation of cells encapsulated in the PMBV/PVA hydrogel. The encapsulated cells (L929) did not proliferate with the excessive proliferation seen on the TCPS. The encapsulated cells were recovered from the PMBV/PVA

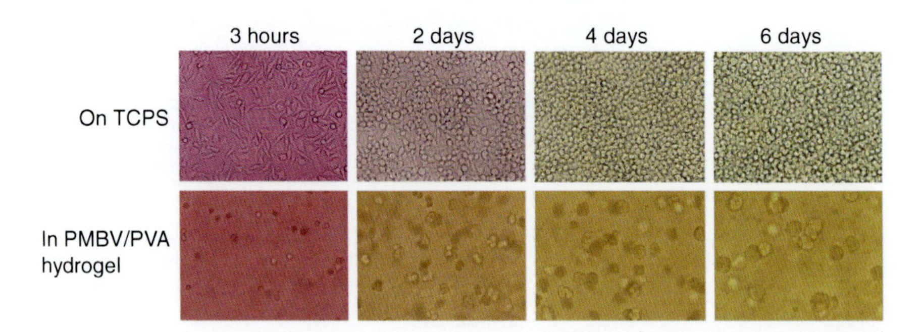

Fig. 7 Phase contrast microscope images of fibroblast (L929) cells on TCPS and in the PMBV/PVA hydrogel

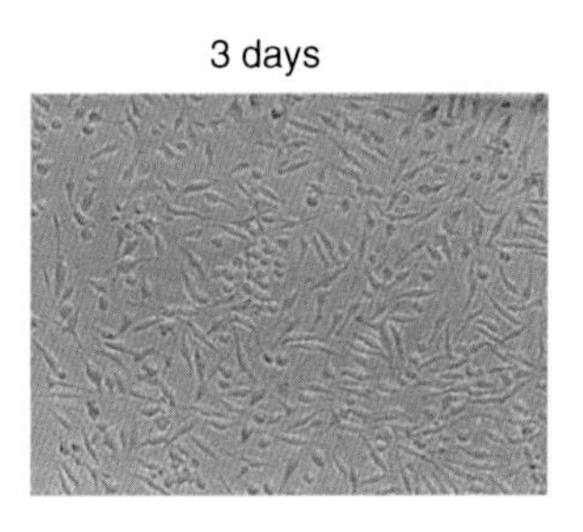

Fig. 8 Phase contrast image of L929 cells recovered from the PMBV/PVA hydrogel after 3 days of encapsulation

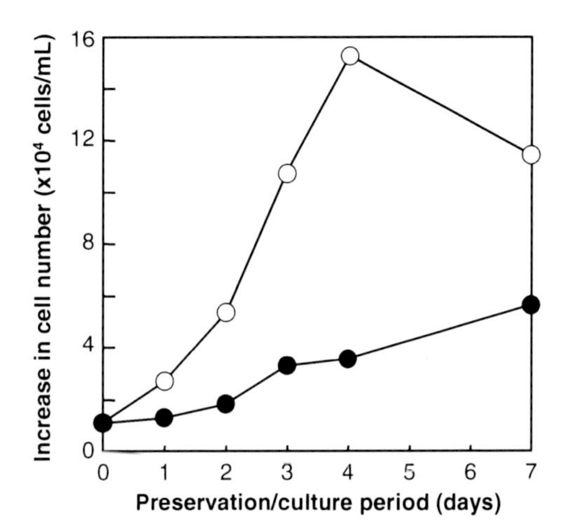

Fig. 9 Proliferation of fibroblast (L929) cells in the PMBV/PVA hydrogel (*filled circles*) and on TCPS (*circles*)

hydrogel after dissociation by the addition of D-fructose. The cell survivability in the hydrogel was 96.5% ± 1.1% after 72 h. The maximum L929 cell proliferation was exhibited after 4 days on the TCPS. On the other hand, the proliferation of L929 cells was maintained for 11 days in the PMBV/PVA hydrogel. The viability of the L929 cells in the PMBV/PVA hydrogel was maintained at more than 90% during the 11 days. The viability of the L929 cells on the TCPS was 60% after 11 days.

4.4 Control of Cell Cycle in the Hydrogel

It is hypothesized that cells proliferate uniformly and that the distribution of cell cycle phases is synchronized in the G0/G1 phase in the PMBV/PVA hydrogel. It has also been reported that cells synchronized at G0/G1 phase express a high level of cell-specific functions [59]. Thus, it is expected that the PMBV/PVA hydrogel can avoid a reduction in activity of entrapped cells and preserve cells with high functionality.

The PMBV/PVA hydrogel containing L929 cells was prepared and incubated for 7 days. After recovery from the PMBV/PVA hydrogel, L929 cells were fixed

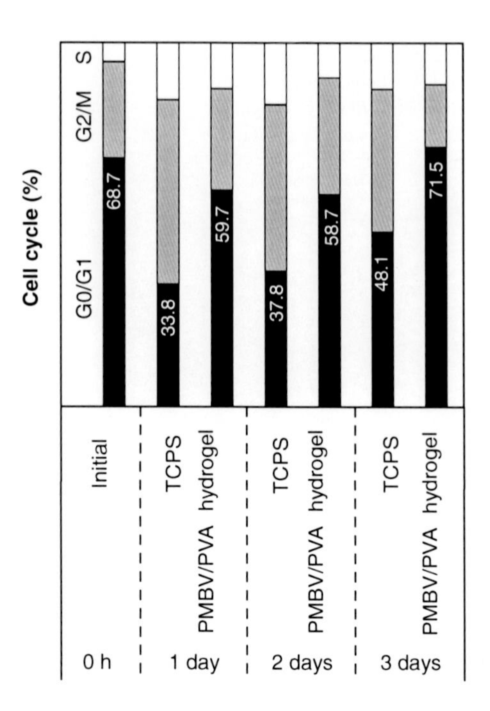

Fig. 10 Cell cycle distribution of fibroblast (L929) cells depends on the elapsed time after encapsulated in the PMBV/PVA hydrogel and on culture time on TCPS

in cooled 70% ethanol for 2 h. The cells were incubated in a solution containing 1 mg/mL RNase and 20 mg/mL propidium iodide (PI) for 30 min. Cell cycle analysis was performed by flow cytometry.

Figure 10 shows the distribution of cell cycle phases of L929 cells encapsulated in the PMBV/PVA hydrogel and those cultured on the TCPS. The percentage of cells in G0/G1 phase was 71% in the PMBV/PVA hydrogel after 3 days. By contrast, the amounts in G0/G1 phase were 30–50% in conventional cultivation on the TCPS during the 3 days. This means that L929 cells divided and proliferated more uniformly in the PMBV/PVA hydrogel than on the TCPS. As a result, the distribution of cell cycle phases of L929 cells cultured in the PMBV/PVA hydrogel was more uniform than that on the TCPS. In the PMBV/PVA hydrogel, 94% of L929 cells were synchronized to G0/G1 phase at 7 days. On the TCPS, 97% of L929 cells were also synchronized to G0/G1 phase at 7 days. However, the cell numbers had been decreasing from 4 days so that L929 cells had entered into the death phase after 7 days on TCPS. L929 cells became confluent and cell death was induced.

5 Cell Functions in the PMBV/PVA Hydrogel

Liver cells, HepG2, could also be encapsulated in the PMBV/PVA hydrogel, and the cell-specific functions were evaluated. HepG2 cells secrete albumin during cell culture. Culture supernatants were collected and albumin content measured using

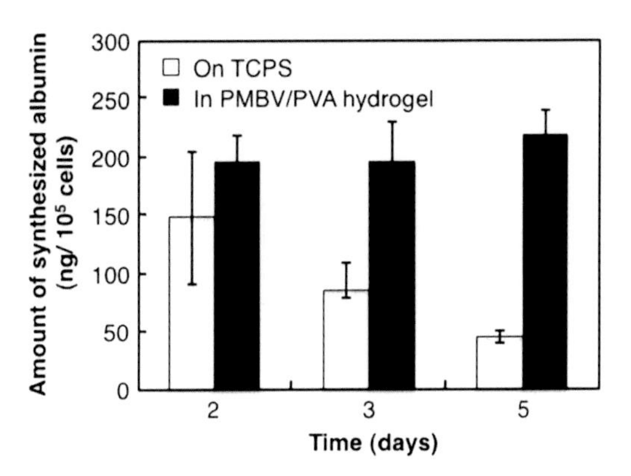

Fig. 11 Amount of secreted albumin from HepG2 cells encapsulated in the PMBV/PVA hydrogel and on TCPS

enzyme-linked immunosorbent assay. The HepG2 cells encapsulated into the PMBV/PVA hydrogel maintained a high level of albumin synthetic function during encapsulation for 5 days (Fig. 11). On the TCPS, the HepG2 activity decreased with culture time. It has been reported that human hepatocellular cells decrease their liver-specific functions in 2D culture due to monolayer failure to accurately mimic the native microenvironment [60]. The results in Fig. 11 show that the PMBV/PVA hydrogel did not induce a reduction in cell-specific function. It has been reported that the functions and proliferation of cells is related to the morphology of cells [61]. A strong cell adhesion and spreading often lead to proliferation, while round-shaped cell morphology is required for cell-specific functions. Thus, cell-specific functions can be preserved in the PMBV/PVA hydrogel.

6 Gene Expression of Cells Encapsulated in PMBV/PVA Hydrogel

Gene expression is an important parameter for estimating the cell activity. In this study, a mouse mesenchymal stem cell line (C3H10T1/2) was encapsulated in the PMBV/PVA hydrogel. The gene expression analysis was performed using the quantified polymerase chain reaction (qPCR) method after 1 day of culture. In this study, four kinds of genes were evaluated, beta-2 microglobulin (*B2m*), elongation factor 1-gamma (*Eef1g*), succinate dehydrogenase complex subunit flavoprotein variant A (*Sdha*), and TATA-binding protein (*Tbp*).

Figure 12 shows the result of qPCR analysis. Since formation of the PMBV/PVA hydrogel is reversible, the encapsulated cells can be recovered from the gel without any damage after treatment with the dissociation reagents. Housekeeping genes are typically constitutive genes that are required for the maintenance of basic cellular function, and are found in all cells of an organism. In the PMBV/PVA hydrogel, all housekeeping genes were expressed. This result indicated that the encapsulated cells maintained their basic cellular function in the PMBV/PVA hydrogels.

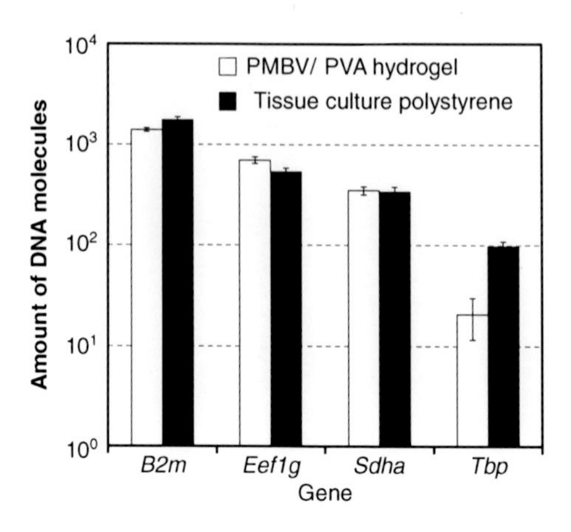

Fig. 12 Amount of DNA molecules of various genes, as measured by qPCR assay. The qPCR was carried out after 1 day of culture in the PMBV/PVA hydrogel or on TCPS. See text for gene descriptions

7 Encapsulation of Stem Cells and Their Undifferentiated Character

Mouse ES cells (129/SvEv, passage 11) were suspended in 5 wt% PMBV solution dissolved in the ES cell culture medium. The PMBV solution containing ES cells was mixed with 2.5 wt% PVA solutions. The mixture was gently shaken to form the PMBV/PVA hydrogel. After confirmation of gelation, the PMBV/PVA hydrogel containing ES cells was stored in an incubator. The cell morphology was observed with a phase contrast microscope.

The ES cells were encapsulated in the PMBV/PVA hydrogel by the same method used for the L929 cells. In general, ES cells form a cell aggregate called an embryoid body in suspension culture. However, it was observed that the encapsulated ES cells in the hydrogel did not form any embryoid body for 72 h.

Figure 13 shows the phase contrast microscope images of the encapsulated ES cells in the PMBV/PVA hydrogel, and those cultured on the water-insoluble phospholipid polymer, poly(MPC-*co*-BMA) (PMB30, MPC unit mole fraction 0.30). The ES cells were uniformly suspended in the PMBV/PVA hydrogel. On the other hand, the ES cells cultured in suspension culture (cell adhesion was completely inhibited on the PMB30 surface) were aggregated and formed an embryoid body. It should be noted that the PMBV/PVA hydrogel did not affect direct cell–cell interaction between the ES cells. This is important because increased cell–cell interaction is a trigger for ES cell differentiation.

The PMBV/PVA hydrogel containing ES cells was dissociated by the addition of 0.2 M D-fructose solution after 3 days. After dissociation of the PMBV/PVA hydrogel, the recovered ES cells were cultured on gelatin-coated TCPS in the culture medium as usual, and the differentiation characters of recovered ES cells were estimated by alkaline phosphatase (ALP) staining (Fig. 14).

ALP staining was performed to estimate the functionality of recovered ES cells, and it was found that the ES cells with undifferentiated character were well stained

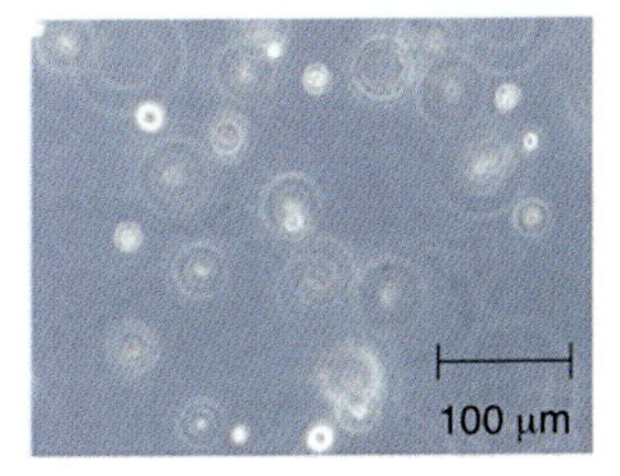

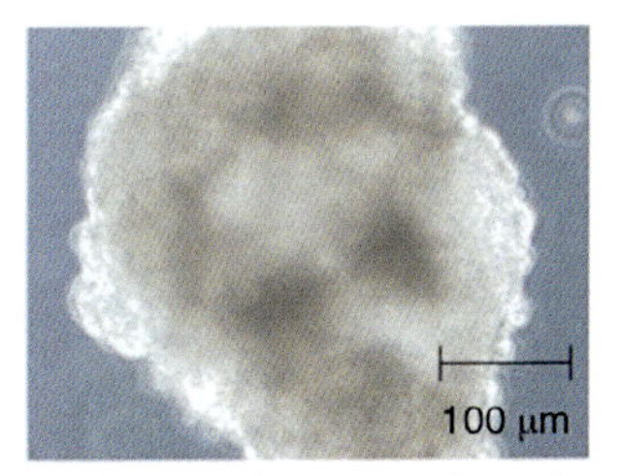

Fig. 13 Phase contrast microscope images of mouse embryonic stem cells in the PMBV/PVA hydrogel (*left*) and on TCPS (*right*)

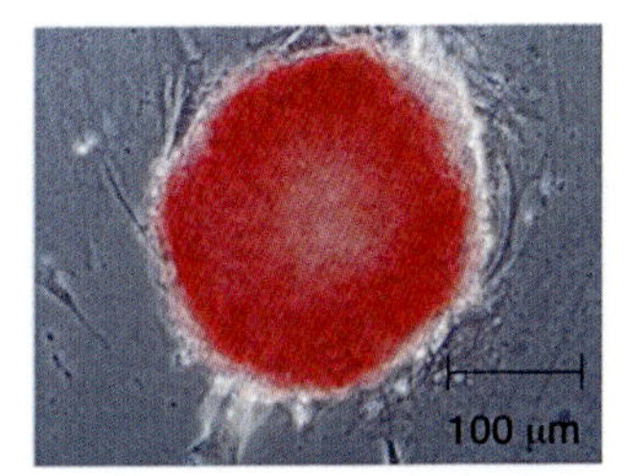

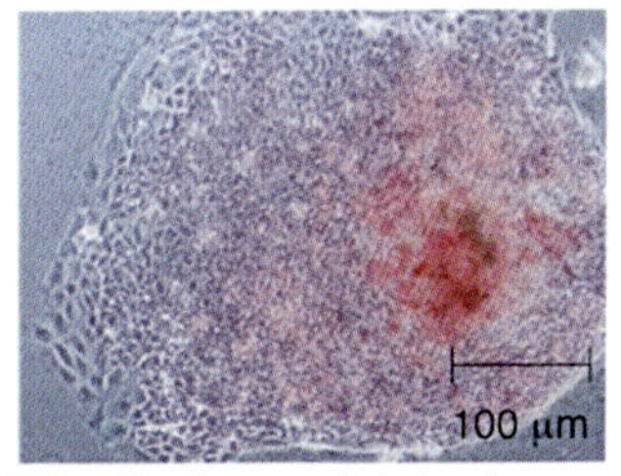

100 µm
ES cell colony and derived (differentiated) cells which were recovered from suspension culture condition on PMB30.

Fig. 14 Alkaline phosphatase staining of recovered ES cells after dissociation of PMBV/PVA hydrogel (*left*), and of the ES cells cultured on PMB30 (*right*). The undifferentiated ES cells were well stained. ES cells encapsulated in the PMBV/PVA hydrogel maintained their undifferentiated character during the 3 days of encapsulation

by ALP staining. Indeed, this is an indicator of the differentiation of ES cells. It should be noted that the colonies were also well stained by ALP staining. This result indicated that the ES cells maintained their undifferentiated character during the immobilization period in the PMBV/PVA hydrogel. It can be seen that the PMBV/PVA hydrogel supported the cell immobilization space in their network without any stimulation to change the cell function.

8 Cell-Based Biochips Fabricated Using a PMBV/PVA Hydrogel

8.1 Microfluidics for Miniaturized Cell Operation

Microfluidics evolved from micro-analytical methods in capillary format such as capillary electrophoresis, high-performance liquid chromatography, and gas chromatography, and has successfully revolutionized chemical and biochemical applications towards miniaturized and integrated on-chip formats, with high speed, high sensitivity, high resolution, and minute sample consumption [62]. Recently, microfluidic chips have been applied to manipulate and analyze cells and to interrogate molecules inside the cell [63–65]. These cell-based microfluidic chips contribute to and advance genomic, proteomic, and cellomic research at the micro- and nanoscale levels [66, 67]. As the development of cell encapsulation and tissue engineering technologies progresses, hydrogels have been utilized to encapsulate cells in microfluidic chips.

8.2 Fabrication and Operation of Cell-Based Biochips

Integration of the PMBV/PVA hydrogel system and microfluidics for the encapsulation and preservation of cells on-chip has been investigated [68, 69]. The PMBV/PVA hydrogel spontaneously encapsulated cells on-chip without inflicting any adverse physical effects (such as seen with thermal- and phototreatments), and the cells encapsulated in the hydrogel in the chip exhibited high viability and low proliferation over a period of days and weeks under cell-based assay conditions. Hence, the research not only presents a promising hydrogel material for long-term cell preservation without perfusion culture on-chip, which has not been realized before, but also expands the application of the PMBV/PVA hydrogel system for cell-based assays from bulk scale to microscale.

The microfluidic chip system for preparing a miniaturized PMBV/PVA hydrogel consists of a two-chamber chip, an aluminum custom-made chip holder, Teflon capillaries, microtubes, and syringes equipped with a microsyringe pump (Fig. 15). The two-chamber chip was fabricated by a photolithographic wet etching technique. Whereas both channels and chambers (200 μm in depth) were fabricated on the top plate, only chambers (200 μm in depth) were fabricated on the bottom plate.

To encapsulate cells in the chip, first, 5 μL of 2.5 wt% PVA solution was delivered into the cell-container chamber through the introducing microchannel using a microsyringe pump with a withdraw mode. Then, 15 μL of cell suspension in a culture medium containing PMBV (5.0 wt%) was introduced. After that, PMBV/PVA hydrogel spontaneously formed and encapsulated the cells in the cell-container chamber. The viability of the cells encapsulated in the miniaturized hydrogel in the chip was investigated using fluorescence-based LIVE/DEAD assays.

8.3 Long-Term Viability of Cells in the Cell-Based Chip

Two mammalian cell lines, mouse fibroblasts (L929 cells) and human arterials endothelial cells (HAECs) were used in the investigation of cell encapsulation. Fibroblast cells and endothelial cells are model cells widely used in many cell biology applications such as cytotoxicity assays and tissue engineering. Cell-encapsulating PMBV/PVA hydrogels with L929 cells and HAECs were

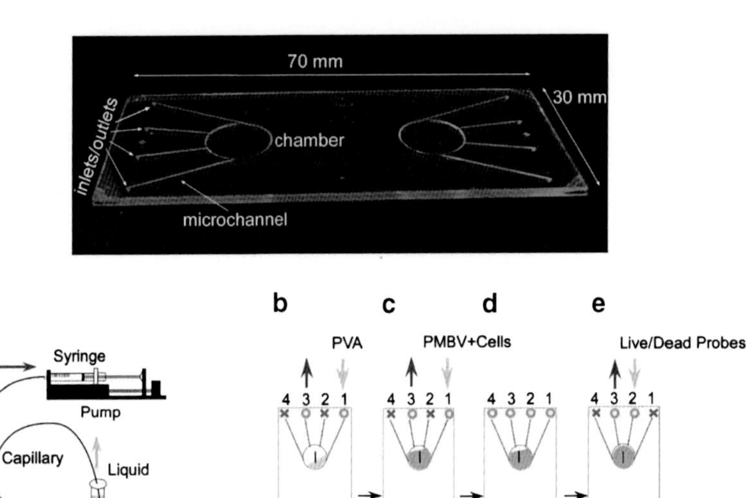

Fig. 15 *Above*: Photo of a two-chamber chip fabricated on glass substrates for preparing miniaturized cell-encapsulating hydrogel. *Below*: Setup and operation of the two-chamber chip. (*a*) The system setup. (*b*) PVA solution is introduced into chamber *I* through inlets *1* and *3*. (*c*) The PMBV/cell suspension is introduced through the same inlet/outlet. Immediately, cell-encapsulating PMBV/PVA hydrogel spontaneously forms in chamber *I*. After that, cells are encapsulated in chamber *II* through the same operation. (*d*) With all inlets/outlets open, the chip is incubated at 37 °C in a cell culture incubator. (*e*) After encapsulation for several days, LIVE/DEAD agents are introduced through inlet *2* and outlet *3* to chamber *I* and through inlet *7* and outlet *6* to chamber *II*, respectively

prepared in both the microplate (bulk hydrogel) and the chip (miniaturized hydrogel).

Figure 16 demonstrates representative fluorescence images of LIVE/DEAD assays of L929 cells after being encapsulated for 4 days in the bulk hydrogel and in the miniaturized hydrogel. In both hydrogel formats, only few dead cells (indicated as red fluorescence) were found, indicating that most cells were live after 4 days of encapsulation in both hydrogel formats. Accordingly, almost equal cell viabilities were calculated (at 4 days), i.e., about 88.7% in the miniaturized hydrogel and about 87.8% in the bulk hydrogel (Fig. 17). This indicates that, after 4 days of encapsulation, the viability of cells in the miniaturized hydrogel was highly uniform with that of cells in the bulk hydrogel. After 8 days of encapsulation, the viability of L929 cells in the miniaturized hydrogel was as high as its viability after 4 days, whereas the viability of L929 cells in the bulk hydrogel decreased slightly to

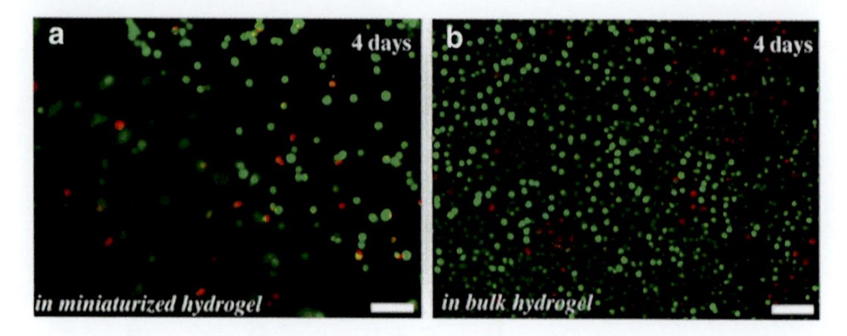

Fig. 16 Fluorescence images of LIVE/DEAD assays of the L929 cells encapsulated for 4 days: (**a**) in the miniaturized PMBV/PVA hydrogel formed in the microfluidic chip, and (**b**) in the bulk PMBV/PVA hydrogel formed in the 96-well microplate. *Green* fluorescence indicates live cells and *red* fluorescence indicates dead cells. *Scale bar*: 100 μm

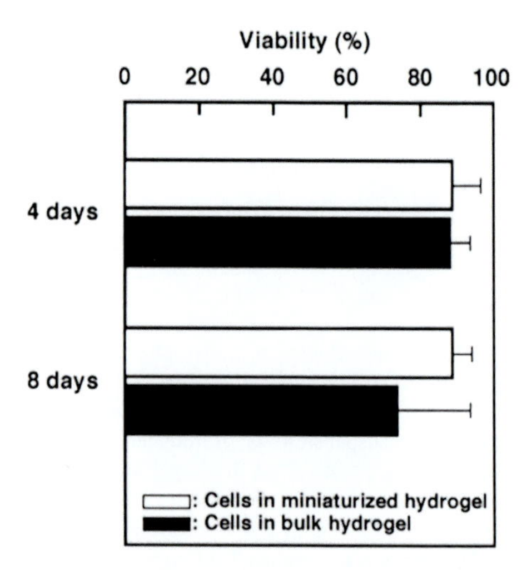

Fig. 17 Viability of L929 cells after 4 and 8 days of encapsulation in miniaturized PMBV/PVA formed in a microfluidic chip and in bulk PMBV/PVA hydrogel formed in a 96-well microplate

73.5%. Similarly, the 7-day viability of HAECs (initial density: 1.6×10^5 cells/mL) in the bulk hydrogel was also slightly lower than that in the miniaturized hydrogel. It is considered that this slight difference may simply result because the bulk hydrogel in the microplate (open system) evaporates faster than the miniaturized hydrogel in the chip (semi-closed system) and this does not appear to be an intrinsic property of the hydrogel. Therefore, to obtain substantial and reliable results, the following comparative investigations were performed on the cells in both hydrogel formats after the cells were encapsulated for 4 days.

8.4 Cytotoxicity Assay in the Cell-Based Chip

Cell cytotoxicity assays with the encapsulated L929 cells were performed. Methanol and $CoCl_2$ solutions were used as toxins. Whereas methanol is a strong cytotoxic agent at a high concentration, $CoCl_2$ is a noncytotoxic or weak cytotoxic agent [70]. Therefore, a $CoCl_2$ solution with a high concentration (50 mM) and methanol solutions with low (7%, v/v) and a high (7%, v/v) concentrations were prepared in PBS for toxin exposure experiments. The cytotoxic sensitivity of cells in the miniaturized hydrogel (initial density 1.0×10^6 cells/mL, 4-day encapsulation) was compared with that of cells in the bulk hydrogel (initial density 1.0×10^6 cells/mL, 4-day encapsulation).

As shown in Fig. 18, cytotoxicity responses expressed as the percentage of dead cells were in accordance with the cytotoxicity of the toxins. Exposure to noncytotoxic (or weakly cytotoxic) solutions (50 mM $CoCl_2$ and 7% v/v methanol) resulted in a very low percentage of dead cells, whereas a cell death of 100% was observed in exposures to the strongly cytotoxic solution (70% v/v methanol). This indicates that cells encapsulated in both hydrogel formats exhibited high cytotoxic sensitivities. Most importantly, for each toxin solution, the cytotoxicity response of cells in the miniaturized hydrogel was not only as uniform as that of cells in the bulk hydrogel, but also as uniform as that of cells cultured in medium. This reveals that, after 4 days of encapsulation, cells in the miniaturized hydrogel maintained a high degree of correlation in cytotoxic sensitivity with the cells in the bulk hydrogel and as well as in the conventional medium culture.

A comparative investigation on the performance of the PMBV/PVA hydrogel in a miniaturized format and in a bulk format was performed. Aspects of hydrogel formation were studied, i.e., cell encapsulation, long-term cell viability, and cell cytotoxicity. Cell encapsulations in the miniaturized hydrogel and in the bulk hydrogel were prepared in a glass microfluidic chip and in a standard 96-well microplate, respectively. The high viability of the cells in the miniaturized hydrogel was maintained, as in the bulk hydrogel, after long-term (4 days and 8 days) encapsulation. Remarkably, not only did the cells in both hydrogel formats exhibit high cytotoxic sensitivities against different toxins with different cytotoxicity after 4 days of encapsulation, but also the cells in the miniaturized hydrogel maintained a

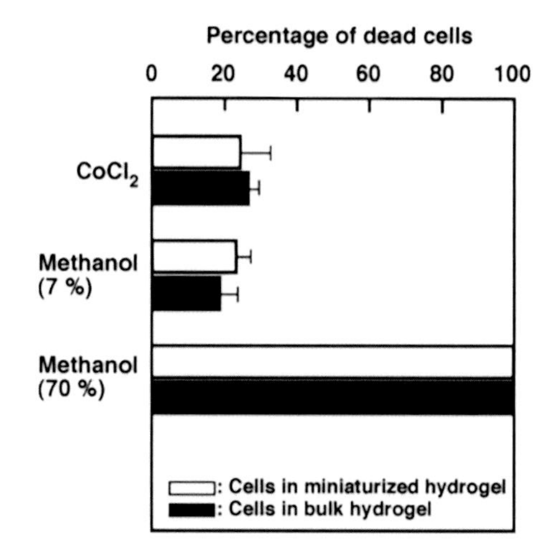

Fig. 18 Cytotoxic response of L929 cells after 4 days of encapsulation in miniaturized PMBV/PVA hydrogel formed in a microfluidic chip and in bulk PMBV/PVA hydrogel formed in a 96-well microplate

high degree of correlation in cytotoxic sensitivity with the cells in the bulk hydrogel and as well as with cells in the conventional medium culture. Therefore, the PMBV/PVA hydrogel behaved as uniformly in the microscale as in the bulk, which is important, useful, and meaningful for use of the PMBV/PVA hydrogel system in cell-based applications from a bulk level to a microscale level.

9 Conclusion and Future Perspectives

A spontaneously forming hydrogel composed of MPC polymer with phenyl boronic acid group and PVA can encapsulate cells under ordinary conditions. The PMBV/PVA hydrogel dissociates again by addition of sugar, and a cell suspension is obtained. During this encapsulation-recovery process, cells do not lose their activity and functions. That is, the PMBV/PVA hydrogel can regulate excessive proliferation, provide normalized cells with uniform cell cycle phases, and maintain cell-specific functions when the cells are encapsulated in the hydrogel. In the case of ES cell encapsulation, the cells maintain their undifferentiated characteristics during the preservation period in the PMBV/PVA hydrogel. Thus, we conclude that the PMBV/PVA hydrogel is a novel and powerful soft device for controlling cell functions in cell engineering fields.

It is usual during differentiation of stem cells for them to form an embryoid body on suspension culture. However, the cell clusters formed on suspension culture are aggregations derived from various different cells. Individual cells from the same tissue may actually differ from each other and have different roles. Furthermore, it has been recently revealed that the absolute amounts of complementary DNA expression differ from cell to cell even if the cells derive from the same cell line [71]. Thus, inducing homogeneous cell clusters derived by expansion of single cells, and not derived by aggregation of various different cells, may maintain high cell-specific functions and highly efficient differentiation of stem cells.

Acknowledgments The authors appreciate Prof. Madoka Takai, Dr. Ryosuke Matsuno, Dr. Yuuki Inoue and Prof. Takehiko Kitamori at The University of Tokyo for helpful discussions during preparation of the manuscript. One of the authors, Dr. Xu Yan, moved to Osaka Prefecture University, Osaka, Japan in April 2011.

References

1. Lutolf MP, Hubbell JA (2005) Synthetic biomaterials as instructive extracellular microenvironments for morphogenesis in tissue engineering. Nat Biotechnol 23:47–55
2. Lutolf MP (2009) Spotlight on hydrogels. Nat Mater 8:451–453
3. Lutolf MP (2009) Integration column: artificial ECM: expanding the cell biology toolbox in 3D. Integr Biol 1:235–241

4. Peppas NA, Hilt JZ, Khademhosseini A et al (2006) Hydrogels in biology and medicine: from molecular principles to bionanotechnology. Adv Mater 18:1345–1360
5. Slaughter BV, Khurshid SS, Fisher OZ et al (2009) Hydrogels in regenerative medicine. Adv Mater 21:3307–3329
6. Hoffman AS (2002) Hydrogels for biomedical applications. Adv Drug Deliv Rev 54(1):3–12
7. Lee KY, Mooney DJ (2001) Hydrogels for tissue engineering. Chem Rev 101:1869–1880
8. Takahashi K, Tanabe K, Yamanaka S et al (2007) Induction of pluripotent stem cells from adult human fibroblasts by defined factors. Cell 131:861–872
9. Takahashi K, Yamanaka S (2006) Induction of pluripotent stem cells from mouse embryonic and adult fibroblast cultures by defined factors. Cell 126:663–676
10. Evans MJ, Kaufman MH (1981) Establishment in culture of pluripotential cells from mouse embryos. Nature 292:154–156
11. Tsang VL, Bhatia N (2004) Three-dimensional tissue fabrication. Adv Drug Deliv Rev 56:1635–1647
12. Liu C, Xia Z, Czernuszka JT (2007) Design and development of three-dimensional scaffolds for tissue engineering. Trans IChemE, Part A, Chem Eng Res Des 85:1051–1064
13. Stoop R (2008) Smart biomaterials for tissue engineering of cartilage. Inj Int J Care Injured 3951:577–587
14. Drury JL, Mooney DJ (2003) Hydrogels for tissue engineering: scaffold design variables and applications. Biomaterials 24:4337–4351
15. Peppas NA, Bures P, Leobandung W et al (2000) Hydrogels in pharmaceutical formulations. Eur J Pharm Biopharm 50:27–46
16. Zhu J (2010) Bioactive modification of poly(ethylene glycol) hydrogels for tissue engineering. Biomaterials 31:4639–4656
17. Van Tomme SR, Storm G, Hennink WE (2008) In situ gelling hydrogels for pharmaceutical and biomedical applications. Int J Pharm 355:1–18
18. Grinnell F (2003) Fibroblast biology in three-dimensional collagen matrices. Trends Cell Biol 13:264–269
19. Tabata Y, Hijikata S, Ikada Y (1994) Enhanced vascularization and tissue granulation by basic fibroblast growth factor impregnated in gelatin hydrogels. J Control Rel 31:189–199
20. Khademhosseini A, Suh KY, Yang JM et al (2004) Layer-by-layer deposition of hyaluronic acid and poly-L-lysine for patterned cell co-cultures. Biomaterials 25:3583–3592
21. Watanabe J, Nederberg F, Atthoff B et al (2007) Cytocompatible biointerface on poly(lactic acid) by enrichment with phosphorylcholine groups for cell engineering. Mater Sci Eng 27:227–231
22. Watanabe J, Eriguchi T, Ishihara K (2002) Stereocomplex formation by enantiomeric poly (lactic acid) graft-type phospholipid polymers for tissue engineering. Biomacromolecules 3:1109–1114
23. Watanabe J, Eriguchi T, Ishihara K (2002) Cell adhesion and morphology in porous scaffold based on enantiomeric poly(lactic acid) graft-type phospholipid polymers. Biomacromolecules 3:1375–1383
24. Watanabe J, Ishihara K (2005) Cell engineering biointerface focusing on cytocompatibility using phospholipid polymer with an isomeric oligo(lactic acid) segment. Biomacromolecules 6:1797–1802
25. Watanabe J, Ishihara K (2003) Phosphorylcholine and Poly(D, L-lactic acid) containing copolymers as substrates for cell adhesion. Artif Organs 27:242–248
26. Jeong B, Choi YK, Bae YH et al (1999) New biodegradable polymers for injectable drug delivery systems. J Control Release 62:109–114
27. Li Z, Ramay HR, Hauch KD et al (2005) Chitosan-alginate hybrid scaffolds for bone tissue engineering. Biomaterials 26:3919–3928
28. Ling Y, Rubin J, Deng Y et al (2007) A cell-laden microfluidic hydrogel. Lab Chip 7:756–762
29. Chenite A, Chaput C, Wang D et al (2000) Novel injectable neutral solutions of chitosan form biodegradable gels in situ. Biomaterials 21:2155–2161

30. Jeong B, Kim SW, Bae YH (2002) Thermosensitive sol-gel reversible hydrogels. Adv Drug Deliv Rev 54:37–51
31. Nuttelman CR, Mortisen DJ, Henry SM et al (2001) Attachment of fibronectin to poly(vinyl alcohol) hydrogels promotes NIH3T3 cell adhesion, proliferation and migration. J Biomed Mater Res 57:217–223
32. Kitano S, Kataoka K, Koyama Y et al (1991) Glucose-responsive complex formation between poly(vinyl alcohol) and poly(*N*-vinyl-2-pyrrolidone) with pendent phenylboronic acid moieties. Makromol Chem Rapid Commun 12:227–233
33. Lutolf MP, Raeber GP, Zisch AH (2003) Cell-responsive synthetic hydrogels. Adv Mater 15:888–892
34. Lutolf MP, Hubbell JA (2003) Synthesis and physicochemical characterization of end-linked poly(ethylene glycol)-co-peptide hydrogels formed by michael-type addition. Biomacromolecules 4:713–722
35. Kimura M, Fukumoto K, Watanabe J et al (2005) Spontaneously forming hydrogel from water-soluble random- and block-type phospholipid polymers. Biomaterials 26:6853–6862
36. Kimura M, Fukumoto K, Watanabe J et al (2004) Hydrogen-bonding-driven spontaneous gelation of water-soluble phospholipid polymers in aqueous medium. J Biomater Sci Polym Edn 15:631–644
37. Nam KW, Watanabe J, Ishihara K (2002) Characterization of the spontaneously forming hydrogels composed of water-soluble phospholipid polymers. Biomacromolecules 3:100–105
38. Nam KW, Watanabe J, Ishihara K (2004) Modeling of swelling and drug release behavior of spontaneously forming hydrogels composed of phospholipid polymers. Int J Pharm 275:259–269
39. Ishiyama N, Moro T, Ishihara K et al (2010) The prevention of peritendinous adhesions by a phospholipid polymer hydrogel formed in situ by spontaneous intermolecular interactions. Biomaterials 31:4009–4016
40. Konno T, Ishihara K (2007) Temporal and spatially controllable cell encapsulation using a water-soluble phospholipid polymer with phenylboronic acid moiety. Biomaterials 28:1770–1777
41. Choi J, Konno T, Matsuno R et al (2008) Surface immobilization of biocompatible phospholipid polymer multilayered hydrogel on titanium alloy. Colloids Surf B Biointerfaces 67:216–223
42. Matthew JE, Nazario YL, Roberts SC et al (2002) Effect of mammalian cell culture medium on the gelation properties of Pluronic F127. Biomaterials 23:4615–4619
43. Ishihara K, Ueda T, Nakabayashi N (1990) Preparation of phospholipid polymers and their properties as polymer hydrogel membrane. Polym J 22:355–360
44. Ishihara K (1997) Novel polymeric materials for obtaining blood compatible surfaces. Trends in Polym Sci 5:401–407
45. Lewis AL (2000) Phosphorylcholine-based polymers and their use in the prevention of biofouling. Colloid Surf B Biointerfaces 18:261–275
46. Ishihara K (2000) Bioinspired phospholipid polymer biomaterials for making high performance artificial organs. Sci Technol Adv Mater 1:131–138
47. Iwsaki Y, Ishihara K (2005) Phosphorylcholine-containing polymers for biomedical applications. Ann Bioanal Chem 381:534–546
48. Ishihara K, Takai M (2009) Bioinspired interfaces for nanobiodevices based on phospholipid polymer chemistry. J R Soc Interface 6:S279–S291
49. Ishihara K, Oshida H, Endo Y et al (1992) Hemocompatibility of human whole blood on polymers with a phospholipid polar group and its mechanism. J Biomed Mater Res 26:1543–1552
50. Ishihara K, Nomura H, Mihara T et al (1998) Why do phospholipid polymers reduce protein adsorption? J Biomed Mater Res 39:323–330

51. Myers GJ, Johnstone DR, Swyer WJ et al (2003) Evaluation of mimesys phosphorylcholine (PC)-coated oxygenators during cardiopulmonary bypass in adults. J Extra Corpor Technol 35:6–12
52. Bakhai A, Booth J, Delahunty N et al (2005) The SV stent study: a prospective, multicentre, angiographic evaluation of the BiodivYasio phosphorylcholine coated small vessel stent in small coronary vessels. Int J Cardiol 102:95–102
53. Snyder TA, Tsukui H, Kihara S et al (2007) Preclinical biocompatibility assessment of the EVAHEART ventricular assist device: coating comparison and platelet activation. J Biomed Mater Res A 81:85–92
54. Goda T, Ishihara K (2006) Soft contact lens biomaterials from bioinspired phospholipid polymers. Expert Rev Med Devices 3:167–174
55. Moro T, Takatori Y, Ishihara K et al (2004) Surface grafting of artificial joints with a biocompatible polymer for preventing periprosthetic osteolysis. Nat Mater 3:829–836
56. Kikuchi A, Suzuki K, Okabayashi O et al (1996) Glucose-sensing electrode coated with polymer complex gel containing phenylboronic acid. Anal Chem 68:823–828
57. Shiino D, Murata Y, Kubo A et al (1995) Amine containing phenylboronic acid gel for glucose-responsive insulin release under physiological pH. J Control Release 37:269–276
58. Lorand JP, Edwards JO (1959) Polyol complexes and structure of the benzeneboronate ion. J Org Chem 24:769–774
59. Zhang E, Li X, Zhang S et al (2005) Cell cycle synchronization of embryonic stem cells: effect of serum deprivation on the differentiation of embryonic bodies in vitro. Biochem Biophys Res Commun 333:1171–1177
60. Chang T, Hughes-Fulford (2009) Monolayer and spheroid culture of human liver hepatocellular carcinoma cell line cells demonstrate distinct global gene expression patterns and functional phenotypes. Tissue Eng Part A 15:559–567
61. Mooney D, Hansen L, Vaccanti J et al (1992) Switching from differentiation to growth in hepatocytes: control by extracellular matrix. J Cell Physiol 151:497–505
62. Ohno K, Tachikawa K, Manz A (2008) Microfluidics: applications for analytical purposes in chemistry and biochemistry. Electrophoresis 29:4443–4453
63. El-Ali J, Sorger PK, Jensen KF (2006) Cells on chips. Nature 442:403–411
64. Griffith LG, Swartz MA (2006) Capturing complex 3D tissue physiology in vitro. Nat Rev Mol Cell Biol 7:211–224
65. Sato K, Mawatari K, Kitamori T (2008) Microchip-based cell analysis and clinical diagnosis system. Lab Chip 8:1992–1998
66. Lion N, Rohner TC, Dayon L et al (2003) Microfluidic systems in proteomics. Electrophoresis 24:3533–3562
67. Andersson H, van den Berg A (2003) Microfluidic devices for cellomics: a review. Sensor Actuat B-Chem 92:315–325
68. Xu Y, Sato K, Konno T et al (2009) An on-chip living cell bank. Proc in Micro-TAS 2009: W82F
69. Xu Y, Sato K, Mawatari K et al (2010) A microfluidic hydrogel capable of cell preservation without perfusion culture under cell-based assay conditions. Adv Mater 22: 3017–3021
70. Wang Z, Kim MC, Marquez M et al (2007) High-density microfluidic arrays for cell cytotoxicity analysis. Lab Chip 7:740–745
71. de Souza N (2010) Single-cell methods. Nat Methods 7:35

Adv Polym Sci (2012) 247: 167–200
DOI: 10.1007/12_2011_161

Published online: 3 November 2011

Design of Biointerfaces for Regenerative Medicine

Yusuke Arima, Koichi Kato, Yuji Teramura, and Hiroo Iwata

Abstract Understanding and controlling biological responses against artificial materials is important for the development of medical devices and therapies. Self-assembled monolayers (SAMs) of alkanethiols provide well-defined surfaces that can be manipulated by varying the terminal functional groups. Thus, SAMs have been extensively used as a platform for studying how artificial materials affect biological responses. Here, we review cell adhesion behavior in response to SAMs with various surface properties and the effects that adsorbed proteins have on subsequent cell adhesion. We also describe an application for SAMs as a substrate for culturing neural stem cells (NSCs). Substrates that induced the correct orientation of immobilized growth factors, like epidermal growth factor, improved the selection of a pure NSC population during cell expansion. In addition, we review new methodologies for using amphiphilic polymers to modify the surfaces of cells and tissues. Coating the cell surface with amphiphilic polymers that can capture and immobilize bioactive substances or cells represents a promising approach for clinical applications, particularly cellular therapies.

Keywords Amphiphilic polymer · Cell adhesion · Chimeric protein · Islet of Langerhans · Self-assembled monolayer · Stem cell

---

Y. Arima, K. Kato, Y. Teramura, and H. Iwata (✉)
Institute for Frontier Medical Sciences, Kyoto University, 53 Kawahara-cho, Sakyo-ku, Shogoin, Kyoto 606-8507, Japan
e-mail: iwata@frontier.kyoto-u.ac.jp

K. Kato
Present address: Department of Biomaterials Science, Graduate School of Biomedical Sciences, Hiroshima University, 1-2-3, Kasumi, Minami-ku, Hiroshima 734-8553, Japan

Y. Teramura
Present address: Department of Immunology,, Genetics and Pathology (IGP), Uppsala University, Rudbecklab C5, 3rd floor, Dag Hammarskjoldsv 20, 751 85 Uppsala, Sweden

Contents

1 Introduction

Much effort has been devoted to understanding biological responses to artificial materials [1, 2] in order to facilitate the development of medical devices and artificial organs. However, many issues remain to be fully understood. Furthermore, these studies require overcoming various difficulties because biological responses are affected by many factors, including surface energy, surface electrostatic properties, macro- and microsurface morphology, surface heterogeneity, different functional groups, and the mobility of functional groups on surfaces. Systematic studies of biological responses to artificial materials require surfaces with well-controlled properties; however, there is a lack of methods for systematically controlling surface properties. Surface chemistry approaches have employed the use of model surfaces, like self-assembled monolayers (SAMs) of alkanethiols, $HS(CH_2)_nX$, where X denotes various functional groups [3–6]. SAMs are also suitable for studying correlations between biological responses and surface properties.

It has been shown that cell adhesion highly depends on the outermost functional groups on SAMs; however, cells do not directly interact with the SAMs. Instead, they interact with proteins adsorbed on SAMs. Cell adherence requires an interaction between integrin molecules in the cell membrane and glycoproteins specialized for cell adhesion, like fibronectin (Fn) and vitronectin (Vn), which are adsorbed on the artificial material. Thus, the presence of glycoproteins in serum plays a crucial role in cell adherence to artificial materials. In the first part of this review (Sect. 2), we will briefly survey recent studies of cell adhesion on SAMs with different functional groups and discuss the mechanisms involved.

Knowledge gained from cell adhesion studies with SAMs has been used to develop culture substrates with the appropriate cell adhesion glycoproteins for different types of cells [7–10]. Stem cells, capable of self-renewal and differentiation into multiple cell types, are found in embryonic and adult tissues. Pluripotent stem cells, like embryonic stem cells and induced pluripotent stem cells, have been developed in vitro. These cells are expected to provide cell sources for regenerative medicine. Various culture conditions have been developed to enable expansion of these cells without loss of their multi- and pluripotency and to induce differentiation into viable cells with specific functions.

In the last few years, our group has focused on neural stem cells (NSCs). NSCs were discovered by screening rodent CNS cells for responses to epidermal growth factor (EGF) [11]. Integrin and epidermal growth factor receptor (EGFR) coordinately regulate cell migration, survival, and growth by modulating a common set of signaling pathways. Moreover, EGF was shown to be a mitogen for NSCs. Taking these facts into consideration, we hypothesized that NSCs might be selectively trapped on SAM surfaces through EGF–EGFR interactions, and that this interaction might strongly promote NSC proliferation due to EGFR signaling. In the second part of this article (Sect. 3), we will describe our own work on cultured NSC interactions with surface-immobilized EGF.

Regenerative medicine and tissue engineering have opened new therapeutic domains. Stem cells have become therapeutic units for generating functional cells and tissues. One of the more successful endeavors has been the transplantation of islets of Langerhans (islets), which produce and release insulin, as a treatment for patients with type 1 diabetes. However, implantation of living cells into a host induces various undesirable biological responses similar to the responses against artificial materials, such as blood coagulation, complement activation, inflammatory reactions, and immune reactions. Understanding biological responses to artificial materials [2] and surface modification methods for biomaterials gives bases to evade adverse host responses and to improve functions of transplanted cells. For living cells and tissues, however, surface treatment should be carried out under the physiological conditions that do not deteriorate their viability and biological functions. We developed new methods to modify the surfaces of cells and tissues to increase their compatibility with a host environment. In the last part of this review (Sect. 4), we discuss the latest methods for modifying islet surfaces and their effects on islet–host compatibility.

2 Cell Adhesion to a Model Biomaterial Surface

2.1 Self-Assembled Monolayer as a Model Surface

Self-assembled monolayers (SAMs) of alkanethiols, $HS(CH_2)_nX$, where X denotes various functional groups, are frequently used to prepare model surfaces [3–6]. Alkanethiols or alkanedisulfides chemisorb from a solution onto a surface coated with a metal such as gold, silver, or platinum. SAMs are commonly formed on a thin gold layer coated on a glass plate, due to its easy preparation and its stability in the ambient environment. Furthermore, gold is compatible with sensitive methods for analyzing surface phenomena, including surface plasmon resonance (SPR) [12], ellipsometry [13], Fourier-transformed infrared-reflection adsorption spectroscopy (FTIR-RAS) [13], and quartz crystal microbalance (QCM) [14]. The gold–sulfur bond is relatively stable, with a $\Delta H° \approx 28$ kcal/mol [15, 16]. In addition, alkanethiols self-assemble through van der Waals interactions between alkyl chains. Alkanethiols with long alkyl chains ($n > 11$) form closely packed SAMs, with approximately 21.4 $Å^2$ of occupied area per molecule [17, 18]. Due to the thiol anchoring to the gold and the close packing of the alkyl chain, another terminal group, X, can be effectively displayed on the surface of the SAM. Alkanethiols with various functional groups are commercially available, and SAMs with different functional groups are easily prepared.

The surface properties of SAMs can be finely controlled by coadsorbing a mixture of alkanethiols with different functional groups. The composition of a SAM can be determined by spectroscopic methods like FTIR-RAS and X-ray photoelectron spectroscopy (XPS). The fraction of given alkanethiol on the mixed SAM surface reflects its mole fraction in the mixed solution, but it is not a linear relationship. The adsorbed fraction is highly dependent on the chain length of alkanethiol [19] and its terminal functional group [20, 21]. The water contact angle measurement of mixed SAMs gradually changes with the fraction of mixed SAMs on the surface, and the relationship can be approximately expressed with Cassie's equation. Thus, the preparation of mixed SAMs with different alkanethiols allows us to systematically change the surface properties to produce a variety of different model surfaces.

A micropattern can be printed onto the SAM surface. This is achieved by UV light irradiation through a photomask to cause photodegradation of alkanethiols [22–24] or by microcontact printing with a pattern stamp made from poly(dimethylsiloxane) [25, 26]. Micropatterned SAMs have been employed as a high-throughput platform for studies on biomolecular interactions that included arrays of DNAs [27–29], proteins [30–33], and cells [34–37]. Micropatterned SAMs have also been used to examine cell fate after controlling the geometry of cell adhesion on a micrometer scale [38–41].

2.2 Cell Adhesion on Material Surfaces

When a cell suspension is applied to a surface, the events that occur can be conceptually classified into three stages: (1) a cell approaches the surface, (2) the cell attaches to the surface, and (3) the cell adheres, and thus, spreads out on the surface. Most studies of cell adhesion on artificial materials measure the number of adherent cells, the cell morphology, and changes in protein expression. To gain more detailed insight into the biophysical mechanism of cell adhesion requires

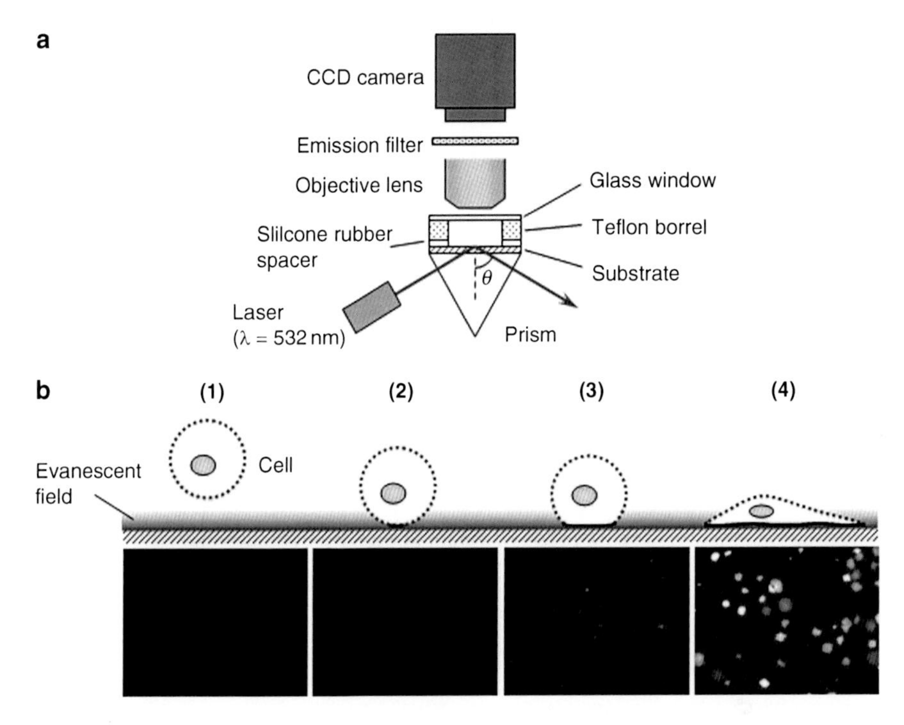

Fig. 1 Real-time tracking of cell adhesion [42]. (**a**) Components of a total internal reflection fluorescent microscope (TIRFM). (**b**) The cell adhesion process; (*1*) a cell approaches the surface, (*2*) the cell lands, (*3*) the cell attaches, and (*4*) the cell spreads out on the surface. The evanescent field was generated by total internal reflection of a laser beam at the glass–water interface. Cells with fluorescently labeled membranes (*dashed lines*) were plated on SAMs. Cell membranes within the evanescent field (*solid line*) were observed by TIRFM. Corresponding TIRFM images are shown *below*

real-time tracking of cell behavior. To observe initial cell adhesion onto SAMs in real time, we employed a total internal reflection fluorescence microscope (TIRFM).

The optical assembly of a TIRFM is schematically shown in Fig. 1a [42]. The TIRFM utilizes an evanescent field, which is generated by laser reflection at a water–glass interface [43, 44]. The intensity of the evanescent field decays exponentially with increasing distance from the interface. The characteristic penetration depth of the evanescent field is approximately 100 nm (depending on wavelength, incident angle, and refractive index of the glass and aqueous solution). Fluorescent dyes are excited in the evanescent field, and the emitted fluorescent light is captured with a charge-couple device (CCD) camera.

We assembled a TIRFM with low magnification to study cell adhesion behavior on SAMs with various functional groups [42]. Figure 1b shows a schematic illustration of the cell adhesion process and the corresponding TIRFM images. A suspension of cells with fluorescently labeled cell membranes is applied onto a substrate (Fig. 1b-1). At first, no bright spots were observed by TIRFM, because the cells were in the medium, outside of the evanescent field. When the cells approached and landed on the surface (Fig. 1b-2), small, bright spots were observed. The cells were round with diameters of approximately 10 μm, but the diameter of the cell membrane visualized in the evanescent field was approximately 2 μm. These spots were difficult to detect by TIRFM with a 10× objective lens. Once cells adhered to the SAM surface (Fig. 1b-3), bright spots were easily detected by TIRFM. When cells spread out on the surface (Fig. 1b-4), the bright spots observed by TIRFM became larger with time. Thus, TIRFM was a useful tool for investigating cell adhesion. Changes observed in labeled cells could be interpreted to infer cell behavior at the solid–liquid interface.

TIRFM was used for time-lapse observations of initial cell adhesion to SAMs with different surface functionalities (Fig. 2). After 10 min of plating a suspension of human umbilical vein endothelial cells (HUVECs), a few bright spots were observed on SAMs with COOH and NH_2 functionalities; this indicated cell adherence. The number of bright spots increased and the spot areas enlarged with incubation time, indicating that HUVECs adhered and spread well on COOH–SAM and NH_2–SAM surfaces. Quantitative analysis of the number of adherent cells and cell adhesion areas

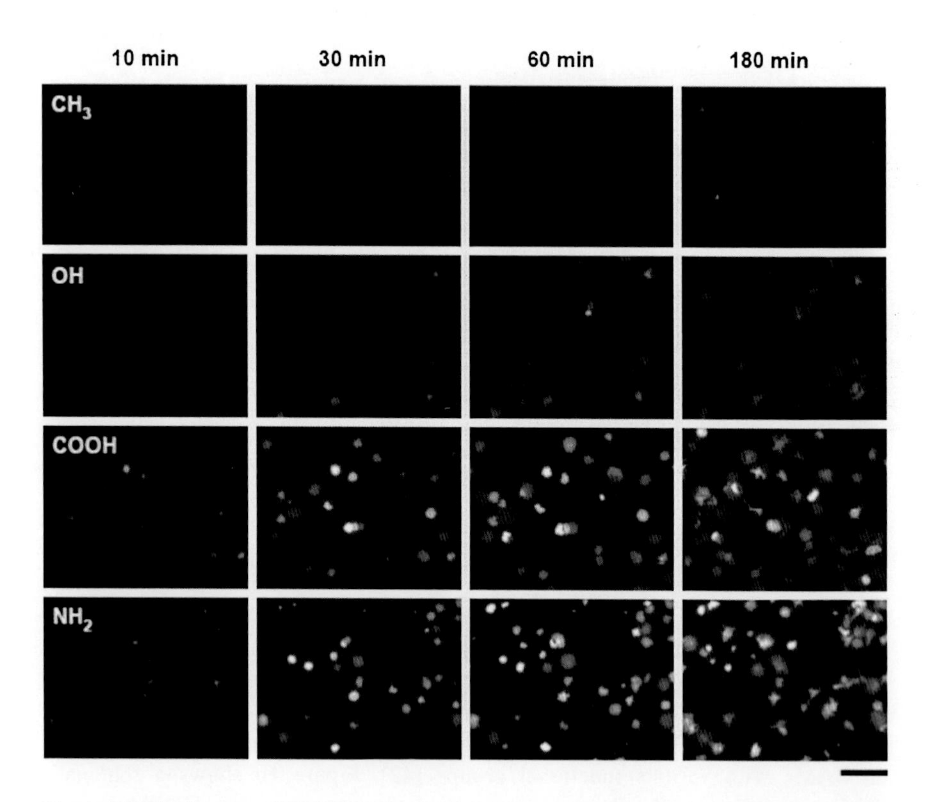

Fig. 2 TIRFM images of HUVEC adhesion behavior on SAMs with four different types of surface functional groups at the indicated times after first applying the cell suspension. Scale bar: 200 μm [42]

on COOH– and NH_2–SAMs showed that cell adhesion reached equilibrium after 60 min of incubation, but the adhesion area increased up to 180 min. In contrast, no spot was observed on SAMs with CH_3 and OH functionalities. The number and total area of bright spots were much lower on OH– and CH_3–SAMs than on COOH– and NH_2–SAMs at all times observed. After 60 min, some bright spots were observed on OH–SAMs, but not on CH_3–SAMs. This indicated that HUVECs adhered poorly to OH–SAM surfaces and worse to CH_3–SAMs. The effects of surface functional groups on SAMs have been extensively studied with various cell lines. Those studies showed that most cell types adhered well to COOH– and NH_2–SAMs and poorly to CH_3– and OH–SAMs [45–50].

Other studies used various polymers to investigate how cell adhesion depended on the wettability of materials. They showed that cells adhered well to moderately wettable materials with water contact angles of 40–60° [51–57]. However, the results were inconclusive. They employed polymeric materials composed of different chemicals that changed wettability concomitantly with changes in the SAM surface functionality. We performed a study with mixed SAMs to determine the effect of contact angles on cell adhesion [21]. SAMs with widely varying wettabilities were prepared by mixing two alkanethiols with different functional groups (CH_3/OH, CH_3/COOH, and CH_3/NH_2). Figure 3 shows that the number of adherent cells depended on the contact angles of SAMs for both HUVECs and human cervical carcinoma (HeLa) cells. The maximum number of adherent cells occurred at different contact angles for all of these mixed SAMs, and it was different for different cell types. Thus, the design of a surface that promotes cell adhesion should take into consideration both the type of surface functional groups and the types of cells targeted.

2.3 *Effect of Protein Adsorption on Cell Adhesion*

When cells are suspended in a biological fluid or culture medium, both serum proteins and cells interact with the surface substrate. Serum protein adsorption behavior on SAMs has been examined with various analytical methods, including SPR [58–61], ellipsometry [13, 62, 63], and quartz QCM [64–66]. These methods allow in situ, highly sensitive detection of protein adsorption without any fluorescence or radioisotope labeling. SPR and QCM are compatible with SAMs that comprise alkanethiols. In our laboratory, we employed SPR to monitor protein adsorption on SAMs.

SPR detects changes in the refractive index near the surface of a metal film. Two optical configurations, Kretchmann (Fig. 4a) and Otto, can be applied with SPR. The former is easily set-up and is suitable for protein adsorption on SAMs. A beam of *p*-polarized light is directed, through a prism, to the back of a sample glass plate, which is coated with a metal thin film. The front of the sample glass plate faces the solution of interest. When the incident angle, θ exceeds the critical angle, total internal reflection occurs. An evanescent wave is generated on the surface facing the solution, which has a lower refractive index than glass. At a specific incident

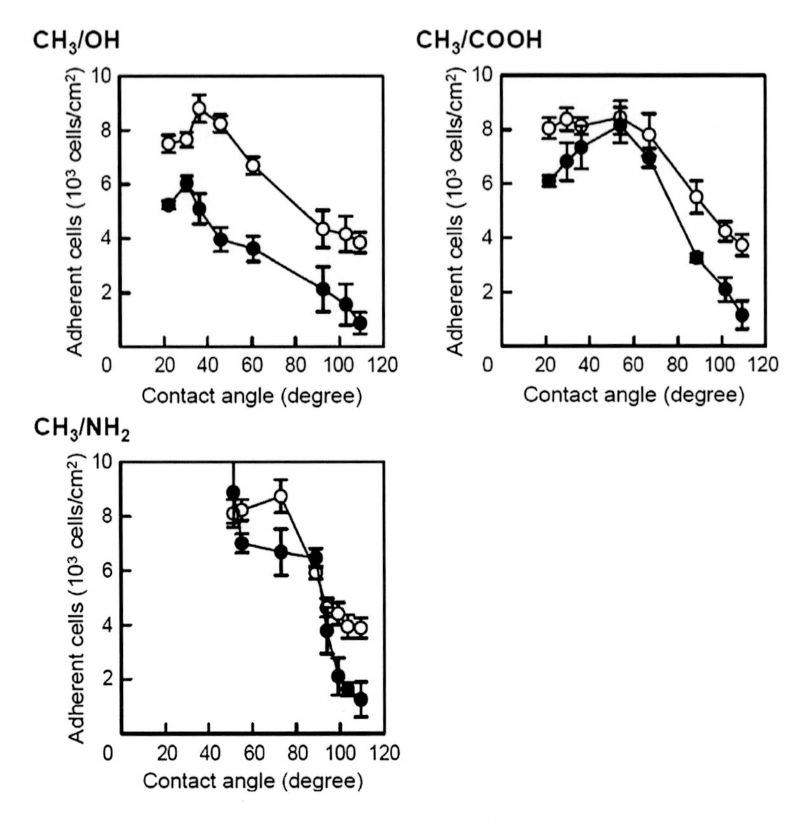

Fig. 3 Relationship between water contact angles on mixed SAMs and the number of adherent cells. HUVECs (*open circles*) and HeLa cells (*filled circles*) were allowed to adhere for 1 h. The averages (± SEM) of five experiments are shown [21]

angle, the evanescent wave of the incoming light is able to couple with the free oscillating electrons (plasmons) in the metal film, and surface plasmon resonance occurs. This resonance causes an energy transfer from the incident light to the plasmons of the metal film, which reduces the intensity of the reflected light (black line in Fig. 4b). Another important point in SPR is that the resonance angle is affected by the refractive index in the vicinity of the metal film (within ~200 nm from the metal surface). Thus, a shift in resonance angle reflects events at the interface, like protein adsorption on the surface. The resonance angle shift ($\Delta\theta$) after protein adsorption is related to the amount of adsorbed protein by the Fresnel relationship [67, 68]. This equation assumes five layers (glass/Au/SAM/protein/water); the refractive index and the density of the protein layer are usually taken as 1.45 and 1 g/cm^3, respectively. For real-time analysis, the change in reflectance (ΔR) is tracked at a fixed incident angle (usually 0.5° lower than the resonance angle) during a measurement, and then it is numerically converted to the resonance angle shift.

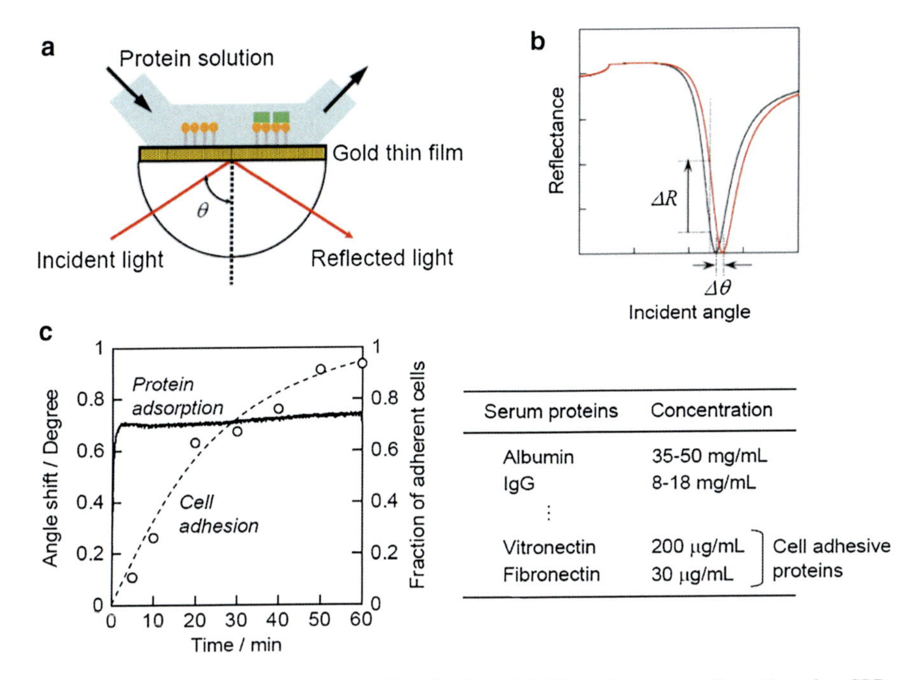

Serum proteins	Concentration	
Albumin	35-50 mg/mL	
IgG	8-18 mg/mL	
⋮		
Vitronectin	200 μg/mL	Cell adhesive proteins
Fibronectin	30 μg/mL	

Fig. 4 The effect of proteins on cell adhesion. (**a**) Kretschmann configuration for SPR. (**b**) Reflectance (R) as a function of incident angle (θ), before (*black*) and after (*red*) the adsorption of substances. (**c**) *Left*: Time course of SPR angle shift during exposure to culture medium supplemented with 2% FBS (*solid line*) and the fraction of adherent cells determined by TIRFM (*circles*) on NH_2-SAM. The *dashed line* is a manual fit to the symbols, included simply as a guide [42]. *Right*: The concentrations of serum proteins in FBS

Figure 4c shows one example of the time course of an SPR angle shift during exposure of a NH_2–SAM to culture medium supplemented with 2% fetal bovine serum (FBS). It also includes the time course of the fraction of adherent cells on the same surface determined by TIRFM observation (Fig. 2). The SPR angle shift rapidly increased, and then leveled off within a few minutes. Cells adhered much more slowly than proteins. Those results indicated that serum proteins in a medium rapidly adsorbed to the surface; then, cells interacted with the adsorbed protein layer, as shown schematically in Fig. 5.

Thus, cell adhesion is determined by nonspecifically adsorbed serum proteins on the surface. Therefore, it is important to consider the characteristics of adsorbed proteins including the amount, composition, and conformation or orientation.

2.3.1 Amount of Protein on a Surface

To suppress cell adhesion on a material surface, one approach is to inhibit the adsorption of proteins. SAMs of alkanethiols that carry oligo(ethylene glycol) (OEG) [69] and phosphorylcholine [46, 70, 71] have been shown to prevent

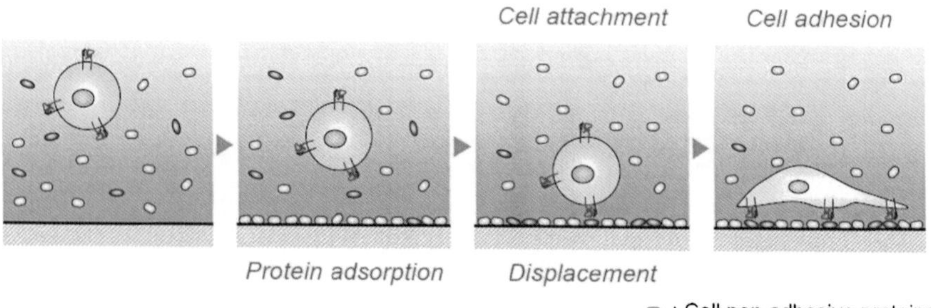

Fig. 5 Cell adhesion to a material surface, including the adsorption of serum proteins

nonspecific adsorption; thus, they effectively prevented cell adhesion. Several studies have described surfaces that prevent nonspecific protein adsorption [59, 72, 73]; those surfaces are also likely to prevent cell adhesion.

We examined protein adsorption to SAMs that carried four different functional groups [42] and mixed SAMs with different wettabilities [21]. Large amounts of serum proteins adsorbed to all these SAMs, but the different surface functional groups greatly affected cell adhesion behavior (Figs. 2 and 3). Thus, the amount of adsorbed proteins did not correlate with the degree of cell adhesion to SAMs.

2.3.2 Protein Composition on a Surface

Serum glycoproteins like Fn and Vn play a crucial role in cell adhesion to artificial materials. These proteins carry peptide motifs, including arginine–glycine–aspartic acid (RGD) [74] and proline–histidine–serine–arginine–asparagine (PHSRN) [75], that specifically interact with integrin receptors on cell membranes. Adsorption of Fn and Vn onto a material surface is required to support cell adhesion. However, the concentrations of Fn and Vn in serum are much lower than that of albumin and IgG (Fig. 4d). The composition of proteins initially adsorbed to a surface reflects the concentrations of each protein in the medium; thus, most proteins on the surface will be albumin and IgG, which cannot support cell adhesion. Controlling the composition of proteins in a solution is important for controlling cell adhesion on artificial materials.

Various methods have been used to examine the composition of proteins adsorbed to SAMs. Overall adsorption patterns can be examined with sodium dodecyl sulfate polyacrylamide gel electrophoresis (SDS-PAGE) [50, 76, 77]. Absorbed proteins are eluted from the surface with surfactant (SDS), and then separated by electrophoresis. The proteins of interest are examined by western blotting [50, 76, 77]. Protein-specific antibodies can be used to detect proteins of interest [19, 62, 76], or radiolabeled proteins can be added to the serum [45, 78]. Lestelius et al. used ellipsometry to examine antibody binding to 12 different plasma proteins that adhered to SAMs exposed to human plasma. Those SAMs carried terminal methyl ($-CH_3$), trifluoromethyl ester ($-OCOCF_3$), sulfate ($-OSO_3H$), carboxyl ($-COOH$), or hydroxyl ($-OH$) groups [62]. Adherence of proteins relevant to coagulation and complement activation depended on the surface functionalities of SAMs. The sulfate and the carboxyl surfaces captured coagulation proteins, like high molecular weight kininogen, factor XII, and prekallikrein; the hydroxyl surface captured low amounts of complement protein, C3c. Tidwell et al. reported the adsorption of radiolabeled Fn to SAMs with terminal methyl, hydroxyl , carboxyl , and methyl ester ($-COOCH_3$) groups in 10% bovine serum [45]. COOH–SAM exhibited high levels of Fn adsorption, which correlated well with the adhesion of endothelial cells.

Figure 4c shows that the amount of adsorbed proteins is rapidly saturated within several minutes of exposing serum-containing medium to a surface. Albumin, the most abundant serum protein, was expected to preferentially adsorb onto the surfaces during early time points. Then, adsorbed albumin was expected to be displaced by cell adhesion proteins. To investigate the effect of preadsorbed albumin displacement on cell adhesion, SAMs were first exposed to albumin; then, HUVECs suspended in a serum-supplemented medium were added [21, 42]. Very few cells adhered to hydrophobic SAMs that had been pretreated with albumin, due to the large interfacial tension between water and the hydrophobic surfactant-like surface. Albumin was infrequently displaced by the cell adhesive proteins Fn and Vn. One the other hand, HUVECs adhered well to hydrophilic SAM surfaces that had been preadsorbed with albumin. In that case, the preadsorbed albumin was readily displaced by cell adhesive proteins.

2.3.3 Protein Conformation and Orientation on a Surface

Proteins undergo conformational (or orientational) changes after adsorption to a material surface, which can influence their subsequent biological functions. Cells adhere to a surface through an interaction between integrin receptors on the cell membrane and the specific amino acid sequences, RGD and PHSRN, of cell adhesion proteins. When cell adhesion proteins undergo a conformational (or orientational) change that hinders the integrin binding site, cells cannot adhere to the surface.

The conformation and orientation of adsorbed proteins has been examined with monoclonal antibodies that recognize a specific site in a protein of interest. Keselowsky et al. examined the conformation of Fn adsorbed to SAMs that carried methyl, hydroxyl, carboxyl, and amine groups [79]. They used monoclonal antibodies that recognized the central cell-binding domain of Fn near the RGD motif. Different SAM functionalities differentially modulated the binding affinities of the monoclonal antibodies ($OH > COOH = NH_2 > CH_3$). The strength of cell adhesion to these

SAMs was correlated to the affinities of the Fn-specific monoclonal antibodies. Although antibody-based measurements could not distinguish between conformational (structural) and orientational changes in the adsorbed proteins, they provided information about the biological activity of adsorbed proteins.

2.3.4 Cell Adhesion in Serum-Free Medium

Cells are sometimes cultured in serum-free medium. In this condition, the surface should carry substituents of serum proteins that can directly interact with cells. SAMs of alkanethiols with bioactive ligands have been used to control interactions between the material surface and cells [80–83]. Several bioactive ligands have been tested, including RGD [80], PHSRN [81], and laminin-derived peptides [82, 83]. These ligands were expected to directly interact with cell surface integrins.

SAMs of alkanethiols that carried RGD peptides were tested to determine the minimum amount of peptide required for cell adhesion to the substrate. Roberts et al. employed SAMs of alkanethiols with mixtures of RGD and oligo(ethylene glycol) moieties that resisted nonspecific protein and cell adsorption [80]. Bovine capillary endothelial cells attached and spread on SAMs that had a mole fraction of RGD, $\chi_{RGD} \geq 10^{-5}$. Cell spreading reached a maximum at $\chi_{RGD} \geq 10^{-3}$. This fraction indicated that the RGD density was on the order of 10^{11} RGD molecules/cm^2, assuming that the RGD occupied an alkanethiol area of ~0.25 nm^2/molecule. Arnold et al. designed a hexagonally close-packed rigid template of cell-adhesive gold nanodots coated with cyclic RGDfK peptide. They used block–copolymer micelle nanolithography to create a patterned surface [84]. The gold nanodots were placed with 28, 58, 73, or 85 nm spacing, based on the molecular weight of the block–copolymer. Adhesion tests with osteoblasts, fibroblasts, and melanocytes showed that cells adhered and spread on the patterned gold nanodots with a spacing of $\leq$58 nm. This result also showed that the RGD density was on the order of 10^{11} molecules/cm^2.

3 Cell Culture Substrates for Specific Cell Proliferation

In standard cell culture methods, cells are plated in a cell culture flask or Petri dish, and they are maintained in medium supplemented with FBS and various growth factors. Cells adhere to the substrate through integrin and cell adhesion glycoproteins that adsorb to the plastic surface of the flask or dish. Primary cells isolated from embryonic and adult tissues are widely cultured with these methods. Nevertheless, it appears that this conventional culture method is inefficient for the production of specific cells in high purity and large quantities. Although different kinds of cells isolated from widely different tissues can adhere and proliferate in cell culture flasks, efficiency may be limited by the heterogeneity of cell populations.

To overcome these limitations, we have developed culture substrates that enable the highly efficient expansion of specific cells in adherent cultures [37, 85–88]. An important characteristic of these substrates is that specifically engineered growth factors are immobilized on the surface. Extensive protein engineering techniques were used to optimize the presentation of growth factors to cells.

3.1 *Strategy for Adherent Cultures of Neural Stem Cells*

NSCs, capable of self-renewal and differentiation into multiple cell types, are found in embryonic and adult tissues of the central nervous system (CNS) [89]. Several studies have demonstrated that NSCs are a potential source for cell replacement therapies in CNS disorders [90, 91]. Those studies have largely relied on the ability to culture NSCs in vitro. To develop culture substrates for the selective expansion of NSCs, we first considered the responsiveness of NSCs to growth factors. Originally, NSCs were discovered as EGF-responsive cells from rodent CNS tissue [11]. Another study [36] showed that the expression of EGFR on rat neurosphere-forming cells was highly correlated to the expression of nestin, an intracellular marker for NSCs. In addition, EGF was shown to be a mitogen for NSCs. Based on these results, we hypothesized that a substrate with surface-immobilized EGF might selectively trap NSCs from a heterogeneous population of cells. Furthermore, the EGF–EGFR interactions that would specifically capture NSCs might also promote NSC proliferation due to EGFR signaling. To test this hypothesis, we focused on surface immobilization of EGF for the selective expansion of rat NSCs.

3.2 *Oriented Immobilization of Engineered EGF*

There are many protein immobilization techniques available. One of the standard techniques uses amine chemistry, where surface-bound nucleophilic groups react with amines, which are abundant in proteins [92]. Although this technique provides covalent immobilization of proteins, the use of amines would cause protein inactivation. In addition, it does not provide control over the orientation of the immobilized protein to ensure efficient recognition by ligands. On the other hand, recombinant DNA technology can overcome these problems. A recombinant protein can be designed that has a specific peptide motif at a given site in the molecule for surface anchoring.

We used recombinant DNA technology to design unique substrates for in vitro expansion of rat NSCs. For example, we fused a small peptide sequence of six consecutive histidine (His) residues to the C-terminus of human EGF (EGF-His). This EGF-His was anchored to the surface of a glass-based substrate by coordination with Ni^{2+} ions, which were fixed on the surface of a SAM of alkanethiol. Strikingly, neither a covalently immobilized EGF-His nor a physically adsorbed

EGF-His could trap cells as efficiently as EGF-His immobilized by surface anchoring through coordination.

To prepare a Ni^{2+}-chelated surface, a thin gold layer was deposited onto a glass plate; then, on the gold surface, a SAM was formed that terminated with trivalent carboxylic acids; finally, Ni^{2+} ions were chelated to the acidic termini. In detail, first, thin chromium and gold layers were deposited onto the surface of a glass plate with a vacuum evaporator. The glass plate was then immersed in ethanol that contained 16-mercapto-1-hexadecanoic acid (COOH-thiol) and (1-mercaptoundec-11-yl) triethylene glycol (TEG-thiol) at various compositions to allow the formation of mixed SAMs. Each glass plate coated with a mixed SAM was immersed in a solution containing *N,N′*-dicyclohexylcarbodiimide (DCC) and *N*-hydroxysuccinimide (NHS), which converted the terminal carboxylic acid to an active ester. Subsequently, a solution of *N*-(5-amino-1-carboxypentyl) iminodiacetic acid (NTA) was plated onto the activated surface to introduce triacetic acid. The glass plate was then immersed in a $NiSO_4$ solution to form the Ni^{2+} chelate. Finally, an EGF-His solution was plated onto the Ni^{2+}-chelated surface to allow immobilization of EGF-His through the coordination of Ni^{2+} with His. A His-tagged protein firmly binds to a Ni^{2+}-chelated substrate; for instance, His-tagged green fluorescent protein (GFP) bound to NTA-Ni^{2+} with a dissociation constant of 4.2×10^{-7} M [93]. The mixed SAM surface was characterized by water contact angle measurements, infrared reflection adsorption spectroscopy, and XPS. These revealed the formation of well-defined surfaces. The expected structure of an EGF-His-immobilized surface is schematically depicted in Fig. 6. The surface density of immobilized EGF-His increased with increases in the COOH-thiol content of the COOH-thiol/TEG-thiol mixture used to prepare the SAM. The maximum EGF-His density was approximately 0.4 $\mu g/cm^2$ with 100% COOH-thiol.

The incorporation of TEG-thiol onto a COOH-thiol SAM elevated the fraction of correctly oriented EGF-His on the surface by preventing nonspecific adsorption of EGF onto the SAM surface. This might be explained by the following findings. The area occupied by a single EGF molecule (2.98 nm^2/molecule) [94] is approximately ten times larger than the area occupied by COOH-thiol (0.25 nm^2/molecule) [95] or TEG-thiol (0.35 nm^2/molecule) [96]. Therefore, on a surface of 100% COOH-thiol, most COOH-thiol molecules would not be expected to be involved in the coordination with EGF-His. Instead, the excess carboxylic acids contained in COOH-thiol and NTA would be expected to trigger nonspecific adsorption of EGF-His. Under conditions where the COOH-thiol content was 10–15%, and 80% of COOH-thiol was converted to Ni^{2+} chelate, the predicted density of Ni^{2+} chelate would be equivalent to that of closely packed EGF-His in a monolayer. However, higher surface densities of COOH-thiol gave rise to larger amounts of immobilized EGF-His, without obvious saturation at 10–15% COOH-thiol. This was probably due to the nonspecific EGF-His binding to residual carboxylic acids present on the surface after NTA derivatization and Ni^{2+} chelation. Furthermore, these nonspecifically bound EGF-His molecules may hinder access to the Ni^{2+} sites for specific coordination.

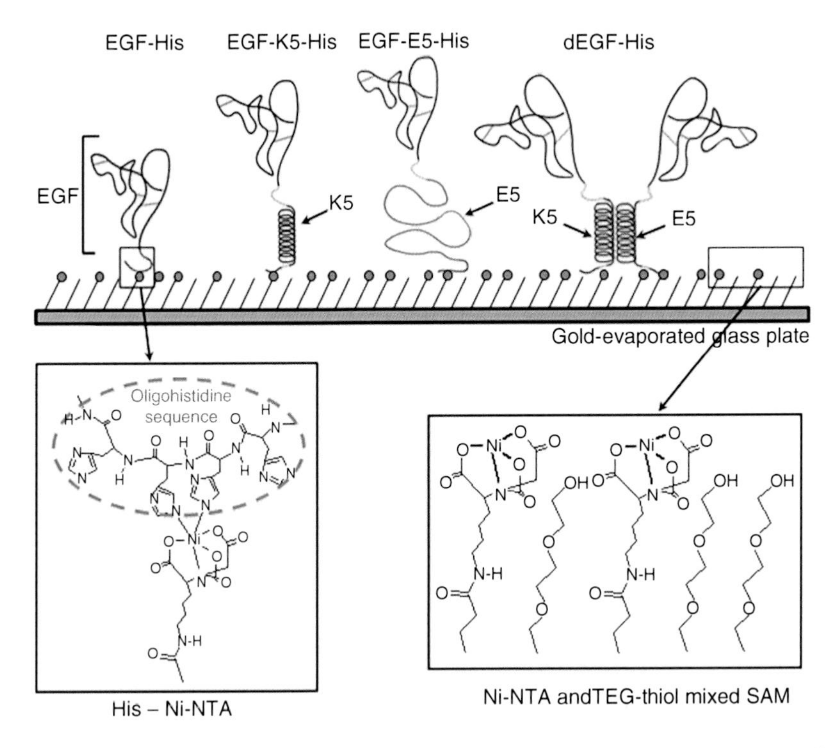

Fig. 6 EGF-containing chimeric proteins anchored to the Ni-chelated surface through coordination. *Bold lines* in the molecular structures represent chelate bonding. *TEG-thiol* triethylene glycol-containing alkanethiol. Reproduced from Nakaji-Hirabayashi et al. [87] with permission from American Chemical Society, copyright 2009

3.3 Proliferation of Rat NSCs on EGF-His-Immobilized Surface

We investigated the efficiency of NSC expansion on surfaces with EGF-His immobilized in the correct orientation. NSCs were obtained from neurosphere cultures prepared from fetal rat striatum harvested on embryonic day 16. NSCs were cultured for 5 days on EGF-His-immobilized substrates prepared with mixed SAMs of different COOH-thiol contents. Cells adhered and formed network structures at a density that increased with the COOH-thiol content of the surface. As a control, cells were seeded onto surfaces without immobilized EGF-His. This resulted in poor cell adhesion during the entire culture period. In addition, when EGF-His adsorbed to SAMs with 100% COOH-thiol or SAMs with NTA-derivatized COOH that lacked Ni^{2+} chelation, we observed poor initial cell adhesion, and the cells formed aggregates within 5 days. Interestingly, the substrate used to covalently immobilize EGF-His with the standard carbodiimide chemistry was not a suitable surface for cell adhesion and proliferation. The control experimental results contrasted markedly with results from EGF-His-chelated surfaces.

As described earlier, the surface density of immobilized EGF-His was shown to be directly correlated to the COOH-thiol content. Therefore, we reasoned that the initial cell attachment must involve an interaction between immobilized EGF-His and an EGFR expressed on the cell membrane. In fact, reverse transcriptase polymerase chain reaction assays revealed that neurosphere cells expressed EGFR mRNA. In addition, cell adhesion to immobilized EGF-His was totally inhibited when soluble EGF was added to the culture medium at 20 ng/mL. Flow cytometry analyses further demonstrated that rat neurosphere cultures could be significantly enriched in NSCs by plating cells onto a glass surface with immobilized EGF-His. All these findings supported our hypothesis that cell adhesion was mediated by an EGF–EGFR interaction.

Table 1 summarizes the results of proliferation assays carried out on EGF-His-immobilized surfaces. The total cell number after 5 days of culture varied with different surfaces. These differences could be attributed principally to differences in the numbers of cells initially attached, demonstrated by the cell numbers observed at 24 h. The doubling time, determined for the 24–72 h period after cell seeding, was similar for all the surfaces, with an average of 16.9 ± 4.7 h. The mitogenic signal of EGF might have been saturated on all surfaces, including the surface with the lowest EGF density. The doubling time determined on the EGF-His-immobilized surfaces was approximately half the doubling time determined for the neurosphere culture (31.2 ± 1.4 h). On the surfaces with 0.01, 0.1, and 1% COOH-thiol content, the growth rate declined slightly after 96 h; this was probably due to a reduction of EGF activity after 4–5 days.

NSC content was assessed in a population of cells expanded on EGF-His-immobilized surfaces for 5 days. Cells were immunocytochemically stained with antibodies specific for nestin, a marker for NSCs, and β-tubulin III (βIII),

Table 1 Proliferation of neurosphere-forming cells in cultures with or without EGF-immobilized surfaces

COOH-thiol content[a] (%)	Surface density of EGF-His[b] (ng/cm^2)	Cell number after 24 h[c] (10^4 $cells/cm^2$)	Cell number after 5 days[c] (10^4 $cells/cm^2$)	Doubling time[d] (h)
0	51	0.7 ± 0.3	11.7 ± 5.1	n.d.
0.01	76	1.2 ± 0.4	38.6 ± 17.6	16.8 ± 5.5
0.1	110	1.4 ± 0.2	46.4 ± 17.2	16.7 ± 5.2
1	215	1.6 ± 0.1	49.6 ± 15.2	17.2 ± 5.4
10	342	1.8 ± 0.2	56.7 ± 10.5	17.3 ± 5.9
100	390	2.5 ± 0.2	64.0 ± 6.6	17.0 ± 4.0
Neurosphere	–	–	–	31.2 ± 1.4[e]

Reproduced from Nakaji-Hirabayashi et al. [85] with permission from Elsevier, copyright 2007
[a]Content in the solution used to prepare the mixed SAMs
[b]Determined with a microBCA assay
[c]Mean ± standard deviation for $n = 5$ assays
[d]Determined from the logarithmic plot of growth curves for the culture period of 24–72 h. Mean ± standard deviation for $n = 4$ assays
[e]Determined for cells during the standard neurosphere culture

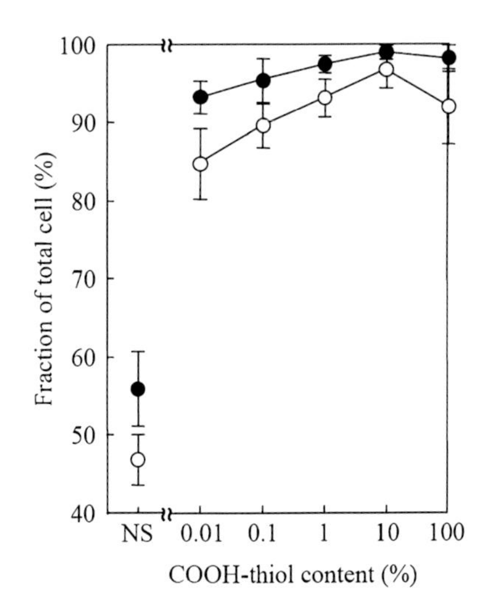

Fig. 7 Quantitative evaluation of NSC enrichments in cultures with different material surfaces. The fraction of NSC phenotypes in the cultures are shown for (*open circles*) nestin$^+$ βIII$^-$ or (*closed circles*) both nestin$^+$ βIII$^-$ and nestin$^+$ βIII$^+$. Data are also shown for neurosphere cultures (*NS*). The averages (± SEM) of five experiments are shown. Reproduced from Nakaji-Hirabayashi et al. [85] with permission from Elsevier, copyright 2007

a marker for differentiated neurons. The number of cells that expressed nestin and βIII was expressed as the percent of the total number of cells in the culture flask (Fig. 7). Cells with a nestin$^+$ βIII$^-$ phenotype represented the most immature population. Both nestin$^+$ βIII$^-$ cells and the total number of nestin$^+$ cells (nestin$^+$ βIII$^-$ plus nestin$^+$ βIII$^+$) were most abundant on the surface with 10% COOH-thiol (97 ± 2.7% and 98 ± 0.8%, respectively). Although the abundance of cells with these phenotypes decreased with decreasing COOH-thiol content, approximately 85% of all cells exhibited the nestin$^+$ βIII$^-$ phenotype on the EGF-His-immobilized surface with 0.01% COOH-thiol.

We also conducted experiments to compare our culture method with the standard neurosphere culture. In the standard neurosphere culture, cell number increased approximately nine times over 5 days. Immunostaining showed that the neurosphere cultures contained 54 ± 5.3% nestin$^+$ cells and 41 ± 7.4% nestin$^+$ βIII$^-$ cells. This demonstrated that the standard neurosphere culturing method was less efficient than EGF-immobilized substrates for selectively expanding NSCs. Thus, the EGF-immobilized substrates prepared from mixed SAMs with 10% COOH-thiol provided the most efficient method for selective NSC expansion.

Cells that were expanded on EGF-immobilized substrates were harvested and cultured on freshly prepared EGF-His-immobilized substrate for another 5 days. These procedures were repeated up to four times. The rate of cell growth was not affected by repeated subculturing; the average doubling time was 18.1 ± 1.8 h. In addition, immunocytochemical staining showed that, after four subcultures, cells retained 93 ± 2.0% nestin$^+$ populations. These results suggested that, in theory, NSCs can be expanded by approximately 10^5- to 10^6-fold within 20 days on the EGF-His-immobilized substrates. Furthermore, these cells retained multipotency. Under appropriate conditions, the cells differentiated to express βIII, GFAP (astrocyte marker), and O4 (oligodendrocyte marker) with marked reductions in nestin expression. The results of the differentiation assays suggested that expanded cells that expressed nestin retained, for the most part, multipotency and could differentiate into neuronal and glial lineages.

Based on those results, we concluded that, when cultured on the EGF-His-immobilized surface prepared from a mixed SAM of 10% COOH-thiol, highly enriched NSC populations could be produced in large quantities. Over a 5-day culture on the substrate, cells were expanded 32 times. These expanded cells consisted of 98% nestin$^+$ cells that retained multipotency for differentiation into neuronal and glial lineages. This suggested that selective expansion could be repeated for large-scale production of highly enriched NSC cells.

3.4 Structural Integrity and Stability of Immobilized EGF-His

We wondered why NSCs proliferated exclusively on surfaces with EGF-His ligands anchored by coordination. We focused on two aspects in particular: the conformational integrity of coordinated EGF-His and the stability of coordinate bonds at the interface. Conformational information was acquired with multiple internal reflection–infrared absorption spectroscopy (MIR-IRAS) [97]. The stability of coordinate bonds was assessed by culturing NSCs on a surface with a small region of EGF-His ligands anchored by coordination. This spatially restricted EGF-His anchoring enabled an intuitive exploration of EGF-His release under cell culture conditions.

The results of these studies showed that EGF-His coordinated to the Ni^{2+}-chelated NTA-surface retained an intact conformation and was firmly anchored to the surface during NSC culturing procedures. Both attributes are essential for establishing adherent cultures and, hence, selective expansion of NSCs. In contrast, with covalent immobilization, the structural integrity of EGF-His was reduced due to multivalent linkages to the surface. In physical adsorption, EGF-His maintained its intact conformation, but was readily released from the surface during cell culturing procedures. We concluded that structural integrity and firm anchorage were optimized with the coordination method.

3.5 Spontaneous Dimerization of EGF

Numerous studies have shown that EGF binding to EGFR triggers receptor dimerization. This is considered a crucial step in intracellular signal transduction [98]. Inspired by this mechanism, we designed EGF chimeric proteins that spontaneously dimerized (dEGF-His). These dimers were terminally anchored to the substrate. We expected that these preformed dimeric EGF structures would facilitate the formation of EGF–EGFR dimer complexes more efficiently than monomeric EGF structures.

Our strategy for the spontaneous dimerization of EGF was to incorporate an α-helical oligopeptide with the ability to participate in forming a coiled-coil structure [99]. We implemented a heterodimerization system in which $(KELASVK)_5$ (K5) and $(EKLASVE)_5$ (E5) peptides could form stable coiled-coil heterodimers. We then synthesized two chimeric proteins that contained EGF attached to either the K5 (EGF-K5-His) or the E5 (EGF-E5-His). Both had the hexahistidine sequence added to the C-terminus for anchoring through coordination with Ni^{2+} ions fixed to a substrate.

Analyses by native polyacrylamide gel electrophoresis and circular dichroism (CD) spectroscopy revealed that spontaneous coiled-coil associations between EGF-E5-His and EGF-K5-His promoted heterodimer (dEGF-His) formation. The CD spectroscopic analysis suggested that the E5 peptide in monomeric EGF-E5-His had a disordered structure. However, the α-helical structure was induced in the E5 peptide when it associated with EGF-K5-His. These findings are shown schematically in Fig. 6.

We tested adhesion and proliferation of rat NSCs on surfaces with immobilized dEGF-His, EGF-K5-His, EGF-E5-His, or EGF-His. The surface with immobilized dEGF-His provided the highest efficiency for selective expansion of NSCs. On this substrate, cells expanded 60-fold over 96 h of culture. Of note, over 98% of the expanded cells expressed nestin, but not βIII.

The monomeric EGF-His appeared to be limited in its capacity for promoting EGFR dimerization, most likely due to the reduced mobility of surface-anchored EGF. In contrast, when dEGF-His was anchored to the surface, EGFR dimerization was most likely enhanced because the EGF domains were paired in close proximity. Presumably, this would promote efficient proliferation of NSCs, as observed on the surface with immobilized dEGF-His. We estimated that the distance between two EGF domains in the dEGF-His dimer was approximately comparable to the distance between the two binding sites in dimerized EGFR [100]. We further observed that both EGF-E5-His and EGF-K5-His, anchored as single components, provided more efficient substrates than EGF-His alone. This might be explained by the E5 and K5 peptides inserted between the EGF domain and the His sequence; this extra component may have increased the mobility of the EGF domain and thus enhanced its accessibility. Consistent with that notion, the effects of these peptides on cell proliferation were not detectable when the chimeric proteins were used as unattached, diffusible factors.

3.6 *Modules for Culturing NSCs in a Closed System*

The method for immobilizing EGF-His on SAMs gave rise to the need for fabricated cultureware that could allow large-scale expansion of pure NSCs. We attempted to construct culture modules with surface areas much larger than the laboratory-scale substrates described above. For uniformly anchoring EGF-His over a large area, we utilized a glass plate with amine functionalities on the surface.

We also found that a gold layer was not required nor advantageous for general use of the technique.

First, the surface of a glass plate (60 mm × 60 mm × 1 mm) was modified with 3-aminopropyltriethoxysilane (APTES) to introduce the amine functionality. The surface amine was reacted with glutaraldehyde to introduce aldehyde, and then it was exposed to a solution of NTA; NTA coupling occurred through the formation of a Schiff base. Subsequently, the glass plate was immersed in a $NiSO_4$ solution to chelate Ni^{2+} ions. Finally, EGF-His was anchored to the Ni^{2+}-chelated surface to obtain an EGF-His-immobilized glass plate. The surface density of coordinated EGF-His was 0.53 ± 0.10 μg/cm^2, measured with the microBCA assay. The MIR-IRAS spectrum of a cognate surface exhibited strong absorption at amide I′ and amide II′ bands (the prime notation designates a deuterated condition), which indicated the presence of EGF-His at the surface. The conformations of characteristic secondary structures were determined from the amide I′ band. The results showed that the content of β-sheet and β-turn structures in surface-anchored EGF-His was similar to those observed in native EGF in solution and in EGF-His anchored to a Ni^{2+}-bound surface created on an alkanethiol SAM.

To fabricate a culture module, we adhered a silicone frame to the edges of the plate to enclose the sides; then, a polystyrene film was adhered to the top of the frame to allow gas exchange. This culture module provided an area of 25 cm^2 and a chamber volume of 2.5 mL. To test the modules, we injected dissociated rat neurosphere-forming cells. After 4 days of culture, 5×10^6 cells were obtained per module. These cells could be subcultured in new modules. We performed six passages by subculturing every 4 days. At each subculture, some of the recovered cells were seeded to a freshly prepared module at a density of 3.0×10^4 cells/cm^2. The average number of cells harvested after each subculture was $5.4 \pm 1.7 \times 10^6$ cells per module. We determined that 95% of the total number of cells had a phenotype of nestin$^+$ βIII$^-$. In addition, the cells obtained after six passages could be differentiated, under appropriate conditions, into three major lineages, including neurons, astrocytes, and oligodendrocytes. This demonstrated that the cells maintained multipotency.

The culture module greatly facilitated homogeneous distribution of seeded cells and cultivation of a large number of cells under identical conditions. In addition, the module required a smaller volume of medium than standard cell culture systems. Importantly, this modular system provides the great advantages of scalability and safety because cell processing can be performed in a closed system. Thus, the modules facilitate the production of cells that are safe for use in cell transplantation therapies.

In a separate study [88], we synthesized EGF fused to a polystyrene-binding peptide [101] (EGF-PSt) that could be immobilized on the surface of a tissue culture polystyrene dish. This surface also permitted efficient expansion of NSCs. Thus, EGF-PSt can be used to produce large quantities of pure NSCs in standard laboratories.

4 Cell Surface Modifications

Cell surfaces can be modified in a number of ways, including covalent conjugation of polymers to amino groups on membrane proteins [102–105], electrostatic interaction between cationic polymers and a negatively charged cell surface [106–109], and hydrophobic interactions that anchor long alkyl chains of amphiphilic polymers to the lipid bilayer of the cell membrane [110–126]. These methods have been used to immobilize various functional groups and bioactive substances onto the cell surface. The covalent conjugation method is expected to impair membrane protein functions because polymers are attached by crosslinking to the amino groups on membrane proteins. Therefore, the extent of this reaction should be carefully controlled. The electrostatic interaction is performed by simply adding a cationic polymer solution to a cell suspension. However, most cationic polymers are cytotoxic; therefore, this treatment causes deterioration or death to most cell types. In contrast, the hydrophobic interaction can be performed by simply adding amphiphilic polymers with long alkyl chains to a cell suspension. This has not caused any major damage to cell function or integrity. Our group has extensively studied surface modifications with amphiphilic polymers [110–126].

4.1 Cell Surface Modifications with Amphiphilic Polymers

Amphiphilic polymers are typically derived by conjugating polyethylene glycol (PEG) to a phospholipid (PEG-lipid) (Fig. 8) [117, 118]. When a PEG-lipid solution is added to a cell suspension, the hydrophobic alkyl chains of the PEG-lipid spontaneously form hydrophobic interactions with the lipid bilayer of the cell membrane (Fig. 8a). This spontaneous anchoring of the PEG-lipid to a supported lipid membrane was confirmed by observation with surface plasmon resonance (SPR) (Fig. 9a). Three kinds of PEG-lipids with different alkyl chain lengths were added to a lipid membrane that had been created on a SPR sensor surface. The SPR angle rapidly increased with the spontaneous anchoring of PEG-lipid into the lipid membrane. In addition, the anchoring rates decreased with increasing alkyl chain lengths. This spontaneous anchoring was also demonstrated with a human cell line derived from T cell leukemia cells (CCRF-CEM). A solution of fluorescein isothiocyanate (FITC)-conjugated PEG-lipid was added to a suspension of CCRF-CEM. Under a confocal laser scanning microscope, the bright fluorescence from FITC was observed at the periphery of all cells (Fig. 9b). This indicated that PEG-lipids had lodged on the cell surface. The retention time of PEG-lipids on cell membranes can be controlled by adjusting the length of the lipid alkyl chain. The dissociation rate of PEG-lipid was much slower with long than with short hydrophobic domains [111].

Proteins can be immobilized on the cell surface with the use of a short, single stranded DNA (ssDNA) attached to the end of a PEG chain (ssDNA–PEG-lipid) [114–116, 119, 125]. First, an ssDNA–PEG-lipid is prepared by conjugating

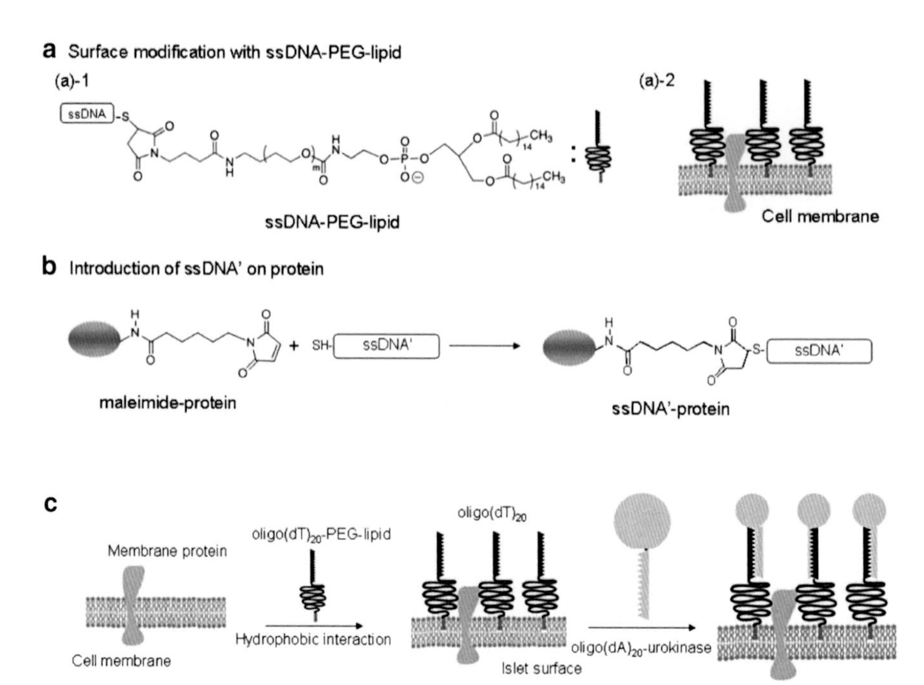

Fig. 8 Immobilization of urokinase on the surfaces of islet cells. (**a**) Surface modification: (*1*) chemical structure of ssDNA–PEG-lipid, and (*2*) ssDNA–PEG-lipid anchoring to the cell membrane. (**b**) Introduction of a complementary ssDNA onto urokinase, which was first modified with a madeimide group by a cross-linker, EMCS. (**c**) Urokinase-immobilization through DNA hybridization

maleimide–PEG-lipid with an ssDNA that carries a thiol group. Then, a protein is modified with a hetero-bifunctional crosslinker, sulfo-EMCS [*N*-(6-maleimidocaproyloxy)sulfosuccinimide]; next, it is treated with a ssDNA′ that is complementary to the ssDNA on the PEG-lipid. Figure 8 shows a schematic of the procedure, where the ssDNA and ssDNA′ are oligo(deoxythymidine) [oligo$(dT)_{20}$] and oligo (deoxyadenine) [oligo$(dA)_{20}$], respectively. The cells with oligo$(dT)_{20}$ attached are exposed to the protein with the oligo$(dA)_{20}$ attached. The protein is immobilized on the cell through hybridization between oligo$(dT)_{20}$ and oligo$(dA)_{20}$.

Figure 9c shows the SPR profiles recorded before and after an oligo$(dT)_{20}$–PEG-lipid had been incorporated into the supported lipid membrane, and its interactions with oligo$(dA)_{20}$–urokinase and oligo$(dT)_{20}$–urokinase on the SPR sensor surface. The first increase in the SPR angle indicated that the oligo$(dT)_{20}$–PEG-lipid had incorporated into the membrane; the second increase showed the binding of oligo $(dA)_{20}$–urokinase to the oligo$(dT)_{20}$ substrate. No increase was observed when oligo$(dT)_{20}$–urokinase was applied to the oligo$(dT)_{20}$ substrate. These results indicated that protein could be specifically conjugated to the cell surface by hybridization between oligo$(dA)_{20}$ and oligo$(dT)_{20}$ moieties.

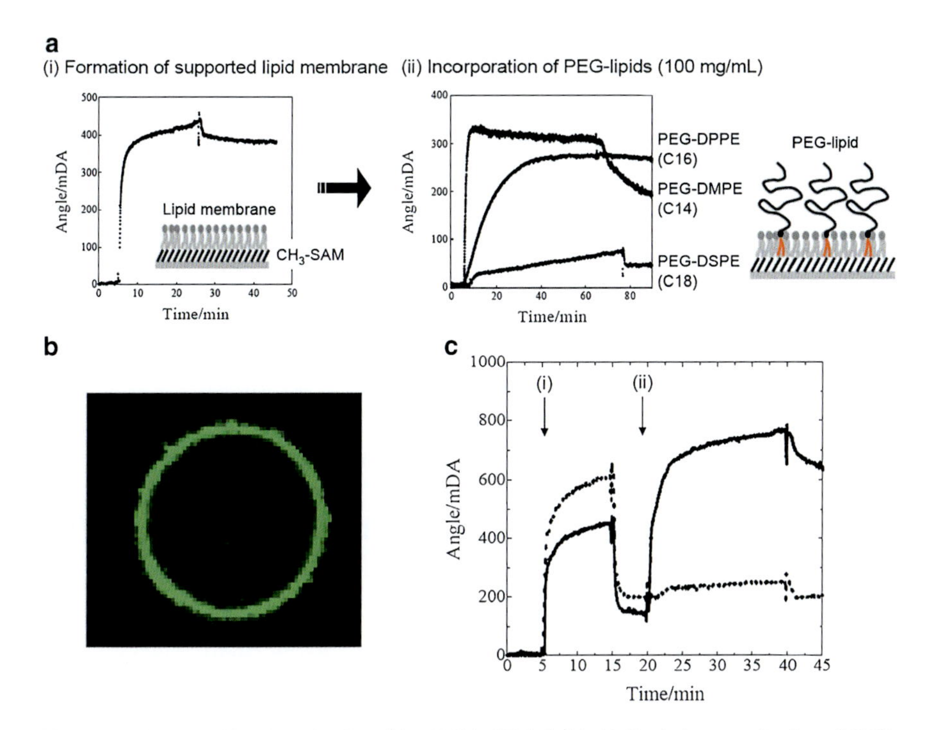

Fig. 9 Surface modification of cells with ssDNA–PEG-lipid. (**a**) Real-time monitoring of PEG-lipid incorporation into a supported lipid membrane by SPR. (*i*) A suspension of small unilamellar vesicles (SUV) of egg yolk lecithin (70 μg/mL) was applied to a CH_3-SAM surface. A PEG-lipid solution (100 μg/mL) was then applied. (*ii*) Three types of PEG-lipids were compared: PEG-DMPE (C14), PEG-DPPE (C16), and PEG-DSPE (C18) with acyl chains of 14, 16, and 18 carbons, respectively. (**b**) Confocal laser scanning microscopic image of an CCRF-CEM cell displays immobilized FITC-oligo$(dA)_{20}$ hybridized to membrane-incorporated oligo$(dT)_{20}$–PEG-lipid. (**c**) SPR sensorigrams of interaction between oligo$(dA)_{20}$–urokinase and the oligo $(dT)_{20}$–PEG-lipid incorporated into the cell surface. (*i*) BSA solution was applied to block nonspecific sites on the oligo$(dT)_{20}$-incorporated substrate. (*ii*) Oligo$(dA)_{20}$–urokinase (*solid line*) or oligo$(dT)_{20}$–urokinase (*dotted line*) was applied

4.2 *Immobilization of Bioactive Substances on an Islet Surface*

Cell transplantation has shown promise as a method for treating serious diseases. Various kinds of pluripotent stem cells have been developed or identified, including embryonic stem cells (ESCs), induced pluripotent stem cells (iPSCs), and mesenchymal stem cells (MSCs). Moreover, the differentiation of stem cells to functional cells has been extensively studied. Previous studies have demonstrated that the transplantation of islet of Langerhans cells (islets) could successfully treat type 1 diabetes. Islets are insulin-secreting cells found in the pancreas. Over 200 patients with type 1 diabetes have been clinically treated with islet transplantation. To cure the disease, a single patient typically requires islets from several donors, due to the destruction of islets just after transplantation. In the clinical setting, islets are infused into the liver through the portal vein. Exposure of islets to the blood activates blood coagulation and complement systems, which induce nonspecific inflammatory reactions or instant blood-mediated inflammatory reactions (IBMIR). These host defense mechanisms destroy donor islets because they are considered foreign bodies [127–129]. Anticoagulants, including aspirin, heparin, and dextran sulfate, are typically administered to inhibit blood coagulation. However, systemic infusion of these drugs increases bleeding. The optimal approach would be to prevent blood coagulation at the islet. Recent studies have been able to immobilize various bioactive substances, like heparin, urokinase, thrombomodulin, and the soluble domain of human complement receptor 1 (sCR1), on islets in attempts to control local activation of the blood coagulation and complement systems [110, 112, 115, 117–119, 121, 125].

As an example, we will describe immobilization of the fibrinolytic enzyme, urokinase (UK), on the islet surface [115, 119]. As shown in Fig. 8, UK could be immobilized on islets through ssDNA hybridization of oligo$(dT)_{20}$–PEG-lipid and oligo$(dA)_{20}$–UK. When the oligo$(dT)_{20}$–PEG-lipid was added to a suspension of islets, the lipid moiety spontaneously anchored to the lipid bilayer of the cell membrane through hydrophobic interactions. The oligo$(dT)_{20}$ segment was exposed on the cell surface, which made it accessible for conjugation with the oligo $(dA)_{20}$ on UK (Fig. 8c).

Figure 10 shows confocal laser-scanning fluorescence images of islets treated with oligo$(dT)_{20}$–PEG-lipid and oligo$(dA)_{20}$–UK (UK-islets). In these experiments, lipids with different alkyl chain lengths were attached to the PEG moiety to test the stability of UK on the islet surface. Both PEG varieties showed clear fluorescence signals (Fig. 10a, b) from UK-islets, which indicated that both facilitated stable UK attachments. On the other hand, fluorescence was nearly undetectable on unmodified islets and islets treated with oligo$(dA)_{20}$–UK in the absence of oligo$(dT)_{20}$–PEG-lipids (Fig. 10c). These results indicated that UK could be immobilized on islets through DNA hybridization. The retention time of the oligo$(dT)_{20}$–PEG-lipid on the cell membrane depended on the chain lengths of the PEG-lipid [111]. At 2 days after conjugation, a strong fluorescence signal was observed for islets treated with oligo$(dT)_{20}$–PEG-lipid (C18), and a weaker fluorescence signal was observed for islets treated with oligo$(dT)_{20}$–PEG-lipid (C16). The longer alkyl chain length conferred longer retention of UK on the islet surface.

UK is a serine protease that activates plasminogen to plasmin. Plasmin dissolves the fibrin in blood clots. The attachment of UK to the islet surface was expected to dissolve blood clots that surrounded the islets in the liver; thus, IBMIR could be inhibited in the initial stages. A fibrin plate-based assay was performed to assess the

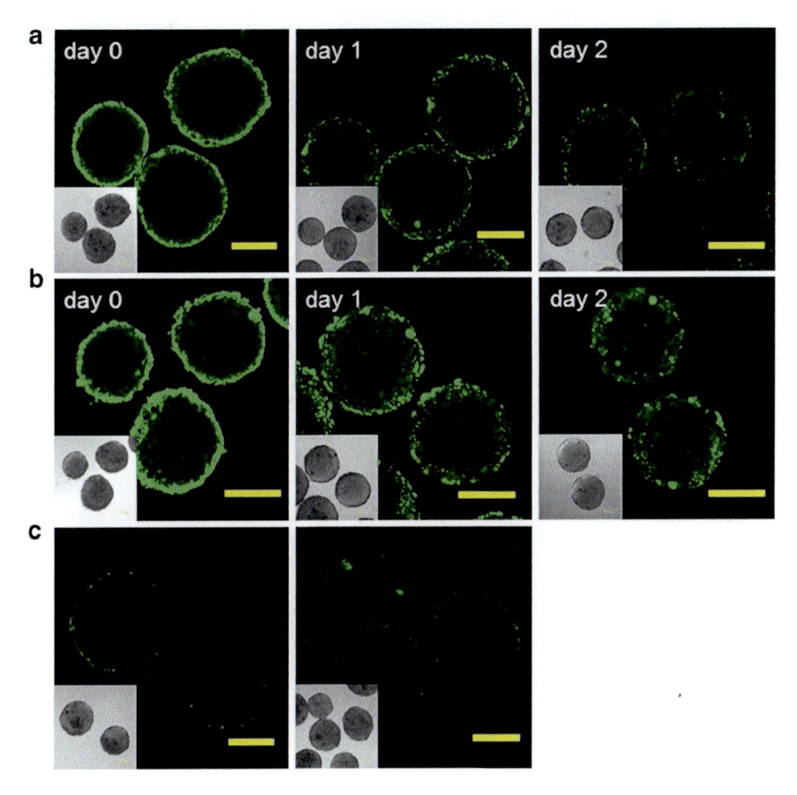

Fig. 10 Confocal laser scanning microscope images of islets with urokinase (UK) immobilized on the membrane. The *green fluorescence* indicates positive immunostaining for UK. (**a**) Islets were modified with oligo(dT)$_{20}$–PEG-lipid (C16) or (**b**) oligo(dT)$_{20}$–PEG-lipid (C18); then, oligo (dA)$_{20}$–UK was added to the media. (**c**) Unmodified islets with (*left*) and without (*right*) oligo (dT)$_{20}$–PEG-lipids added to the solution. *Insets*: Bright field images. Scale bars: 100 μm

function of the UK attached to the islets. Fifty islets with or without immobilization of UK were spotted onto a fibrin gel plate. After incubation, transparent areas around the spots indicated UK dissolution of the fibrin. Figure 11 shows the fibrin plate at 14 h after spotting the islets. Larger transparent areas were observed around the UK-islets compared to those around the unmodified islets (Fig. 11a-1, a-2, a-4). These indicated that the immobilized UK retained its activity on the islets. UK-islets were also tested after 2 days of culture in the presence of serum (Fig. 11a-3, a-5). UK activity rapidly decreased with 2 days in culture. The morphology of all islets after modification with UK was well maintained after 7 days of culture (Fig. 11b). Islets with UK maintained the ability to regulate insulin release in response to changes in glucose concentration (data not shown). We also performed transplantation of UK-islets by transfusion to the liver through the portal vein [119]. The transplantation results indicated that donor islets were rescued from host defenses by attaching UK to their surfaces. It remains to be determined how

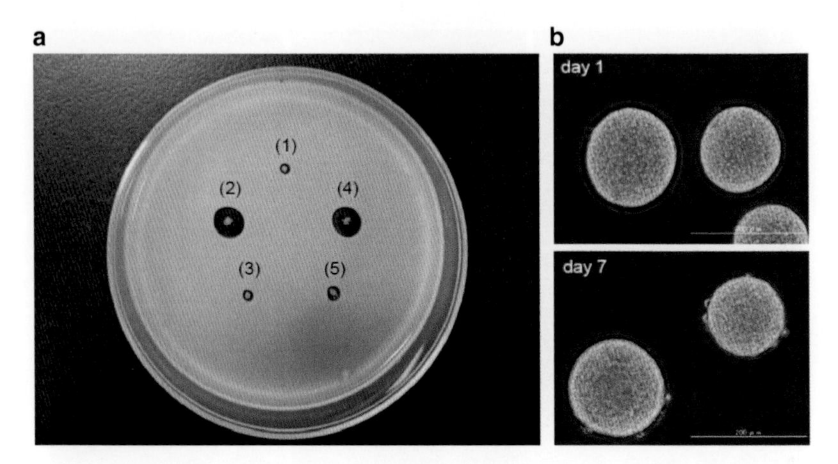

Fig. 11 Islets with immobilized urokinase (UK-islets) were tested for the ability to dissolve fibrin. (**a**) Fibrin in the plate gel medium was dissolved by UK-islets (*clear areas*). Fifty islets were applied to each spot, and the plate was observed after incubation at 37 °C for 14 h. (*1*) untreated islets; (*2*) UK-islets treated with oligo(dT)$_{20}$–PEG-lipid (C16), just after preparation; (*3*) UK-islets treated with oligo(dT)$_{20}$–PEG-lipid (C16) lost activity after 2 days in culture; (*4*) UK-islets treated with oligo(dT)$_{20}$–PEG-lipid (C18), just after preparation; and (*5*) UK-islets treated with oligo (dT)$_{20}$–PEG-lipid (C16) lost activity after 2 days in culture. (**b**) Morphology of UK-islets after 1 and 7 days of culture

long UK-islets can maintain the inhibition of IBMIR; however, these data suggested that UK immobilization on islets is a promising approach for islet transplantation.

4.3 Encapsulation of Islets with Living Cells

The histocompatibility and blood compatibility of donor islets can be significantly improved by enclosing them inside a capsule made of the patient's vascular endothelial cells. The ssDNA–PEG-lipid method was utilized to enclose islets with living cells [125]. The method is schematically shown in Fig. 12a. Oligo(dT)$_{20}$ was introduced onto the surface of HEK293 cells with an oligo(dT)$_{20}$–PEG-lipid, and oligo(dA)$_{20}$ was introduced onto the surface of islets with an oligo(dA)$_{20}$–PEG-lipid. Then, the oligo(dA)$_{20}$–islets were mixed with the oligo(dT)$_{20}$–HEK293 cells. The HEK293 cells were immobilized on the islet surface through DNA hybridization, as shown in Fig. 12b. Although the HEK293 cells existed as single cells on the islet just after immobilization, the surface of islets were completely covered with a cell layer after 3 days in culture (Fig. 12b). No central necrosis of the islet cells was observed. Immunostaining showed that insulin remained inside the islets after culturing for 3 days (Fig. 12b). Furthermore, after cell encapsulation, insulin secretion in response to glucose stimulation was well maintained (data not shown). This technique will greatly facilitate islet transplantation for treating type 1 diabetes.

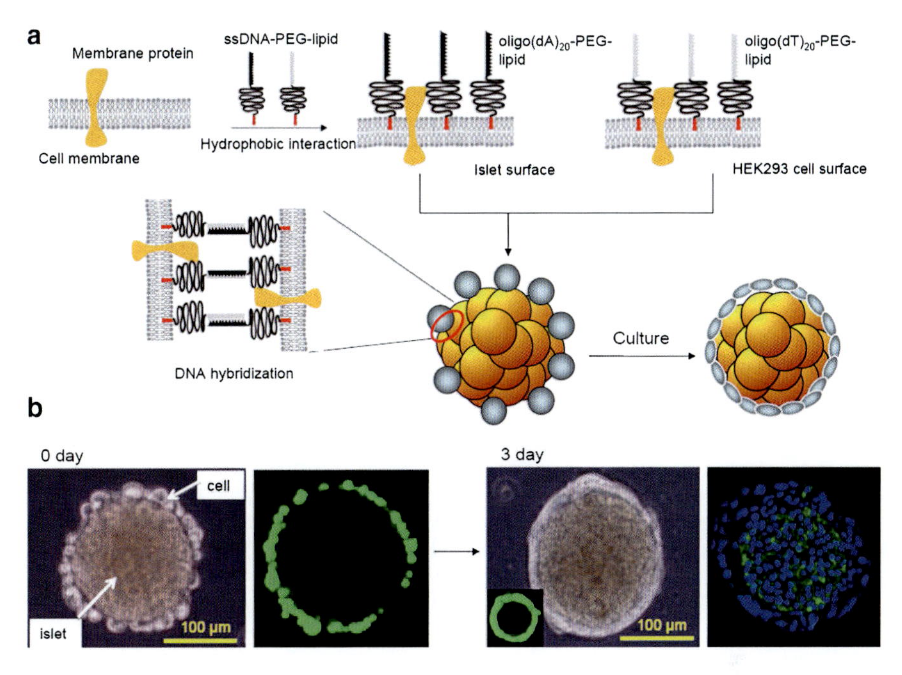

Fig. 12 Islet encapsulation within living HEK293 cells that express GFP. (**a**) Islets are enclosed within a layer of HEK293 cells (that express GFP) by introducing surface modifications of complementary ssDNAs. Islets modified with oligo(dT)$_{20}$–PEG-lipid are combined with HEK293 cells that have oligo(dA)$_{20}$–PEG-lipid immobilized on the surface. DNA hybridization immobilizes the HEK293 cells to the surface of the islets. After 3 days in culture, islets are completely encapsulated within HEK293 cells. (**b**) Phase contrast (*left panels*) and fluorescence images (*right panels*) of islets with attached HEK293 cells at 0 and 3 days after immobilization. GFP–HEK293 cells immobilized to islets were observed with a confocal laser-scanning microscope. Frozen sections of islets with attached GFP–HEK293 cells were stained with AlexaFluor 488-labeled anti-insulin antibody (*green*) (3-day samples). The 3-day samples were also stained with Hoechst 33342 dye (*blue*) for nuclear staining

5 Summary

Understanding biological responses against foreign substances has led to the development of various medical devices, artificial organs, and regenerative medicine approaches. SAMs of alkanethiols formed on a thin gold layer coated on a glass plate provide a platform for studying protein adsorption and cell adhesion to artificial materials. Although many questions remain to be answered before we fully understand the mechanisms of cell adhesion to artificial materials, rational studies can be conducted with SAMs. SAMs were used for developing a culture substrate for selective proliferation of NSCs. Based on the findings, we constructed a NSC culture module to facilitate the large-scale production of high purity NSCs.

The ssDNA–PEG-lipid provides versatility in cell surface modifications. It enables the immobilization of a broad spectrum of proteins and low molecular weight drugs on living cell surfaces. The encapsulation of islets inside a cell layer illustrates the versatility of this approach. The concept of controlling the host defense response against a living cell graft is likely to lead to important steps forward in cellular therapies. Various types of functional cells can be derived from pluripotent stem cells, including ESCs, iPSCs, and MSCs for use in cell replacement therapies. Our method is likely to be a key technology for the control of both acute and chronic host reactions and facilitation of graft adaptation in patients.

References

1. Park JB (1984) Tissue response to implants (biocompatibility). In: Biomaterials science and engineering. Plenum, New York
2. Anderson JM, Cook G, Costerton B, Hanson SR, Hensten-Pettersen A, Jacobsen N et al (2004) Host reactions to biomaterials and their evaluation. In: Ratner BD, Hoffman AS, Schoen EJ, Lemons JE (eds) Biomaterials science: an introduction to materials in medicine, 2nd edn. Elsevier Academic, Amsterdam
3. Ulman A (1996) Formation and structure of self-assembled monolayers. Chem Rev 96:1533–1554
4. Ostuni E, Yan L, Whitesides GM (1999) The interaction of proteins and cells with self-assembled monolayers of alkanethiolates on gold and silver. Colloids Surf B 15:3–30
5. Love JC, Estroff LA, Kriebel JK, Nuzzo RG, Whitesides GM (2005) Self-assembled monolayers of thiolates on metals as a form of nanotechnology. Chem Rev 105:1103–1169
6. Senaratne W, Andruzzi L, Ober CK (2005) Self-assembled monolayers and polymer brushes in biotechnology: Current applications and future perspectives. Biomacromolecules 6:2427–2448
7. Xu C, Inokuma MS, Denham J, Golds K, Kundu P, Gold JD, Carpenter MK (2001) Feeder-free growth of undifferentiated human embryonic stem cells. Nat Biotechnol 19:971–974
8. Flaim CJ, Chien S, Bhatia SN (2005) An extracellular matrix microarray for probing cellular differentiation. Nat Methods 2:119–125
9. Soen Y, Mori A, Palmer TD, Brown PO (2006) Exploring the regulation of human neural precursor cell differentiation using arrays of signaling microenvironments. Mol Syst Biol. doi:10.1038/msb4100076
10. Nakajima M, Ishimuro T, Kato K, Ko IK, Hirata I, Arima Y, Iwata H (2007) Combinatorial protein display for the cell-based screening of biomaterials that direct neural stem cell differentiation. Biomaterials 28:1048–1060
11. Reynolds BA, Tetzlaff W, Weiss SA (1992) A multipotent EGF-responsive striatal embryonic progenitor cell produces neurons and astrocytes. J Neurosci 12:4565–4574
12. Green RJ, Frazier RA, Shakesheff KM, Davies MC, Roberts CJ, Tendler SJB (2000) Surface plasmon resonance analysis of dynamic biological interactions with biomaterials. Biomaterials 21:1823–1835
13. Tengvall P, Lundström I, Liedberg B (1998) Protein adsorption studies on model organic surfaces: an ellipsometric and infrared spectroscopic approach. Biomaterials 19:407–422
14. Reimhult E, Larsson C, Kasemo B, Höök F (2004) Simultaneous SPR and QCM-D monitoring measurements of biomolecular adsorption events involving structural transformations and variations in coupled water. Anal Chem 76:7211–7220
15. Nuzzo RG, Zegarski BR, Dubois LH (1987) Fundamental studies of the chemisorption of organosulfur compounds on Au(111). Implications for molecular self-assembly on gold surfaces. J Am Chem Soc 109:733–740

16. Nuzzo RG, Dubois LH, Allara DL (1990) Fundamental studies of microscopic wetting on organic surfaces. 1. Formation and structural characterization of a self-consistent series of polyfunctional organic monolayers. J Am Chem Soc 112:558–569
17. Porter MD, Bright TB, Allara DL, Chidsey CED (1987) Spontaneously organized molecular assemblies. 4. Structural characterization of n-alkyl thiol monolayers on gold by optical ellipsometry, infrared spectroscopy, and electrochemistry. J Am Chem Soc 109:3559–3568
18. Strong L, Whitesides GM (1988) Structures of self-assembled monolayer films of organosulfur compounds adsorbed on gold single crystals: electron diffraction studies. Langmuir 4:546–558
19. Hirata I, Hioki Y, Toda M, Kitazawa T, Murakami Y, Kitano E, Kitamura H, Ikada Y, Iwata H (2003) Deposition of complement protein C3b on mixed self-assembled monolayers carrying surface hydroxyl and methyl groups studied by surface plasmon resonance. J Biomed Mater Res 66A:669–676
20. Bain CD, Evall J, Whitesides GM (1989) Formation of monolayers by the coadsorption of thiols on gold: variation in the head group, tail group, and solvent. J Am Chem Soc 111:7155–7164
21. Arima Y, Iwata H (2007) Effect of wettability and surface functional groups on protein adsorption and cell adhesion using well-defined mixed self-assembled monolayers. Biomaterials 28:3074–3082
22. Tarlov MJ, Burgess DRF Jr, Gillen G (1993) UV photopatterning of alkanethiolate monolayers self-assembled on gold and silver. J Am Chem Soc 115:5305–5306
23. Huang J, Dahlgren DA, Hemminger JC (1994) Photopatterning of self-assembled alkanethiolate monolayers on gold: a simple monolayer photoresist utilizing aqueous chemistry. Langmuir 10:626–628
24. Ryan D, Parviz BA, Linder V, Semetey V, Sia SK, Su J, Mrksich M, Whitesides GM (2004) Patterning multiple aligned self-assembled monolayers using light. Langmuir 20:9080–9088
25. Kumar A, Biebuyck HA, Whitesides GM (1994) Patterning self-assembled monolayers: applications in materials science. Langmuir 10:1498–1511
26. Xia Y, Whitesides GM (1998) Soft lithography. Angew Chem Int Ed Engl 37:550–575
27. Brockman JM, Frutos AG, Corn RM (1999) A multistep chemical modification procedure to create DNA arrays on gold surfaces for the study of protein-DNA interactions with surface plasmon resonance imaging. J Am Chem Soc 121:8044–8051
28. Liebermann T, Knoll W (2003) Parallel multispot detection of target hybridization to surface-bound probe oligonucleotides of different base mismatch by surface-plasmon field-enhanced fluorescence microscopy. Langmuir 9:1567–1572
29. Shumaker-Parry JS, Aebersold R, Campbell CT (2004) Parallel, quantitative measurement of protein binding to a 120-element double-stranded DNA array in real time using surface plasmon resonance microscopy. Anal Chem 76:2071–2082
30. Houseman BT, Huh JH, Kron SJ, Mrksich M (2002) Peptide chips for the quantitative evaluation of protein kinase activity. Nat Biotechnol 20:270–274
31. Wegner GJ, Lee NJ, Marriott G, Corn RM (2003) Fabrication of histidine-tagged fusion protein arrays for surface plasmon resonance imaging studies of protein-protein and protein-DNA interactions. Anal Chem 75:4740–4746
32. Houseman BT, Gawalt ES, Mrksich M (2003) Maleimide-functionalized self-assembled monolayers for the preparation of peptide and carbohydrate biochips. Langmuir 19:1522–1531
33. Kanda V, Kariuki JK, Harrison DJ, McDermott MT (2004) Label-free reading of microarray-based immunoassays with surface plasmon resonance imaging. Anal Chem 76:7257–7262
34. Yamauchi F, Kato K, Iwata H (2004) Micropatterned, self-assembled monolayers for fabrication of transfected cell microarrays. Biochim Biophys Acta 1672:138–147
35. Yamazoe H, Iwata H (2005) Cell microarray for screening feeder cells for differentiation of embryonic stem cells. J Biosci Bioeng 100:292–296

36. Ko I-K, Kato K, Iwata H (2005) Parallel analysis of multiple surface markers expressed on rat neural stem cells using antibody microarrays. Biomaterials 26:4882–4891
37. Kato K, Sato H, Iwata H (2005) Immobilization of histidine-tagged recombinant proteins onto micropatterned surfaces for cell-based functional assays. Langmuir 21:7071–7075
38. Chen CS, Mrksich M, Huang S, Whitesides GM, Ingber DE (1997) Geometric control of cell life and death. Science 276:1425–1428
39. Brock A, Chang E, Ho CC, LeDuc P, Jiang XY, Whitesides GM, Ingber DE (2003) Geometric determinants of directional cell motility revealed using microcontact printing. Langmuir 19:1611–1617
40. Lehnert D, Wehrle-Haller B, David C, Weiland U, Ballestrem C, Imhof BA, Bastmeyer M (2004) Cell behaviour on micropatterned substrata: limits of extracellular matrix geometry for spreading and adhesion. J Cell Sci 117:41–52
41. McBeath R, Pirone DM, Nelson CM, Bhadriraju K, Chen CS (2004) Cell shape, cytoskeletal tension, and RhoA regulate stem cell lineage commitment. Dev Cell 6:483–95
42. Arima Y, Iwata H (2007) Effects of surface functional groups on protein adsorption and subsequent cell adhesion using self-assembled monolayers. J Mater Chem 17:4079–4087
43. Axelrod D, Hellen EH, Fulbright RM (1992) Total internal reflection fluorescence. In: Lakowicz JR (ed) Topics in fluorescence microscopy. Plenum, New York
44. Burmeister JS, Olivier LA, Reichert WM, Truskey GA (1998) Application of total internal reflection fluorescence microscopy to study cell adhesion to biomaterials. Biomaterials 19:307–325
45. Tidwell CD, Ertel SI, Ratner BD (1997) Endothelial cell growth and protein adsorption on terminally functionalized, self-assembled monolayers of alkanethiolates on gold. Langmuir 13:3404–3413
46. Tegoulia VA, Cooper SL (2000) Leukocyte adhesion on model surfaces under flow: effects of surface chemistry, protein adsorption, and shear rate. J Biomed Mater Res 50:291–301
47. McClary KB, Ugarova T, Grainger DW (2000) Modulating fibroblast adhesion, spreading, and proliferation using self-assembled monolayer films of alkylthiolates on gold. J Biomed Mater Res 50:428–439
48. Franco M, Nealey PF, Campbell S, Teixeira AI, Murphy CJ (2000) Adhesion and proliferation of corneal epithelial cells on self-assembled monolayers. J Biomed Mater Res 52:261–269
49. Scotchford CA, Gilmore CP, Cooper E, Leggett GJ, Downes S (2002) Protein adsorption and human osteoblast-like cell attachment and growth on alkylthiol on gold self-assembled monolayers. J Biomed Mater Res 59:84–99
50. Faucheux N, Schweiss R, Lützow K, Werner C, Groth T (2004) Self-assembled monolayers with different terminating groups as model substrates for cell adhesion studies. Biomaterials 25:2721–2730
51. van Wachem PB, Beugeling T, Feijen J, Bantjes A, Detmers JP, van Aken WG (1985) Interaction of cultured human endothelial cells with polymeric surfaces of different wettabilities. Biomaterials 6:403–408
52. van Wachem PB, Hogt AH, Beugeling T, Feijen J, Bantjes A, Detmers JP, van Aken WG (1987) Adhesion of cultured human endothelial cells onto methacrylate polymers with varying surface wettability and charge. Biomaterials 8:323–328
53. Tamada Y, Ikada Y (1986) Cell attachment to various polymer surfaces. In: Chiellini E, Giusti P, Migliaresi C, Nicolais L (eds) Polymers in medicine II. Plenum, New York
54. Tamada Y, Ikada Y (1993) Effect of preadsorbed proteins on cell adhesion to polymer surfaces. J Colloid Interface Sci 155:334–339
55. Tamada Y, Ikada Y (1993) Cell adhesion to plasma-treated polymer surfaces. Polymer 34:2208–2212
56. Lee JH, Khang G, Lee JW, Lee HB (1998) Interaction of different types of cells on polymer surfaces with wettability gradient. J Colloid Interface Sci 205:323–330

57. Lee JH, Lee JW, Khang G, Lee HB (1997) Interaction of cells on chargeable functional group gradient surfaces. Biomaterials 18:351–358
58. Sigal GB, Mrksich M, Whitesides GM (1998) Effect of surface wettability on the adsorption of proteins and detergents. J Am Chem Soc 120:3464–3473
59. Ostuni E, Chapman RG, Holmlin RE, Takayama S, Whitesides GM (2001) A survey of structure-property relationships of surfaces that resist the adsorption of protein. Langmuir 17:5605
60. Michael KE, Vernekar VN, Keselowsky BG, Meredith JC, Latour RA, García AJ (2003) Adsorption-induced conformational changes in fibronectin due to interactions with well-defined surface chemistries. Langmuir 19:8033–8040
61. Arima Y, Toda M, Iwata H (2011) Surface plasmon resonance in monitoring of complement activation on biomaterials. Adv Drug Delivery Rev 63(12): 988–999
62. Lestelius M, Liedberg B, Tengvall P (1997) In vitro plasma protein adsorption on ω-functionalized alkanethiolate self-assembled monolayers. Langmuir 13:5900–5908
63. Prime KL, Whitesides GM (1993) Adsorption of proteins onto surfaces containing end-attached oligo(ethylene oxide): A model system using self-assembled monolayers. J Am Chem Soc 115:10714–10721
64. Evans-Nguyen KM, Schoenfisch MH (2005) Fibrin proliferation at model surfaces: influence of surface properties. Langmuir 21:1691–1694
65. Rodahl M, Höök F, Fredriksson C, Keller CA, Krozer A, Brzezinski P, Voinova M, Kasemo B (1997) Simultaneous frequency and dissipation factor QCM measurements of biomolecular adsorption and cell adhesion. Faraday Discuss 107:229–246
66. Sellborn A, Andersson M, Fant C, Gretzer C, Elwing H (2003) Methods for research on immune complement activation on modified sensor surfaces. Colloids Surf B 27:295–301
67. Azzam RMA, Bashara NM (1977) Ellipsometry and polarized light. North-Holland, Amsterdam
68. Knoll W (1991) Polymer thin films and interfaces characterized with evanescent light. Makromol Chem 192:2827–2856
69. López GP, Albers MW, Schreiber SL, Carroll R, Peralta E, Whitesides GM (1993) Convenient methods for patterning the adhesion of mammalian cells to surfaces using self-assembled monolayers of alkanethiolates on gold. J Am Chem Soc 115:5877–5878
70. Tegoulia VA, Rao W, Kalambur AT, Rabolt JF, Cooper SL (2001) Surface properties, fibrinogen adsorption, and cellular interactions of a novel phosphorylcholine-containing self-assembled monolayer on gold. Langmuir 17:4396–4404
71. Chung YC, Chiu YH, Wu YW, Tao YT (2005) Self-assembled biomimetic monolayers using phospholipid-containing disulfides. Biomaterials 26:2313–2324
72. Holmlin RE, Chen X, Chapman RG, Takayama S, Whitesides GM (2001) Zwitterionic SAMs that resist nonspecific adsorption of protein from aqueous buffer. Langmuir 17:2841–50
73. Chen S, Yu F, Yu Q, He Y, Jiang S (2006) Strong resistance of a thin crystalline layer of balanced charged groups to protein adsorption. Langmuir 22:8186–91
74. Pierschbacher MD, Ruoslahti E (1984) Cell attachment activity of fibronectin can be duplicated by small synthetic fragments of the molecule. Nature 309:30–33
75. Aota S, Nomizu M, Yamada KM (1994) The short amino acid sequence Pro-His-Ser-Arg-Asn in human fibronectin enhances cell-adhesive function. J Biol Chem 269:24756–24761
76. Toda M, Kitazawa T, Hirata I, Hirano Y, Iwata H (2008) Complement activation on surfaces carrying amino groups. Biomaterials 29:407–417
77. Cornelius RM, Shankar SP, Brash JL, Babensee JE (2011) Immunoblot analysis of proteins associated with self-assembled monolayer surfaces of defined chemistries. J Biomed Mater Res 98A:7–18
78. Barrias CC, Martins MCL, Almeida-Porada G, Barbosa MA, Granja PL (2009) The correlation between the adsorption of adhesive proteins and cell behaviour on hydroxyl-methyl mixed self-assembled monolayers. Biomaterials 30:307–316

79. Keselowsky BG, Collard DM, García AJ (2003) Surface chemistry modulates fibronectin conformation and directs integrin binding and specificity to control cell adhesion. J Biomed Mater Res 66A:247–259
80. Roberts C, Chen CS, Mrksich M, Martichonok V, Ingber DE, Whitesides GE (1998) Using mixed self-assembled monolayers presenting RGD and $(EG)_3OH$ groups to characterize long-term attachment of bovine capillary endothelial cells to surfaces. J Am Chem Soc 120:6548–6555
81. Feng Y, Mrksich M (2004) The synergy peptide PHSRN and the adhesion peptide RGD mediate cell adhesion through a common mechanism. Biochemistry 43:15811–15821
82. Orner BP, Derda R, Lewis RL, Thomson JA, Kiessling LL (2004) Arrays for the combinatorial exploration of cell adhesion. J Am Chem Soc 126:10808–10809
83. Derda R, Li L, Orner BP, Lewis RL, Thomson JA, Kiessling LL (2007) Defined substrates for human embryonic stem cell growth identified from surface arrays. ACS Chem Biol 2:347–55
84. Arnold M, Cavalcanti-Adam EA, Glass R, Blümmel J, Eck W, Kantlehner M, Kessler H, Spatz JP (2004) Activation of integrin function by nanopatterned adhesive interfaces. ChemPhysChem 5:383–388
85. Nakaji-Hirabayashi T, Kato K, Arima Y, Iwata H (2007) Oriented immobilization of epidermal growth factor onto culture substrates for the selective expansion of neural stem cells. Biomaterials 28:3517–3529
86. Nakaji-Hirabayashi T, Kato K, Iwata H (2008) Essential role of structural integrity and firm attachment of surface-anchored epidermal growth factor in adherent culture of neural stem cells. Biomaterials 29:4403–4408
87. Nakaji-Hirabayashi T, Kato K, Iwata H (2009) Surface-anchoring of spontaneously dimerized epidermal growth factor for highly selective expansion of neural stem cells. Bioconjug Chem 20:102–110
88. Konagaya S, Kato K, Nakaji-Hirabayashi T, Iwata H (2011) Design of culture substrates for large-scale expansion of neural stem cells. Biomaterials 32:992–1001
89. Temple S (2001) The development of neural stem cells. Nature 414:112–117
90. Björklund A, Lindvall O (2000) Cell replacement therapies for central nervous system disorders. Nat Neurosci 3:537–544
91. Ronaghi M, Erceg S, Moreno-Manzano V, Stojkovic M (2010) Challenges of stem cell therapy for spinal cord injury: Human embryonic stem cells, endogenous neural stem cells, or induced pluripotent stem cells? Stem Cells 28:93–99
92. Hermanson GT, Mallia AK, Smith PK (1992) Immobilized affinity ligand techniques. Academic, San Diego, pp 57–63
93. Lauer SA, Nolan JP (2002) Development and characterization of Ni-NTA-bearing microspheres. Cytometry 48:136–145
94. Lu HS, Chai JJ, Li M, Huang BR, He CH, Bi RC (2001) Crystal structure of human epidermal growth factor and its dimerization. J Biol Chem 276:34913–34917
95. Laibinis PE, Whitesides GM, Allara DL, Tao Y-T, Parikh AN, Nuzzo RG (1991) Comparison of the structures and wetting properties of self-assembled monolayers of n-alkanethiols on the coinage metal surfaces, copper, silver, and gold. J Am Chem Soc 113:7152–7167
96. Grosdemange GP, Simon ES, Prime KL, Whitesides GM (1991) Formation of self-assembled monolayers by chemisorption of derivatives of oligo(ethylene glycol) of structure HS $(CH_2)_{11}(OCH_2CH_2)_mOH$ on gold. J Am Chem Soc 113:12–20
97. Miyamoto K, Yamada P, Yamaguchi RT, Muto T, Hirano A, Kimura Y, Niwano M, Isoda H (2007) In situ observation of a cell adhesion and metabolism using surface infrared spectroscopy. Cytotechnology 55:143–149
98. Ullrich A, Schlessinger J (1990) Signal transduction by receptors with tyrosine kinase activity. Cell 61:203–212
99. Tripet B, Wagschal K, Lavigne P, Mant CT, Hodges RS (2000) Effects of side-chain characteristics on stability and oligomerization state of a *de Novo*-designed model coiled-coil: 20 amino acid substitutions in position "d". J Mol Biol 300:377–402

100. Ogiso H, Ishitani R, Nureki O, Fukai S, Yamanaka M, Kim J-H, Saito K, Sakamoto A, Inoue M, Shirouzu M, Yokoyama S (2002) Crystal structure of the complex of human epidermal growth factor and receptor extracellular domains. Cell 110:775–787
101. Kumada Y, Shiritani Y, Hamasaki K, Ohse T, Kishimoto M (2009) High biological activity of a recombinant protein immobilized onto polystyrene. Biotechnol J 4:1178–1189
102. JI C, Xie D, Mays J et al (2004) A novel approach to xenotransplantation combining surface engineering and genetic modification of isolated adult porcine islets. Surgery 136:537–547
103. Hashemi-Najafabadi S, Vasheghani-Farahani E, Shojaosadati SA et al (2006) A method to optimize PEG-coating of red blood cells. Bioconjug Chem 17:1288–1293
104. Nacharaju P, Boctor FN, Manjula BN et al (2005) Surface decoration of red blood cells with maleimidophenyl-polyethylene glycol facilitated by thiolation with iminothiolane: an approach to mask A, B, and D antigens to generate universal red blood cells. Transfusion 45:374–383
105. Scott MD, Murad KL, Koumpouras F et al (1997) Chemical camouflage of antigenic determinants: stealth erythrocytes. Proc Natl Acad Sci USA 94:7566–7571
106. Elbert DL, Herbert CB, Hubbell JA (1999) Thin polymer layers formed by polyelectrolyte multilayer techniques on biological surfaces. Langmuir 15:5355–5362
107. Germain M, Balaguer P, Jc N et al (2006) Protection of mammalian cell used in biosensors by coating with a polyelectrolyte shell. Biosens Bioelectron 21:1566–1573
108. Krol S, Del Guerra S, Grupillo M et al (2006) Multilayer nanoencapsulation. New approach for immune protection of human pancreatic islets. Nano Lett 6:1933–1939
109. Ng V, Pl G, Ss S-C et al (2007) Nanoencapsulation of stem cells within polyelectrolyte multilayer shells. Macromol Biosci 7:877–882
110. Chen H, Teramura Y, Iwata H (2011) Co-immobilization of urokinase and thrombomodulin on islet surfaces by poly(ethylene glycol)-conjugated phospholipid. J Control Release 150:229–234
111. Inui O, Teramura Y, Iwata H (2010) Retention dynamics of amphiphilic polymers PEG-lipids and PVA-Alkyl on the cell surface. ACS Appl Mater Interfaces 2:1514–1520
112. Luan NM, Teramura Y, Iwata H (2011) Immobilization of soluble complement receptor 1 on islets. Biomaterials 32:4539–4545
113. Miura S, Teramura Y, Iwata H (2006) Encapsulation of islets with ultra-thin polyion complex membrane through poly(ethylene glycol)-phospholipids anchored to cell membrane. Biomaterials 27:5828–5835
114. Sakurai K, Teramura Y, Iwata H (2011) Cells immobilized on patterns printed in DNA by an inkjet printer. Biomaterials 32:3596–3602
115. Takemoto N, Teramura Y, Iwata H (2011) Islet surface modification with urokinase through DNA hybridization. Bioconjug Chem 22:673–678
116. Teramura Y, Chen H, Kawamoto T et al (2010) Control of cell attachment through polyDNA hybridization. Biomaterials 31:2229–2235
117. Teramura Y, Iwata H (2010) Bioartificial pancreas microencapsulation and conformal coating of islet of Langerhans. Adv Drug Deliv Rev 62:827–840
118. Teramura Y, Iwata H (2010) Cell surface modification with polymers for biomedical studies. Soft Matter 6:1081–1091
119. Teramura Y, Iwata H (2011) Improvement of graft survival by surface modification with poly (ethylene glycol)-lipid and urokinase in intraportal islet transplantation. Transplantation 91:271–278
120. Teramura Y, Iwata H (2009) Islet encapsulation with living cells for improvement of biocompatibility. Biomaterials 30:2270–2275
121. Teramura Y, Iwata H (2008) Islets surface modification prevents blood-mediated inflammatory responses. Bioconjug Chem 19:1389–1395
122. Teramura Y, Iwata H (2009) Surface modification of islets with PEG-lipid for improvement of graft survival in intraportal transplantation. Transplantation 88:624–630
123. Teramura Y, Kaneda Y, Iwata H (2007) Islet-encapsulation in ultra-thin layer-by-layer membranes of poly(vinyl alcohol) anchored to poly(ethylene glycol)-lipids in the cell membrane. Biomaterials 28:4818–4825

124. Teramura Y, Kaneda Y, Totani T et al (2008) Behavior of synthetic polymers immobilized on a cell membrane. Biomaterials 29:1345–1355
125. Teramura Y, Ln M, Kawamoto T et al (2010) Microencapsulation of islets with living cells using polyDNA-PEG-lipid conjugate. Bioconjug Chem 21:792–796
126. Totani T, Teramura Y, Iwata H (2008) Immobilization of urokinase on the islet surface by amphiphilic poly(vinyl alcohol) that carries alkyl side chains. Biomaterials 29:2878–2883
127. Bennet W, Sundberg B, Cg G et al (1999) Incompatibility between human blood and isolated islets of Langerhans: a finding with implications for clinical intraportal islet transplantation? Diabetes 48:1907–1914
128. Moberg L, Johansson H, Lukinius A et al (2002) Production of tissue factor by pancreatic islet cells as a trigger of detrimental thrombotic reactions in clinical islet transplantation. Lancet 360:2039–2045
129. Nilsson B, Korsgren O, Jd L et al (2010) Can cells and biomaterials in therapeutic medicine be shielded from innate immune recognition? Trends Immunol 31:32–38

Adv Polym Sci (2012) 247: 201–232
DOI: 10.1007/12_2011_149

Published online: 21 September 2011

Advances in Tissue Engineering Approaches to Treatment of Intervertebral Disc Degeneration: Cells and Polymeric Scaffolds for *Nucleus Pulposus* Regeneration

Jeremy J. Mercuri and Dan T. Simionescu

Abstract Synthetic polymers and biopolymers are extensively used within the field of tissue engineering. Some common examples of these materials include polylactic acid, polyglycolic acid, collagen, elastin, and various forms of polysaccharides. In terms of application, these materials are primarily used in the construction of scaffolds that aid in the local delivery of cells and growth factors, and in many cases fulfill a mechanical role in supporting physiologic loads that would otherwise be supported by a healthy tissue. In this review we will examine the development of scaffolds derived from biopolymers and their use with various cell types in the context of tissue engineering the *nucleus pulposus* of the intervertebral disc.

Keywords Biomaterials · Biopolymers · Intervertebral disc · Intervertebral disc degeneration · Nucleus pulposus · Tissue engineering

Contents

J.J. Mercuri
Department of Bioengineering, Clemson University, 311 Rhodes Center, Clemson, SC 29634, USA

D.T. Simionescu (✉)
Department of Bioengineering, Clemson University, 304 Rhodes Annex, Clemson, SC 29634, USA
e-mail: dsimion@clemson.edu

1 Introduction

Nearly 5.7 million individuals are diagnosed with intervertebral disc (IVD) disorders annually [1]. In the USA alone, approximately 350,000 individuals underwent surgical procedures in 2005, 75% of which were performed to treat degenerative changes within the spine, including those changes associated with intervertebral disc degeneration (IDD) [2, 3]. Some suggest that there are currently between 1.5 and 4 million adults in the USA with disc related low back pain (LBP) [4]. Alarmingly, 20% of teenagers have lumbar IVDs that exhibit signs of mild degeneration; these numbers increase dramatically with age as 10% of all discs in 50-year-olds and 60% of discs in those over the age of 70 exhibit severe degeneration [5].

From a clinician's perspective, patients with severe IDD may present with debilitating LBP. Although a direct causal relationship between IDD and LBP has been hard to establish due to the nearly 35–40% of asymptomatic individuals whom exhibit radiographic evidence of IDD [6], there is mounting evidence that suggests that a relationship does in fact exist [7–10]. IDD has also been associated with sciatica and disc herniation, which in the long-term can lead to spinal stenosis, a major source of LBP and disability [11]. This suggests that at least a portion of the socio-economic impact related to LBP can be attributed to IDD, making it one of the most expensive healthcare burdens to date. When extrapolating these effects to a global scale, one quickly comes to appreciate the potential for LBP and IDD to reach epidemic proportions.

These staggering statistics thus provide the impetus for research into the etiology of IDD and the development of new treatment strategies, including surgical techniques and tissue engineering approaches to regeneration of the IVD. As a preamble to reviewing these strategies, a short description of IVD biology, physiology, pathology, and current treatment options utilizing polymeric biomaterials is provided.

2 Anatomy of the Intervertebral Disc

The human spine is an exquisite and complex structure from both an anatomical and functional standpoint. The spine is composed of approximately 24 boney elements, termed vertebral bodies, which are separated by 23 cartilaginous cushions or "shock absorbers" termed IVDs. The spine serves three functional roles: (1) to protect the spinal cord, (2) to support and transmit axial loads arising from the head and torso to the hips and lower extremities, and (3) to allow for mobility and stability such that bending and torsional motion of the spine can occur. Without IVDs, however, the latter two roles of the spine could not be effectively fulfilled.

The IVD is composed of three morphologically distinct regions known as the *nucleus pulposus* (NP), the *annulus fibrosus* (AF), and cartilaginous endplates (CEP) (Fig. 1). The NP is a hydrophilic gel-like material located within the central region of the disc. Its material properties are largely due to its primary constituent components: a loose network of type II collagen and the large proteoglycan aggrecan (Fig. 1) [12, 13]. The NP is sequestered by the fibro-cartilaginous AF, which is comprised of approximately 15–25 concentric lamellae of type I collagen that are well adapted to resist torsional and tensile loading experienced during bending [13]. The CEP is a thin layer (~0.6–1 mm) of tissue resembling articular cartilage and lies at the interface of the IVD and vertebral bodies [14]. Early in life, the CEP is thought to function as a growth plate for the vertebral bodies and later functions to prevent the NP from extruding into the vertebral body while potentially limiting the loss of fragmented osmotically active proteoglycan from the IVD [14, 15]. The CEP is also sufficiently permeable such that diffusion of nutrients can occur between the vascular buds in the vertebral bodies and the largely avascular IVD.

3 Intervertebral Disc Biochemistry and Cell Biology

The biochemical composition of the NP and AF, summarized in Table 1, reflects the specialized function for which each region of the IVD is adapted. The IVD is a highly hydrated structure containing approximately 70–90% water. The major biochemical constituents of the IVD include collagens, proteoglycans, elastin, and non-collagenous proteins, which are differentially located throughout the IVD. The primary biochemical component found in the NP is the proteoglycan aggrecan. Aggrecan molecules are composed of a core protein with greater than 100 covalently attached glycosaminoglycan chains (chondroitin sulfate and keratin sulfate). These molecules complex together with hyaluronic acid via a link protein, and together form a polyanionic bottle-brush structure. Collagen type II is the predominant form of collagen found within the NP and forms a loose network with other minor collagens, including types IX and XI, that function as cross-bridges between type II molecules and other extracellular matrix (ECM) components. Together, they form an enhanced ECM network providing resistance to swelling [16–19]. Other ECM components found within the human NP include minor collagens such as types III and VI (found in the pericellular environment), the glycoprotein fibronectin (which links matrix fibers to cell surface integrins), elastin, and the small

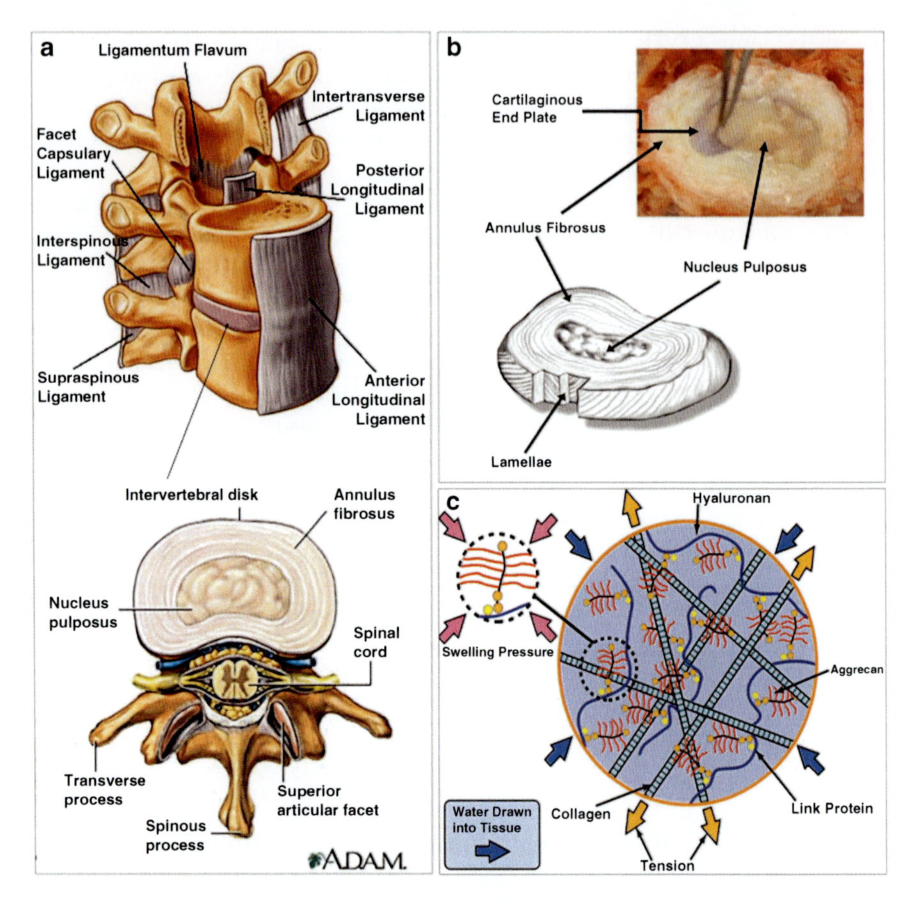

Fig. 1 (**a**) Function spinal unit composed of cartilagenous, boney, ligamentous, and neural structures Available from http://www.spinediagrams.com/ and http://www.healthcentral.com/osteoarthritis/h/c4-c5-intervertebral-disc-degenerative-disease.html. Last viewed 26 AUG 2011. (**b**) Image taken during the dissection of a porcine IVD that clearly indicates the delineation of the outer annulus fibrosus and central gelatinous nucleus pulposus. (**c**) Nucleus pulposus extracellular matrix composed of collagen type II (*green*), aggrecan (red bottle-brushes), hyaluronan (*blue*) and link protein (*yellow*) which endow the tissue with its physico-chemical properties including its characteristic swelling pressure [12]

Table 1 Summary of the regional differences in the biochemical composition of the intervertebral disc

Component	Nucleus pulposus	Annulus fibrosus
Water (%)	70–90	50–60
Collagen (%)	15 (collagen II)	65 (collagen I)
Proteoglycan (%)	60	15
Non-collagen proteins/elastin (%)	25	20
Cell morphology	Chondrocyte-like	Fibroblast-like

leucine-rich proteoglycans fibromodulin, decorin, biglycan, and lumican (which are believed to modulate collagen fibrillogenesis) [16, 20–22].

The AF is comprised primarily of sheets of aligned type I collagen, although types II and III are also present but to a lesser extent than in the NP [22]. Additionally, 2% of the dry weight of the AF is comprised of elastin, which can be found running parallel and perpendicular between lamellae and may prove advantageous for returning the layers to their original orientation following bending [11, 16, 22]. Proteoglycans are also found within the AF, but quantities here are lower than those found within the NP. The presence of these molecules may play a limited role in resisting compression in the AF; however, it is more likely to contribute to the dissipation of shear between lamellae during bending and torsion.

Maintenance of the distinct morphological structure of the IVD is a daunting task for the cells of the IVD, especially considering that the IVD is the largest avascular structure in the body, resulting in a relatively hypoxic and hypo-cellular environment. Two phenotypically distinct cell types are found to inhabit the adult IVD and account for between 0.25 and 1% of the total disc volume (Fig. 2) [14, 20]. Chondrocyte-like cells of an approximate density of 4×10^6 cells/cm^3 are found within the human NP. These cells appear rounded or oval in shape, some of which appear to reside within a lacunae similar to articular chondrocytes. Cells within the AF appear more like fibroblasts and are found in greater numbers (9×10^6 cells/cm^3) extended along collagen fibrils within lamellae. The chondrocyte-like NP cells of the IVD differ from those of normal articular cartilage not only in cell number, but also in ECM production. These cells primarily produce aggrecan, which results in an ECM containing a glycosaminoglycan to hydroxyproline ratio of 27:1, this is considerably different from the 2:1 ratio found in normal articular cartilage [23]. AF cells have been shown to produce predominantly type I collagen and versican; however, they have also been shown to produce minor amounts of aggrecan and collagen type II [24]. A third population of cells can also be found in the IVD: so-called notochordal cells (Fig. 2), which are large in size and highly vacuolated.

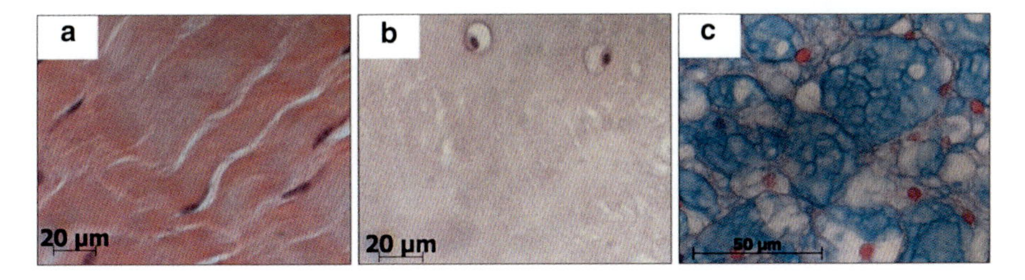

Fig. 2 Representative histological images of (**a**) bovine annulus fibrosus (H&E staining), (**b**) bovine nucleus pulposus (H&E staining), and (**c**) porcine notochordal cells (Alcian blue staining for GAGs with fast red nuclei stain)

These cells populate the central NP at birth but quickly disappear during early childhood [25]. The function of these cells is still under investigation, but it is believed that they produce matrix and serve to coordinate matrix production of the NP cells in the vicinity.

In the normal healthy IVD, the cells not only produce matrix macromolecules and growth factors, they also produce a myriad of proteases [26, 27]. Included in this list of proteases are the matrix metalloproteinases (MMPs) and aggrecanolytic members of the disintegrin and metalloproteinase with thrombospondin motif family (ADAMTS) as well as their respective inhibitors. It is the maintenance of this critical balance that results in a healthy IVD ECM that is subsequently well adapted for its physiologic and biomechanical function.

4 Intervertebral Disc Function and Mechanical Loading

IVDs are subjected to complex loads generated during compression, flexion–extension, lateral bending, and axial rotations. The mechanical function of the disc relies largely upon the synergistic relationship between NP and AF to absorb and distribute spinal loads. Nearly 75% of the compressive load observed by the IVD is borne by the NP itself, with the remaining 25% supported by the AF. Load management in the AF and NP clearly relates to specific biochemical components, their spatial relationships, and the resulting biphasic viscoelastic properties. Under equilibrium conditions, compressive loads on the IVD are borne predominantly by the swelling pressure (Donnan osmotic pressure) and solid matrix of the NP. More specifically, the fixed charge density of the aggrecan molecules found in the NP generate a swelling pressure of about 0.1–0.3 MPa [28]. When loads on the spine increase, for instance when bending to pick up a heavy box, the low hydraulic permeability of the NP tissue causes fluid pressurization. As a result of this, the NP not only supports compressive loading, but can convert some of the compressive stresses into tensile stresses in the AF, which aids in stabilizing the spine. Compressive loads typically observed within the human lumbar spine have been estimated to be in the range of 150–250 N when lying prone, 500–800 N during relaxed standing, and 1,900 N when lifting 10 kg with a rounded back [29]. Corresponding intradiscal pressures within the lumbar NP have been measured to be 0.1–0.24 MPa lying prone, 0.5 MPa standing relaxed, and 2.3 MPa lifting 20 kg weight with flexed back [30]. Application of such loads result in the exudation of nearly 20–25% of disc water, resulting in a fluctuation in spine length of approximately 1–2 cm throughout the course of a day. As the volume fraction of water in the NP decreases concomitant with an increase in fixed charge density, osmotic pressure rises, allowing for the influx of fluid back into the IVD when the spine is unloaded during diurnal rest periods.

5 Intervertebral Disc Degeneration

IDD is an aberrant cell-mediated process driven by a multitude of contributing factors including disc nutrition, mechanical forces, and genetics. The interplay of these elements culminates in detrimental changes within the disc's ECM. Histological evidence suggests that alterations in IVD ECM biochemistry appear to initiate within the end-plates and NP as early as 11–16 years of age [31]. Together, these changes eventuate in the discs' structural and functional demise and result in downstream effects on adjacent spinal structures, spinal biomechanics, and pain generation.

During degeneration, a number of changes in the biochemistry of the IVD are observed: disc water content and aggregating proteoglycan content decrease, non-enzymatic crosslinking of type II collagen occurs inhibiting matrix turnover and repair, and the production of proteoglycan decreases concomitant with a shift from chondroitin sulfate-rich aggrecan to a keratin sulfate-substituted version [14, 16, 21, 32–34]. Additionally, overall matrix degradation, in particular aggrecan and collagen type II, has been shown to increase in the degenerating IVD [35]. Increased fragmentation of aggrecan has been shown to occur, which may ultimately lead to glycosaminoglycan leaching and reduced osmotic pressure within the NP [32, 36]. Although disc collagen content does not significantly change with degeneration, fibrillar collagens (especially type II collagen) become more fragmented and their spatial distribution changes. This is particularly evident in the NP, which becomes infiltrated with collagen type I with increasing degeneration [35, 37]. IVD fibronectin content increases and becomes more fragmented with degeneration as well [38]. Interestingly, fragments of both collagen type II and fibronectin have exhibited the ability to perpetuate the degenerative cascade by upregulating MMP activity, causing further degradation of type II collagen and aggrecan [38–40].

Current evidence suggests that the cells of the IVD are themselves the primary culprits responsible for the destruction of the IVD ECM via the increased production of numerous proteinases (Table 2) [26, 27, 41, 42]. In most cases, production of tissue inhibitors of MMPs (TIMPs) increases in parallel with MMP synthesis; however, TIMP-3 (an inhibitor of ADAMTS-4) may not, thus potentially

Table 2 Summary of the matrix metalloproteinases involved in intervertebral disc degeneration [26, 41]

Enzyme name	Synonym	Major substrate
MMP-1	Collagenase I (interstitial)	Fibrillar collagen (prefers type III)
MMP-2	72-kDa gelatinase A	Gelatins, fibronectins, elastins
MMP-3	Stromelysin-1	Aggrecan, fibronectins, gelatins
MMP-7	Matrilysin	Aggrecan, fibronectins, gelatins
MMP-8	Neutrophil collagenase	Fibrillar collagen (prefers type I), aggrecan
MMP-9	92-kDa gelatinase B	Gelatins
MMP-13	Collagenase III	Fibrillar collagens (prefers type I)
ADAMTS-4	Aggrecanase	Aggrecan core protein
ADAMTS-5	Aggrecanase	Aggrecan core protein

implicating ADAMTS-4 as the major source of aggrecan destruction in the degenerative process. In addition, ECM synthesis can be dramatically hampered due to cell senescence, apoptosis, necrosis, and altered cell phenotype within degenerated IVDs [43].

Taken together, these biochemical and cellular transformations within the degenerating IVD result in macroscopic changes manifest as a dehydrated and fibrous NP, disorganized AF lamellae, and an inconspicuous boundary between the NP and AF. Concentric and/or radial annular tears, and NP fissure formations are also observed within the severely degenerated IVD, and an overall reduction in the central disc height may occur [11, 31]. Additionally, blood vessels and nerve endings have been found to infiltrate into the center of the disc, which may harbor inflammatory cells and their mediators.

6 Functional Consequences of IDD

Degenerative changes within the IVD, and in particular the NP, result in functional changes to the disc and adjacent spinal structures. The NP has been shown to transition from a fluid or gel-like material to that of a solid as indicated by an increase in shear modulus (range: 5–60 kPa) and a diminished capacity to dissipate energy [44]. As proteoglycan content of the NP decreases, the water content, swelling capacity and thus the hydrostatic nature of the NP is lost [45]. This in turn affects the IVD's ability to distribute stress evenly across the disc and causes stress concentrations to be most prevalent in the postero-lateral AF. As a result, the AF directly supports compressive spinal loads, which together with loss of hydrostatic pressure of the NP may lead to increased shear stresses between AF lamellae and thus contribute to their structural failure, increasing the risk of disc herniation [46, 47]. Concomitantly, circumferential and radial strains have been observed to be greater in the postero-lateral region of degenerated IVD, resulting in disc bulging and reduction in disc height [48].

7 Nucleus Pulposus Replacement with Polymeric Materials: A Potential Treatment Option for Symptomatic IDD

Surgical options commonly employed to combat the effects of IDD range in invasiveness and depend on the extent and indication of the degenerative process. While approximately two-thirds of patients undergo successful pain relief using conservative treatments such as bed rest, anti-inflammatory drugs, chiropractic adjustment, and physical therapy, the remainder will require more invasive surgical intervention. These procedures include the removal or obliteration of herniated or degenerate NP material, or complete removal of the disc, which requires subsequent spinal fusion or total disc arthroplasty [11, 49]. Despite the availability of these treatments, they are palliative solutions at best. Fusion and disc arthroplasty are considered to be last resort options for patients with severe symptomatic IDD. Patients must wait until the degeneration and resulting pain and instability is severe enough to warrant the removal of the disc, only to be replaced with stiff immobile metallic fusion hardware or articulating joints. Consequences arising from these treatments include spinal stenosis, accelerated degeneration in adjacent spinal segments, altered range of motion and intradiscal pressure in adjacent levels, subsidence, wear debris generation, and implant migration [50–55]. From this, it is clear that current surgical options have serious limitations and that a gap in treatment options exists [56].

An alternative surgical solution, nucleus pulposus replacement (NPR), is being investigated as a minimally invasive, early-stage intervention for patients with mild to moderate IDD. As degeneration first manifests within the NP, it is believed that removal of the degenerated NP and its subsequent replacement with an alternative mimetic material can mitigate the progression of IDD, while relieving pain and restoring anatomic and biomechanical function in select patients [57]. To date, most NPRs have been developed from in situ forming or pre-formed synthetic and biologic polymers, which are currently in different stages of development or clinical evaluation. We will briefly describe select devices and the polymeric materials from which they have been developed. For a more in depth review of commercial device designs, and pre-clinical, and clinical outcomes the readers are directed to the following resource: "Motion Preservation of the Spine: Advanced Techniques and Controversies" [58].

In situ curing materials offer the advantageous characteristics of being implanted using minimally invasive techniques and allow for complete filling of the NP region of the IVD. Detrimental characteristics of this class of materials include the potential for exothermic reactions during curing which can elicit native tissue/ cellular destruction and migration of these flowable materials away from the NP region prior to curing. One copolymer material developed in consideration of these characteristics is a methylenediphenyldiisocyanate-polyurethane mixture, which is injected into an expandable polycarbonate urethane containment balloon tailored to the central region of the patient's IVD [59]. Once cured, this material forms a resilient and incompressible material that can re-establish pre-degeneration disc heights.

Copolymers of polyethylene glycol dimethacrylate (PEGDM) and poly (*N*-isopropylacrylamide) (PNIPAAM) have also been investigated as potential injectable materials for NPR [60]. Coupling together the thermoresponsive nature of PNIPAAM with the hydrophilicity of PEGDM, investigators were able to form hydrogels that exhibited the capacity to gel at 32 °C, which allows for gelation within the human body. These hydrogels exhibit mechanical properties that can be tailored and an elastic characteristic that allows for the recovery of 85–98% of hydrogel original height by 55 min following stress relaxation testing [60].

As an alternative to purely synthetic injectable materials, an NPR system composed of a recombinant protein block copolymer of silk and elastin has been developed [61]. Genetically altered *Escherichia coli* bacteria produce the copolymer, which is subsequently modified such that a diisocyante-based crosslinker can be used to stabilize the system. The result is a cured material that has similar water, protein, and mechanical properties to the native NP [61]. An in situ polymerizing glutaraldehyde crosslinked albumin-based hydrogel has also been investigated as protein-based NPR; however, information on this particular material is limited [62].

Pre-formed NPRs should exhibit the ability to be compressed or dehydrated to minimize their profile for minimally invasive insertion and for preservation of surrounding anatomic structures. Moreover, the dimensional constraints of the patient's NP must be considered pre-operatively in order to determine appropriate implant sizing and optimal fit. Theoretically, as these materials swell, the pressure generated aids in restoring/maintaining IVD height while the application of compressive loading during daily activities would result in partial load transmission to the AF by the implant. One of the most widely investigated pre-formed devices is composed of a hydrogel pellet core of polyacrylonitrile and polyacrylamide copolymer that is surrounded by a woven jacket of ultrahigh molecular weight polyethylene (UHMWPE). As the hydrophilic core swells and absorbs body fluid, the surrounding jacket prevents excessive expansion that could damage the adjacent endplates. This device has exhibited excellent compressive fatigue-resistant properties and optimistic clinical outcomes [63]. Similarly, an acrylic copolymer layered with dacron mesh that prohibits excessive lateral expansion has been developed and exhibits the innate ability to absorb water while concomitantly generating an axial lift force of up to 400 N [64].

Polyvinyl alcohol (PVA)/polyvinyl pyrrolidone (PVP) copolymers have also been investigated for use as pre-formed NPRs. This material can be formed by successive freeze–thaw cycles that create physical crosslinks via interchain hydrogel bonding between the two components [65–67]. It was theorized that this material can be partially dehydrated for implantation and that, following insertion and rehydration, it would recapitulate the intradiscal pressure observed within the normal NP [66]. Unconfined compressive fatigue testing of the copolymer showed that there was no change in polymer content, indicating the stability of the material [65]. However, after 10 million testing cycles a permanent set was indicated by an un-recoverable decrease in sample height and an increase in sample diameter compared with original values [65]. Hydrogels composed of PVA alone have also been developed that exhibit osmotic swelling pressure profiles similar to those of human NP [68].

In consideration of the clinical importance of being able to visualize any implant material within the body, radio-opaque hydrogels for NPR have been formulated [69]. Copolymers of iodobenzoyl-oxo-ethyl methacrylate (4IEMA) and hydrophilic PVP or hydroxylethyl methacrylate (HEMA) exhibit appropriate swelling characteristics, viscoelastic mechanical properties, and excellent cytocompatibility [69]. Moreover, inclusion of the covalently attached iodine molecules allowed for hydrogel visualization via X-ray in a porcine cadaveric spine model [69].

Although many designs and materials to date have shown promise, some limitations are becoming evident. Pre-formed synthetics fail to fill the entire void space following removal of the NP, leaving open the potential for device migration and expulsion [63, 70]. Synthetic materials may also generate wear debris or toxic leachables due to incomplete in situ curing, which in the long term may adversely affect adjacent disc and neurologic tissue. Additionally, none of the synthetic implants attempt to regenerate healthy host NP tissue.

8 Tissue Engineering

The field of tissue engineering (TE) has been developed in part in response to chronic organ and tissue donor shortages. TE has been broadly defined by Laurencin as "the application of biological, chemical, and engineering principles toward the repair, restoration, or regeneration of living tissue by using biomaterials, cells, and factors alone or in combination." [71]. Nerem elaborates that TE utilizes "living cells, manipulated through their extracellular environment or even genetically, to develop biological substitutes for implantation into the body and/or to foster the remodeling of tissue in some other active manner. The purpose is to either either repair, replace, maintain, or enhance the function of a particular tissue or organ" [72]. In general, the foundation of any TE strategy lies within the ability of healthy cells to induce the de novo production and subsequent maintenance of an ECM. Clinically speaking, this strategy aims to produce healthy tissue with which to restore damaged or diseased tissues in hopes of recapitulating normal structure and function. Although TE does not always allege to address underlying pathology, one can speculate that the regeneration of a living tissue surrogate may be advantageous compared to implanting synthetic materials alone in that the tissue surrogate may possess the ability to remodel, integrate, and respond to the physiologic environment in which it is placed.

In the last 20 years, significant interest has been raised in utilizing TE strategies to mitigate the progression of IDD. Theoretical advantages of such approaches as outlined by Halloran et al. include "the use of degradable polymers, less invasive surgical procedures, preservation of native tissue, non-preclusion of future spinal surgery and multi-level disc treatments" [73]. Effects of IDD are initially manifest within the centrally located, hydrophilic NP, which when left unaddressed may progress to affect the remainder of the IVD and adjacent spinal structures. In response to this issue, researchers are developing TE strategies that aim to replace the degenerate NP in symptomatic patients with mild to moderate IDD in an attempt to mitigate its progression. The following sections provide details on the use of cells and biopolymeric scaffolds in attempts at IVD regeneration.

8.1 Cell Sourcing for Tissue Engineering of the NP

For IVD applications, a suitable cell source must be identified for the following reasons. First, dramatic reductions in viable cell number are typically observed in degenerated IVDs. Second, any remaining viable cells may exhibit altered phenotypes or varying degrees of senescence. The NP cell population found in the degenerative IVD is no exception to this fact [43]. Finally, as shown below, use of scaffolds without cells has provided only limited successes. A variety of cell sources have been described in the literature, including NP cells, chondrocytes, notochordal cells, and stem cells. A short survey of the pros and cons of these cell sources follows.

8.1.1 Autologous and Allogenic NP Cells

Theoretically, autogenic NPs would prove immunologically ideal. However, Richardson et al. suggest that use of these cells may not be not clinically feasible for several reasons: (1) NP cells are sparse and do not provide adequate quantity for self repair, (2) harvest of NP cells from a healthy adjacent IVD will lead to degeneration of that IVD, (3) harvest of NP cells from a degenerating IVD will accelerate the degenerative process, (4) NP cells from degenerated discs exhibit altered phenotype, increased senescence, increased expression of catabolic cytokines and degradative enzymes as well as a diminished capacity for matrix production [74]. Cadaveric NP cells from non-degenerated allogenic discs may also prove to be a feasible option; however, optimal cryopreservation techniques are still being developed, [75] and thus warrant further investigation. Although animal and human studies utilizing autogenic and allogenic NP cell transplantation provide some evidence of success in mitigating the progression of degeneration, these results do not appear to indicate that implantation of these cells alone (without scaffolds) have the capability to fully restore a degenerated disc to a normal healthy status [76].

8.1.2 Chondrocytes

NP cells are often described as "chondrocyte-like" owing to their round morphology and their ability to synthesize an ECM rich in collagen type II and the proteoglycan aggrecan. In a study performed in New Zealand white rabbits, autologous auricular chondrocytes were expanded and re-inserted into nucleotomized IVDs, which resulted in neo-formation of hyaline cartilage within the IVDs [77]. Although these results appear intriguing, it must be kept in mind that a striking difference between ECMs produced by chondrocytes and NP cells does exist [23]. More recently, use of human fetal chondrocytes (hFCs) to mitigate IDD in a lapine model was presented at the 23rd annual meeting of the North American Spine Society [78]. Discs receiving hFC injection retained 87.1% of their pre-operative disc height, and 60% of these discs had brighter MRI signal intensities compared with untreated degenerative controls. Moreover, the authors state that the injected human cells did not elicit immunological reactions, possibly due to the hFCs lacking major histocompatibility complex (MHC) class II molecules and to the immune privileged nature of the IVD [79, 80]. Thus, these results suggest that hFCs may be an appropriate cells source for use to mitigate IDD.

8.1.3 Notochordal Cells

Notochordal cells (NC) can be found in the IVD of many species throughout their entire lifespan [25]; however, in humans these cells disappear after the first decade of life and this precludes them from being used directly as an alternative cell source for TE strategies [81]. However, research over the last decade has shown that that the soluble factors produced by NCs possess anabolic characteristics that could be crucial to the maintenance of a healthy NP [81–85]. In addition, NC conditioned media has been shown to stimulate the differentiation of human mesenchymal stem cells (MSCs) towards an early NP cell-like phenotype in 3D pellet culture [86]. Our group has also observed a similar phenomenon in which MSCs from human adipose tissue were cultured in monolayer in the presence of conditioned media derived from a mixed population of porcine NP cells and NCs. The stem cells began to exhibit gene and protein expression profiles, including the upregulation of aggrecan, sox-9, and collagen type II, which could be indicative of differentiation towards an NP cell-like phenotype [87].

Due to the transitory nature of the NC population within the human IVD, the use of this cell population directly for NP TE applications might not be clinically feasible. Although harvest and expansion of these cells from young cadavers or the use of transgenic animal donors has been suggested as a possible alternative [25], further investigations of such methods are warranted.

8.1.4 Adult Mesenchymal Stem Cells

One of the most widely investigated and promising alternative cell sources for TE of the human NP is that of adult MSCs, in particular bone marrow-derived and adipose-derived stem cells (BMSCs and ADSCs, respectively). Stem cells are aptly defined as having the capacity for unlimited self-renewal paralleled with the ability to develop into progenitor cells, which include osteogenic, chondrogenic, and adipogenic lineages [88]. MSCs can be expanded in number in vitro and they appear to possess the ability to suppress or alter allogeneic immune response [89, 90]. Coupled with the fact that MSCs can differentiate into a chondrogenic lineage, many researchers are investigating the ability to direct the differentiation of MSCs towards an NP cell phenotype utilizing various inductive mechanisms. If successful differentiation of MSCs into an NP-like cell can be realized, MSCs could serve as an ideal alternative cell source for TE of the NP. Differentiation of human BMSCs towards an IVD cell-like phenotype can be accomplished using several different approaches: high cell density 3D spheroid cultures in the presence of transforming growth factor beta-3 (TGF-β3) [91], hypoxia with cell culture media containing TGF-β_1 and ascorbic acid [92], NP-cell conditioned media [93], and 3D scaffolds composed of a thermoresponsive 3% chitosan–0.5 M glycerophosphate crosslinked hydrogel [74]. Others have shown that direct injection of BMSCs into the lumbar IVDs of New Zealand white rabbits resulted in an initial increase proteoglycan content [94]. However, by 24 weeks, injected BMSCs had migrated to the transition zone between the NP and AF and had developed spindle-shaped morphologies reminiscent of AF cells. These interesting aspects definitely necessitate further studies. In a similar approach, Sakai et al. injected allogenic BMSCs encapsulated in atellocollagen type II hydrogels into rabbit degenerated IVDs with outstanding results [95]. Similar to studies on BMSCs, several reports have documented the potential of ADSCs to differentiate into NP-like cells using indirect co-culture systems with NP cells [96, 97].

8.2 *Biopolymeric Scaffolds for Tissue Engineering of the NP*

Although it has been shown that there are many feasible cell sources for TE of the human NP, it is clear that a suitable scaffold to support these cells is required for successful tissue regeneration. In the event that the NP is severely degenerated, creating an environment rich in destructive enzymes [27], one could speculate that removal of the degraded NP could prove advantageous in allowing for the best possible outcome for any alternative cell source to rescue the NP. Accordingly, the use of a scaffold on which to seed cells would be advantageous not only for aiding in the restoration of the mechanical function of the NP, but also in facilitating the maintenance or re-establishment of a proper NP cell phenotype. An ideal NP scaffold would include the use of a material that could (1) protect cells from loads experienced by the IVD, (2) supports either the re-differentiation of cell sources that may de-differentiate during monolayer expansion or induce the outright differentiation of cells, (3) is non-cytotoxic, and (4) ideally degrades into non-cytotoxic components [74, 98–100]. To this list one could add that the chosen scaffold material should be able to be implanted using a minimally invasive approach and should mimic the biochemical and mechanical properties found in the native NP while allowing for cell proliferation and infiltration. To date, numerous scaffolds have been developed from biopolymers in attempts to support cell-driven regeneration of a healthy NP.

8.2.1 Collagen–Glycosaminoglycan-Based Scaffolds

One of the more recent attempts at developing a scaffold that mimics the native ECM of human NP was investigated by Halloran et al. [73]. Using a blend of atellocollagen II, aggrecan, and hyaluronic acid crosslinked with a calcium-dependent enzyme (microbial transglutaminase), the authors' were able to form an in situ curing hydrogel. The authors demonstrated that bovine NP cells remained viable throughout the 7 day study and were producing sulfated glycosaminoglycan (sGAG); however, these molecules did not appear to be retained within the scaffold itself.

In a similar fashion, Yang et al. developed a 70% porous hydrogel scaffold with average pore diameter of 100 μm using a formulation of gelatin, chondroitin-6-sulfate, and hyaluronic acid. Hydrogels were assembled via multiple freeze-dry cycles and crosslinked with glutaraldehyde [101, 102]. Studies using these gels indicated that human NP cells remained viable over a 4-week culture period and appeared to proliferate significantly for between 2 and 4 weeks. Immunohistochemistry for collagen type II stained positive within the scaffold, and gene expression illustrated the upregulation of NP phenotype markers including collagen type II, aggrecan, and Sox-9 in comparison to NP cells cultured in monolayer. Interestingly, a downregulation in interleukin 1 (IL-1), which has been implicated in increasing MMP activity and proteoglycan degradation, was observed along with an upregulation of TIMP-1. In general, the scaffolds developed by Yang appear to be conducive to repopulation, with NP cells maintaining their viability, proliferative, and synthetic capacity. However, both control scaffolds (those without cells) and scaffolds with NP cells were only able to maintain approximately half their initial GAG content at best.

Li et al. have developed a porcine collagen type II, hyaluronate, chondroitin-6-sulfate tri-copolymer sponge [103]. Unlike Yang et al. who utilized glutaraldehyde, which does not have the ability to crosslink GAGs, Li utilized a 1-ethyl-3-(3-dimethylaminopropyl)-carbodiimide (EDC) and *N*-hydroxysuccinimide (NHS) crosslinking system to stabilize GAGs within their scaffolds. Evaluation of hydrogel cytotoxicity carried out using rabbit NP cells indicated no significant difference in DNA synthesis of NP cells in monolayer and scaffold. Histocompatibility following implant in Sprague-Dawley rats indicated a foreign-body reaction, with inflammatory cell infiltration at day 3 within and around the implant followed by a gradual replacement of these cells with fibroblasts at day 14. The scaffold was almost completely degraded after 84 days and fully replaced with vascularized granulation tissue. There was no difference in circulating antibodies towards the porcine collagen, as noted via ELISA between implant, untreated, and sham-operated groups.

In an attempt to mimic the physiological ratio of collagen type II to hyaluronan in the healthy human NP, Calderon et al. constructed hydrogels scaffolds composed of these two elements in a 9:1 (w/w) ratio [104]. Scaffolds were crosslinked with various concentrations of EDC/NHS, but it was found that 8 mM EDC/NHS resulted in a confined compressive modulus on the order of the native NP, while allowing for optimal rat mesenchymal stem cell (rMSC) viability and proliferation. Additionally, real time PCR results from rMSCs seeded on the scaffolds for 21 days indicated that the scaffolds promoted increased aggrecan expression and inhibited collagen type I expression compared to rMSCs cultured on monolayers.

Sakai et al. investigated the influence of three different atelocollagen scaffolds (3% atelocollagen I, 0.3% atelocollagen I, 0.3% atellocollagen II) on human NP cell proliferation and proteoglycan synthesis compared to 1.2% alginate hydrogels [105]. In general, the results indicate varied influence of scaffold material on NP cells. Cells in atelocollagen I scaffolds showed significantly more [3H]-thymidine uptake compared to alginate scaffolds, indicative of increased proliferation. Interestingly, higher concentration gels (3% atelocollagen I and 1.2% alginate) had significantly increased [^{35}S] sulfate uptake compared to gels of low concentration. However, normalized GAG content within the gels indicated that the alginate gels had significantly more GAG than any of the atelocollagen gels (there were no significant differences in GAG content between the atelocollagen gels). Additionally, the assessment of in vitro NP tissue formation on the three atelocollagen gels over a 4 week period indicated that the 3% atelocollagen gel contained significantly more sGAG per milligram dry tissue weight, which may indicate that scaffolds comprised of higher concentrations of atelocollagen are able to physically entrap and retain more GAG.

8.2.2 Elastin-Based Scaffolds

Elastin is a biopolymer that exhibits the unique ability to deform in response to applied load and to subsequently recoil to its original molecular orientation when that load is removed. This molecule has been shown to be present in the NP and, although not a major component by weight, elastin is thought to play a crucial role in aiding in the restoration of IVD matrix deformation following bending [16, 106]. Under this premise, we have developed an elastin-based hydrogel scaffold for NP tissue engineering. We hypothesized that a hydrogel scaffold that could deform and expel water upon compression, and recover both its original shape and water content once unloaded, could mimic the diurnal function of the native NP. Accordingly, we developed an elastin–glycosaminoglycan–collagen (EGC) hydrogel scaffold (Fig. 3) composed of soluble elastin, chondroitin-6-sulfate, hyaluronic acid, and collagen. Scaffolds were formed via gelation of the soluble mixture at 37 °C followed by lyophilization. The EGC scaffolds were subsequently crosslinked using a carbodiimide-based fixative to allow for glycosaminoglycan incorporation, and treated with penta-galloyl glucose, which is a polyphenolic tannin with proven ability to stabilize both collagen and elastin [107, 108]. Scaffolds were then treated with a mixture of enzymes to relieve some of the crosslinking, yielding a highly deformable and resilient biomaterial. Characterization of the EGC hydrogel

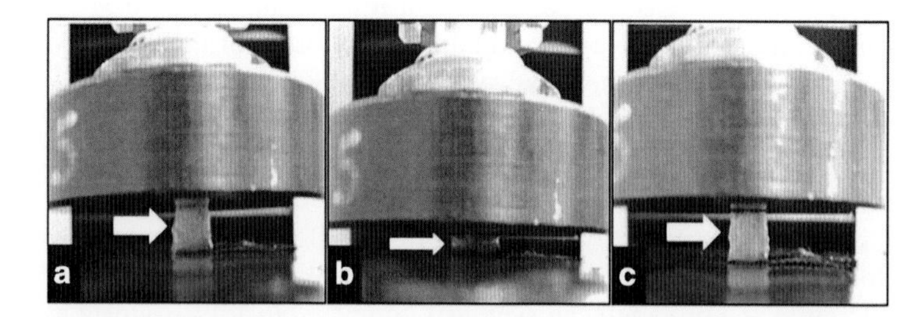

Fig. 3 Image frames captured from a video of an EGC hydrogel scaffold (**a**) prior to, (**b**) during, and (**c**) immediately following compression, illustrating its elastic shape-memory properties

indicated its ability to withstand unconfined compression while concurrently expressing water during loading [109]. Following load removal, EGC hydrogels exhibited the capacity to immediately re-establish their pre-compression water content and recover their initial height and shape (Fig. 3) [109]. We define these combined characteristics as a shape-memory sponge effect, which to the best of our knowledge has yet to be reported in the literature. The unconfined equilibrium compressive modulus of ECG hydrogels is similar to that of the human NP, and long-term cell culture studies utilizing porcine NP cells indicate that the material is cytocompatible [109]. Although others have attempted to develop biomaterials composed of these elements, their formulations were predominantly composed of collagen and the material applications was not specifically tailored for NP regeneration [110]. Thus, our results demonstrate the formation of a novel biopolymeric material for NP replacement that has physical characteristics similar to the native NP and may allow for adaptation to a patients' anatomy following minimally invasive surgical introduction.

8.2.3 Tissue-Based Scaffolds

In our research laboratory, we have also considered an alternative approach to generation of a scaffold for TE of the human NP. The use of ECM-based scaffolds derived from the decellularization of healthy xenogenic or allogenic tissues has become increasingly popular [111]. It is believed that scaffold development utilizing such techniques may prove advantageous because biologically appropriate scaffolds with an ECM microarchitecture tailored towards specific tissues are formed from the outset by the resident host tissue cells themselves. Recent literature indicates that this approach is suitable for the regeneration of numerous tissues including heart valves, vascular grafts, and corneal tissue [107, 108, 112]. In considering this, we hypothesized that an ideal scaffold for TE of the NP could be generated via the utilization of decellularized xenogenic NP. Theoretical advantages to our approach include the formation of a natural biopolymeric scaffold whose components, including whole aggrecan molecules and minor collagens, may be maintained in relevant ratios while innate ECM molecule interactions are preserved, thus yielding a close match to the native tissue microarchitecture. Using a combination of chemical and physical methodologies to successfully decellularize porcine NP, we were able to create a scaffold (Fig. 4) that closely approximates the chemical, physical, and mechanical nature of the human NP [113]. Our acellular porcine nucleus pulposus (APNP) scaffolds contain many of the major and minor ECM components found in the human NP including aggrecan and collagen types II, IX, and XI [113]. Quantification of the APNP GAGs and hydroxyproline content revealed a 21:1 ratio of these components, respectively. This closely approximates the 27:1 ratio found in the healthy human NP [23, 113]. Porcine cell remnants including DNA and the antigenic epitope alpha-Gal are absent within this material. The unconfined compressive material properties of the APNP approach values reported by others for the human NP [113–115].

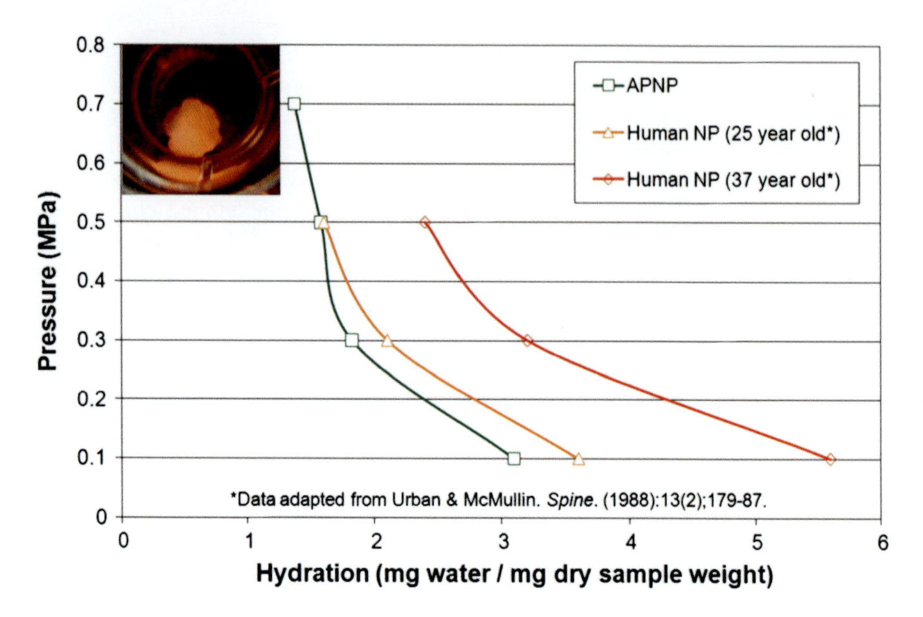

Fig. 4 *Insert*: Representative macroscopic image of an APNP hydrogel scaffold seeded with human adipose-derived stem cells following 7 days of culture in a 12-well plate. *Graph*: Osmotic swelling pressure profile of APNP hydrogel scaffolds (*green-squares*) compared to human NP (*orange-triangles* and *red-diamonds*) values found in the literature [28]

We have also shown that these scaffolds generate an osmotic swelling pressure profile (Fig. 4) that mirrors that of a 25-year-old human and that they are conducive to repopulation with human ADSC [28, 113].

Le Visage et al. investigated the use of small intestine submucosa, an ECM composed predominantly of type I collagen, GAGs, and growth factors, which they hypothesized would provide an attractive substrate for disc cells [116]. Following 3 months in standard cell culture conditions, human NP cells (obtained from degenerated discs) attached and migrated into the scaffolds and remained metabolically active, as determined by histology. Dimethyl-methylene blue (DMMB) assay indicated that control scaffolds (without cells) slowly lost their original GAG content, but that scaffolds seeded with NP cells steadily gained significant amounts of GAG with an average increase of 40%. This was confirmed by toluidine blue staining, which showed large areas of sGAG not observed in control scaffolds. Ultimately however, cells began to die at the 2 and 3 month time-points, and gene and protein expression for aggrecan, collagen type II, and Sox-9 were not representative of active NP cells.

8.2.4 Alginate-Based Scaffolds

Due to its similar structure to chondroitin sulfate and its ability to absorb 200–300 times its weight in water, many investigators have hypothesized that this brown algae derivative can be used to form NP TE scaffolds. Alginate is a linear copolymer consisting of repeating blocks of (1,4)-linked D-mannuronate and L-glucuronate. Upon submersion in an aqueous environment containing divalent calcium ions, sodium ions from sodium alginate are displaced by calcium allowing for the formation of a calcium crosslink between adjacent carboxylic acid groups at neutral pH, creating a solution to gel transition. However, despite exhibiting the ability to maintain an NP-like phenotype, reports of shortcomings in utilizing sodium alginate include the inability of IVD cells to assemble and maintain a functional ECM, as well as a decrease in the mechanical integrity of the scaffolds that is dependent on the time in culture [117]. It has been suggested that such observations are a result of Na^{+}/Ca^{2+} ion exchange in culture media, increased cell-mediated degradation of the gels, cellular Ca^{2+} metabolism, or a combination thereof [117].

In an attempt to overcome the aforementioned pitfalls, Chou and Nicoll suggested that methacrylate-modified alginate solutions would create scaffolds that have improved mechanical stability, maintain NP cell viability, and allow in-situ free-radical crosslinking via a photo-initiator and UV light exposure [118]. Encapsulated bovine NP cells in 3% alginate hydrogels modified with 2.5% methacrylate were shown to be viable and exhibit a round NP cell morphology at 14 days, concomitant with staining positive with immunohistochemistry for chondroitin-sulfate proteoglycan. The mean equilibrium Young's modulus for these gels was ~1.2 kPa, which is slightly lower than that of native human NP tested in unconfined compression [114]. Although the gel modulus increased with increasing methacrylate incorporation (5–7.5% methacrylation), cell viability decreased. In an extension of this work, Chou et al. implanted 2% alginate hydrogels modified with 3.5% methacrylate seeded with bovine NP cells into a murine subcutaneous pouch model to assess the maintenance of phenotype and matrix production [119]. Results indicated that the gels elicited a mild foreign body reaction, as evidenced by thin fibrous capsule formation; however, the photo-crosslinked gels remained intact. NP cells maintained collagen type II gene expression while showing a significant increase in aggrecan expression between 4 and 8 weeks. Equilibrium Young's modulus significantly increased in cell-encapsulated scaffolds (4.31 ± 1.39 kPA) in comparison to acellular controls (2.34 ± 0.33 kPa), which could be indicative of NP-like tissue formation.

Leone et al. developed a new amidic derivative of alginate crosslinked with 1,3 diaminopropane as they hypothesized that a hydrogel more closely mimicking the physical and chemical properties of hyaluronic acid would be ideal for tissue engineering the NP [120]. A 1% alginate solution resulted in a hydrogel that reached a swelling equilibrium after 1 h and was capable of swelling up to 250% its volume. Moreover, rheological analysis indicated that the alginate gel behaved in similar manner to non-degenerated NP and exhibited a predominantly elastic solid-like behavior under dynamic conditions with increasing stiffness upon increasing angular frequency. When seeded with normal primary chondrocytes from human knee articular cartilage, the gels allowed cell proliferation, maintenance of phenotype, and even appeared to produce low levels of MMP-13, which may be indicative of a cellular attempt to remodel the scaffold.

In an alternative approach to utilization of alginate scaffolds for NP TE, Gaetani investigated the use of semipermeable alginate microcapsules with a liquid core for the co-culture of human NP cells and ADSCs to create NP tissue in vitro [121]. Using a one-step process, a cell suspension in 40 mM barium chloride was created and added drop-wise into a 0.5% (w/v) sodium alginate solution in DMEM. The authors cite the advantages of using barium due to its reduced involvement in biological processes compared to that of calcium and an observed increase in alginate mechanical integrity. Culture for 7 days resulted in a pseudo-tissue structure in which tightly aggregated and elongated spindle-shaped cells resided at the liquid core–alginate interface, with rounded cells evenly distributed throughout the core itself. Interestingly, the authors also found that the matrix surrounding the rounded cells stained positively for aggrecan and collagen type I. The authors theorized that using a liquid-core scaffold may be advantageous because cells are not trapped in a gel matrix, but rather are suspended in liquid media. Careful consideration of this theory, however, is warranted as most cells derived from tissues are anchorage-dependent. The observed production of collagen type I by the cells within the scaffold in this study suggests that the cells are attempting to create their own substrate for attachment; however, the production of collagen type I would probably not result in an appropriate NP-like tissue. Additionally, since the authors used NP cells harvested from herniated discs, long-term studies with healthy cells are certainly warranted.

8.2.5 Miscellaneous Scaffolds

The use of chitosan, a cationic polysaccharide derived from depolymerization and deacytlation of chitin found in crustacean shells, has been investigated by Roughly as an alternative scaffold for NP TE [122]. The authors hypothesize that the thermoresponsive cationic chitosan hydrogels would be able to trap and retain anionic proteoglycan molecules produced by encapsulated NP cells, an ability significantly lacking in many other hydrogel systems. The authors found that gels made of 1.5% chitosan crosslinked with glycerophosphate in the presence of hydroxyethyl cellulose allowed bovine NP cell proliferation while retaining nearly 75% of synthesized GAG within the constructs after 20 days of in vitro culture. This resulted in a total GAG content that was 8–12% of GAG levels found in the native bovine NP. With the addition of TGF-β to the culture media, total GAG content approached but never exceeded 15% of native NP values, despite increased time in culture. Moreover, the rate of proteoglycan accumulation eventually diminished, which may have been due to a potential inhibitory effect of the surrounding proteoglycan on the cells, as suggested by the authors.

Reza et al. illustrated the feasibility of using methacrylating carboxymethylcellulose (CMC), a water-soluble polysaccharide derivative of cellulose, to yield a photo-crosslinkable hydrogel scaffold for encapsulating NP cells [123]. The authors utilized this material due to the fact that carboxylic acid on the CMC molecule becomes deprotonated at physiologic pH, resulting in a negative charge similar to that found on the native GAG molecules found in the NP. Bovine NP cell-laden scaffolds composed of a 3% solution of 250 kDa CMC exhibited an equilibrium Young's modulus of 3.53 $\pm$ 0.87 kDa and 60% cell viability after 7 days in culture. Additionally, histologic analysis indicated that the NP cells had a round morphology and resided in lacunae surrounded by pericellular and interterritorial deposits of chondroitin sulfate proteoglycan.

Yang et al. developed scaffolds derived from fibrin, citing such advantages as the ability to deliver cells and scaffolds in a minimally invasive fashion while maintaining the ability to form a scaffold that conforms to the nucleotomized space [124]. NP cells were suspended in fibrinogen concentrate derived from centrifuged human plasma, which was subsequently clotted via the addition of calcium. Results suggest a significant twofold increase GAG production, a 326% increase in aggrecan, and a significantly greater amount of DNA in the fibrin clot scaffolds as compared to human NP cells cultured in alginate. However, NP cells cultured in fibrin also expressed increased collagen types I and VI, which could be indicative of fibrotic phenotype.

Seguin et al. developed NP tissue in vitro on a porous calcium phosphate (CPP) substrate surface, which could theoretically allow the construct to be anchored to bone of the vertebral body [115]. CPP substrates were formed by sintering CaP powder, yielding a scaffold with a mean porosity of ~37% and pore size of 27 μm. After 2 weeks in culture on CPP, bovine NP cells were able to form a continuous layer, with subsequent ECM production by 6 weeks resulting in a tissue thickness of approximately 1.8 mm and a dry weight of 2.5 mg. This matrix stained with toluidine blue throughout the ECM, with more intense staining in the pericellular regions indicative of proteoglycan production. The growth rate of newly formed tissue decreased as time passed (0.6 mg/week at 2 weeks to 0.3 mg/week at 6 weeks), despite a steady increase in Hoescht DNA staining indicative of cell proliferation. GAG content within the tissue formed on CPP was not significantly different at 6 weeks from content found in native bovine tissue (218 and 227 μg/mg dry weight, respectively). SDS-PAGE illustrated that collagen II was predominantly found in the formed tissue, whereas unconfined compression testing indicated no significant difference in mechanical properties of in-vitro formed tissue compared to native bovine NP. Although the CPP appears to be an excellent substrates on which to develop NP tissue, further investigation of the implanted construct should be evaluated to determine the potential bone in-growth and possible over-growth into the formed NP tissue which would be detrimental. Additionally, this initial work does not take into consideration that the native NP tissue does not directly interface with bone; instead a cartilaginous endplate would lie between the two.

The latter issue was addressed more recently by Hamilton et al. who developed a trilayer composite consisting of CPP, articular cartilage (CEP), and NP tissue [125]. By sequential seeding of chondrocytes onto the CPP surface followed 2 weeks later by seeding NP cells onto the matrix produced by the chondrocytes, the authors were able to form a tissue composite construct. Although it appeared that the NP cells were able to maintain a rounded morphology, the interfacial shear load required to cause delamination of the in-vitro formed constructs (1.3 ± 0.23 N) was approximately 60 times less than that required of ex vivo samples.

Others have investigated the use of various composite scaffold materials with which to tissue engineer the entire IVD. Brown et al. investigated the effects of gelatin, demineralized bone matrix (DBM), and polylactide on porcine NP cell viability, metabolic activity, and gene expression [126]. Following 1 month of culture, NP cells assumed a flattened polygonal shape on gelatin microspheres, whereas those on DBM appeared elongated and spindle shaped. Interestingly, NP cells did not appear to attach to the polylactide scaffolds after 1 month and exhibited limited metabolic activity and viability. Cells on gelatin scaffolds exhibited higher metabolic activity, significantly higher mRNA expression for collagen type II (3.7-fold increase) and no significant difference in aggrecan expression compared to NP cells seeded on DBM scaffolds. Interestingly, cells on DBM expressed significantly more collagen type I, which the authors attribute to the presence of bone morphogenetic protein within the DBM. Additionally, the authors suggest that surface texture and/or the polymer degradation products could have influenced the observed cell response. From the data the authors suggest that gelatin and DMB should be investigated further for use in a composite cell scaffold for IVD TE.

8.2.6 Scaffolds Without Cells

A limited number of researchers have investigated the utilization of scaffolds alone to recruit neighboring disc cells and MSCs in an attempt to restore native NP integrity. Abbushi et al. implanted a polyglycolic acid felt material coated with hyaluronic acid and rabbit allogenic serum into rabbit IVDs that had been operated on to evacuated the NP [127]. Although MRI, lateral radiographs, and histological results after 6 months indicated mitigation of degeneration concomitant with the presence of NP-like cells secreting proteoglycan-rich ECM, the discs progressively degenerated and were unable to re-establish pre-operation disc height and MRI signal values. Additionally, Revell et al. injected a hyaluronic acid-based hydrogel (HYAFF120) into the lumbar spine of pigs immediately following subtotal nucleotomy [128]. Again, similar results were obtained following 6 weeks of implant in which discs receiving HYAFF120 gels maintained normal disc height and had striking resemblances to native NP tissue containing chondrocytes (suggested by the authors to have originated from neighboring disc tissue or remaining NP tissue). Although these results suggest that implantation of cell-free scaffolds may initially mitigate degeneration of the NP, the authors did not allow for the onset of degeneration before implantation and thus results need to be interpreted with caution.

As seen above, implanting cells without scaffold support or scaffolds without cells may not prove efficacious. NP allograft studies in a lapine disc degeneration model indicate that injection of NP cells alone does not mitigate degeneration as effectively as implanting NP allografts, and can result in the formation of fibrotic

NP tissue [129]. Thus, the use of a healthy alternative human NP cell source in conjunction with an appropriate scaffold may be the most beneficial approach to NP TE.

9 Animal Models of IDD

Animal models make possible the study of etiological mechanisms associated with IDD while allowing for the investigation of the efficacy of potential therapeutic approaches. Numerous animal models have been investigated that reflect the biochemical, radiological, and histological changes observed within degenerating human IVDs. Although these models provide important clues towards developing treatment strategies, the selection of model and interpretation of results must be approached cautiously due to intrinsic differences between animal species and humans (e.g., IVD size and aspect ratio, anatomical variation in adjacent spinal structures, variations in disc cell populations and metabolism, biochemical composition, biomechanics) [130, 131].

Animal models of IDD can be broadly classified into two categories: naturally occurring and experimentally induced. Macaques, beagles, and sand rats are a few examples of animals that appear to naturally develop evidence of IDD over the course of their lives [132–134]. One of the most widely investigated animal models of induced degeneration includes lumbar disc puncture models. Masuda et al. found that using 16G and 18G needles to penetrate through the AF into the NP of New Zealand white rabbits resulted in a mild and reproducible model of disc degeneration, as indicated by reductions in disc height, histological staining for proteoglycan, and MRI signal intensity [135]. Degenerative changes were observed as early as 2 weeks post-puncture with progression of degeneration through the 8 week study. Similar results were attained by Sobajima et al. who observed a progressive loss in NP area and MRI signal intensity as early as 3 weeks, which progressed over the 24-week study [136]. Degenerative findings included a decrease in the number of notochordal cells, fluctuating numbers of NP cells, formation of a fibrocartilagenous NP, osteophyte formation, and end-plate sclerosis [136]. Annular stab models using scalpels have also been performed in Sprague-Dawley rats, sheep, and pigs resulting in similar outcomes [137, 138].

Other forms of induced degeneration include axial overloading of the disc to stimulate degeneration [139, 140], intradiscal injection of an apoptotic (camptothecin) [141], and chemonucleolytic agents such as chondroitinase ABC [142]. Zhou et al. showed that intradiscal injection of bromodeoxyuridine (BrdU), a known mutagen that incorporates into DNA in place of thymidine, created a cellular senescence/aging model of degeneration in sheep IVDs [143].

Choice of the ideal animal model to test tissue engineered strategies has to balance cost and relevance. The naturally occurring macaque model would be a primary choice; however, the costs may be prohibitive, especially for the academic arena. Other large animals such as goats and sheep may also pose cost limitations and appear to exhibit degenerative changes that are milder, less repeatable, and take long periods of time to develop. The rabbit model of induced degeneration via needle puncture may be more appropriate for screening purposes. Although there is debate as to the influence that the notochordal cell population retained within the NP of this species may have on therapeutic intervention, degeneration in rabbits repeatedly occurs and it seems unlikely that the presence of this cell population will favorably alter any particular therapeutic outcome. Regardless, all results from animal studies must be cautiously interpreted in the context of feasibility in humans.

10 Conclusions

The adult IVD is a large avascular structure composed of a proteoglycan-rich gelatinous core (the NP) surrounded by concentric sheets of fibrous collagen (the AF). The biochemical components, specialized architecture, and mechanical properties allow the support of compressive loads while allowing for varying degrees of range of motion. Distinct cell populations residing within the IVD account for the differential distribution of ECM components found in the NP as compared with the AF. Clearly, the ECM homeostasis in the NP is in a very delicate balance. Aging, disc nutrition, mechanical forces, and genetic factors contribute to chronic IDD, characterized by high levels of proteases and dramatic changes in NP ECM biochemistry. Loss of NP proteoglycans and inherent hydration capacity has important functional changes, which may become symptomatic and require intervention. Current treatment options such as fusion and total disc arthroplasty exhibit limitations including adjacent level degeneration and wear debris production. A gap in surgical treatment options exists because current methods are only considered when degeneration has progressed significantly, resulting in spinal instability and persistent pain. NPR may be a feasible option for mitigation of mild to moderate IDD. Existing NPR devices developed from synthetic pre-formed or in situ forming materials are very promising, but have some serious limitations such as failure of pre-formed devices to fill the entire void space following nucleotomy and generation of debris or toxic leachables from synthetic materials. More importantly, none of the synthetic implants actually regenerate healthy host NP tissue, including the appropriate cells and ECM components.

Tissue engineering approaches using cells and scaffolds hold the promise for effective NP regeneration by creating tissue analogs capable of maintaining NP ECM homeostasis. Among the numerous cell types that have been investigated, adult stem cells have shown the most promise for differentiation into NP cells. Several types of scaffolds of biologic and synthetic origins have been developed to mimic the physico-chemical properties of the native NP and also to support seeded cells and encourage NP regeneration. The ideal scaffold should be sufficiently strong to withstand mechanical loads immediately after implantation but should also be slowly degradable to allow for remodeling. Scaffolds should also mimic the physico-chemical of the native NP. Ideally, scaffold degradation products should not elicit local or systemic reactions and should not disturb local pH and gas/nutrient exchange. Use of scaffolds or cells alone does not appear to be sufficient to entirely prevent the progression of IDD. To test feasibility, lumbar disc puncture in rabbits appears to provide a progressive and reproducible animal model of degeneration that closely approximates that observed in humans. For more clinically relevant functional studies of tissue engineered devices, large animal models are more adequate for functional analyses.

Future studies in this field should focus on better understanding the biology of NP cells, the functional interactions between IVD components, and the pathobiology of the IDD process in humans, so that early non-surgical interventions can be tested. Use of adult stem cells pre-differentiated under physiologically relevant conditions on 3D scaffolds prior to implantation by minimally invasive procedures may prove most advantageous.

All in all, the future of IVD tissue engineering is bright and novel developments in scaffold chemistry utilizing both synthetic and biopolymers are bound to should provide excellent alternatives for millions of patients suffering from IDD.

References

1. An HS, Thonar EJ, Masuda K (2003) Biological repair of intervertebral disc. Spine 28(15 Suppl):S86–S92
2. Deyo RA, Nachemson A, Mirza SK (2004) Spinal-fusion surgery - the case for restraint. N Engl J Med 350(7):722–726
3. Frymoyer JW, Cats-Baril WL (1991) An overview of the incidences and costs of low back pain. Orthop Clin North Am 22(2):263–271
4. Masuda K, Lotz JC (2010) New challenges for intervertebral disc treatment using regenerative medicine. Tissue Eng Part B Rev 16(1):147–158
5. Miller JA, Schmatz C, Schultz AB (1988) Lumbar disc degeneration: correlation with age, sex, and spine level in 600 autopsy specimens. Spine (Phila Pa 1976) 13(2):173–178
6. Boden SD (1996) The use of radiographic imaging studies in the evaluation of patients who have degenerative disorders of the lumbar spine. J Bone Joint Surg Am 78(1):114–124
7. Videman T, Nurminen M (2004) The occurrence of anular tears and their relation to lifetime back pain history: a cadaveric study using barium sulfate discography. Spine 29 (23):2668–2676
8. Pye SR et al (2004) Radiographic features of lumbar disc degeneration and self-reported back pain. J Rheumatol 31(4):753–758
9. McNally DS et al (1996) In vivo stress measurement can predict pain on discography. Spine 21(22):2580–2587
10. Schwarzer AC et al (1995) The prevalence and clinical features of internal disc disruption in patients with chronic low back pain. Spine 20(17):1878–1883
11. Raj PP (2008) Intervertebral disc: anatomy-physiology-pathophysiology-treatment. Pain Pract 8(1):18–44
12. Hardingham T (1998) Cartilage: aggrecan – link protein – hyaluronan aggregates. Available from http://www.glycoforum.gr.jp/science/hyaluronan/HA05/HA05E.html. Last accessed 25 July 2011

13. Middleditch A, Oliver J (2005) Functional anatomy of the spine, 2nd edn. Elsevier, New York
14. Roberts S et al (2006) Histology and pathology of the human intervertebral disc. J Bone Joint Surg Am 88(Suppl 2):10–14
15. Roberts S et al (1996) Transport properties of the human cartilage endplate in relation to its composition and calcification. Spine (Phila Pa 1976) 21(4):415–420
16. Roughley PJ (2004) Biology of intervertebral disc aging and degeneration: involvement of the extracellular matrix. Spine 29(23):2691–2699
17. Eyre DR, Matsui Y, Wu JJ (2002) Collagen polymorphisms of the intervertebral disc. Biochem Soc Trans 30:844–848
18. Vaughan L et al (1988) D-periodic distribution of collagen type IX along cartilage fibrils. J Cell Biol 106(3):991–997
19. Eyre DR (1988) Collagens of the disc. In: Ghosh P (ed) The biology of the intervertebral disc. CRC, Boca Raton, pp 171–188
20. Urban JP (2000) The nucleus of the intervertebral disc from development to degeneration. Am Zool 40(1):53–61
21. Le Maitre CL et al (2007) Matrix synthesis and degradation in human intervertebral disc degeneration. Biochem Soc Trans 35(Pt 4):652–655
22. Richardson SM et al (2007) Intervertebral disc biology, degeneration and novel tissue engineering and regenerative medicine therapies. Histol Histopathol 22(9):1033–1041
23. Mwale F, Roughley P, Antoniou J (2004) Distinction between the extracellular matrix of the nucleus pulposus and hyaline cartilage: a requisite for tissue engineering of intervertebral disc. Eur Cell Mater 8:58–63
24. Horner HA et al (2002) Cells from different regions of the intervertebral disc: effect of culture system on matrix expression and cell phenotype. Spine (Phila Pa 1976) 27(10):1018–1028
25. Hunter CJ, Matyas JR, Duncan NA (2003) The notochordal cell in the nucleus pulposus: a review in the context of tissue engineering. Tissue Eng 9(4):667–677
26. Roberts S et al (2000) Matrix metalloproteinases and aggrecanase: their role in disorders of the human intervertebral disc. Spine (Phila Pa 1976) 25(23):3005–3013
27. Rutges JP et al (2008) Increased MMP-2 activity during intervertebral disc degeneration is correlated to MMP-14 levels. J Pathol 214(4):523–530
28. Urban JP, McMullin JF (1988) Swelling pressure of the lumbar intervertebral discs: influence of age, spinal level, composition, and degeneration. Spine 13(2):179–187
29. Sato K, Kikuchi S, Yonezawa T (1999) In vivo intradiscal pressure measurement in healthy individuals and in patients with ongoing back problems. Spine 24(23):2468–2474
30. Wilke HJ et al (1999) New in vivo measurements of pressures in the intervertebral disc in daily life. Spine 24(8):755–762
31. Boos N et al (2002) Classification of age-related changes in lumbar intervertebral discs: 2002 volvo award in basic science. Spine (Phila Pa 1976), 27(23):2631–2644
32. Roughley PJ et al (2006) The structure and degradation of aggrecan in human intervertebral disc. Eur Spine J 15(Suppl 3):S326–S332
33. Walker MH, Anderson DG (2004) Molecular basis of intervertebral disc degeneration. Spine J 4(6 Suppl):158S–166S
34. Urban JP, Roberts S (2003) Degeneration of the intervertebral disc. Arthritis Res Ther 5(3):120–130
35. Antoniou J et al (1996) The human lumbar intervertebral disc: evidence for changes in the biosynthesis and denaturation of the extracellular matrix with growth, maturation, ageing, and degeneration. J Clin Invest 98(4):996–1003
36. Sztrolovics R et al (1997) Aggrecan degradation in human intervertebral disc and articular cartilage. Biochem J 326(Pt 1):235–241
37. Hollander AP et al (1996) Enhanced denaturation of the alpha (II) chains of type-II collagen in normal adult human intervertebral discs compared with femoral articular cartilage. J Orthop Res 14(1):61–66

38. Oegema TR Jr et al (2000) Fibronectin and its fragments increase with degeneration in the human intervertebral disc. Spine (Phila Pa 1976) 25(21):2742–2747
39. Yasuda T, Poole AR (2002) A fibronectin fragment induces type II collagen degradation by collagenase through an interleukin-1-mediated pathway. Arthritis Rheum 46(1):138–148
40. Yasuda T et al (2006) Peptides of type II collagen can induce the cleavage of type II collagen and aggrecan in articular cartilage. Matrix Biol 25(7):419–429
41. Goupille P et al (1998) Matrix metalloproteinases: the clue to intervertebral disc degeneration? Spine (Phila Pa 1976) 23(14):1612–1626
42. Le Maitre CL, Freemont AJ, Hoyland JA (2004) Localization of degradative enzymes and their inhibitors in the degenerate human intervertebral disc. J Pathol 204(1):47–54
43. Zhao CQ et al (2007) The cell biology of intervertebral disc aging and degeneration. Ageing Res Rev 6(3):247–261
44. Iatridis JC et al (1997) Alterations in the mechanical behavior of the human lumbar nucleus pulposus with degeneration and aging. J Orthop Res 15(2):318–322
45. Johannessen W, Elliott DM (2005) Effects of degeneration on the biphasic material properties of human nucleus pulposus in confined compression. Spine (Phila Pa 1976) 30 (24):E724–E729
46. Goel VK et al (1995) Interlaminar shear stresses and laminae separation in a disc: finite element analysis of the L3-L4 motion segment subjected to axial compressive loads. Spine (Phila Pa 1976) 20(6):689–698
47. Niosi CA, Oxland TR (2004) Degenerative mechanics of the lumbar spine. Spine J 4(6 Suppl):202S–208S
48. Tsantrizos A et al (2005) Internal strains in healthy and degenerated lumbar intervertebral discs. Spine (Phila Pa 1976) 30(19):2129–2137
49. Perez-Cruet MJ, Khoo LT, Fessler RG (eds) (2006) An anatomical approach to minimally invasive spine surgery. Quality Medical Publishing, St. Louis
50. Bastian L et al (2001) Evaluation of the mobility of adjacent segments after posterior thoracolumbar fixation: a biomechanical study. Eur Spine J 10(4):295–300
51. Eck JC, Humphreys SC, Hodges SD (1999) Adjacent-segment degeneration after lumbar fusion: a review of clinical, biomechanical, and radiologic studies. Am J Orthop 28(6): 336–340
52. Putzier M et al (2006) Charite total disc replacement–clinical and radiographical results after an average follow-up of 17 years. Eur Spine J 15(2):183–195
53. van Ooij A et al (2007) Polyethylene wear debris and long-term clinical failure of the charite disc prosthesis: a study of 4 patients. Spine 32(2):223–229
54. van Ooij A, Oner FC, Verbout AJ (2003) Complications of artificial disc replacement: a report of 27 patients with the SB charite disc. J Spinal Disord Tech 16(4):369–383
55. Zeh A et al (2007) Release of cobalt and chromium ions into the serum following implantation of the metal-on-metal Maverick-type artificial lumbar disc (Medtronic Sofamor Danek). Spine 32(3):348–352
56. Girardi FP, Viscogliosi Bros (2007) Worldwide orthopedic and spine market. In: Davis RJ, Girardi FP (eds) Nucleus arthroplasty technology in spinal care, vol 1. Raymedica, Minneapolis, pp 21–26. Available at http://www.thesona.com/sona2.swf. Last accessed 25 July 2011
57. Ahrens M, Tsantrizos A, LeHuec J (2008) DASCOR. In: Yue J, Bertagnoli R, McAfee P, An H (eds) Motion preservation surgery of the spine; advanced techniques and controversies. Elsevier, Philadelphia, pp 397–407
58. Yue J, Bertagnoli R, McAfee P, An H (2008) Motion preservation surgery of the spine; advanced techniques and controversies. Elsevier, Philadelphia, pp 397–465
59. Ahrens M et al (2009) Nucleus replacement with the DASCOR disc arthroplasty device: interim two-year efficacy and safety results from two prospective, non-randomized multicenter European studies. Spine (Phila Pa 1976) 34(13):1376–1384

60. Vernengo J et al (2008) Evaluation of novel injectable hydrogels for nucleus pulposus replacement. J Biomed Mater Res B Appl Biomater 84(1):64–69
61. Boyd LM, Carter AJ (2006) Injectable biomaterials and vertebral endplate treatment for repair and regeneration of the intervertebral disc. Eur Spine J 15(Suppl 3):S414–S421
62. Wardlaw D (2008) BioDisc nucleus pulposus replacement. In: Yue J, Bertagnoli R, McAfee P, An H (eds) Motion preservation surgery of the spine: advanced techniques and controversies. Elsevier, Philadelphia, pp 431–441
63. Bertagnoli R, Schonmayr R (2002) Surgical and clinical results with the PDN prosthetic disc-nucleus device. Eur Spine J 11(Suppl 2):S143–S148
64. Bertagnoli R et al (2005) Mechanical testing of a novel hydrogel nucleus replacement implant. Spine J 5(6):672–681
65. Joshi A et al (2006) Functional compressive mechanics of a PVA/PVP nucleus pulposus replacement. Biomaterials 27(2):176–184
66. Thomas J et al (2004) The effect of dehydration history on PVA/PVP hydrogels for nucleus pulposus replacement. J Biomed Mater Res B Appl Biomater 69(2):135–140
67. Thomas J, Lowman A, Marcolongo M (2003) Novel associated hydrogels for nucleus pulposus replacement. J Biomed Mater Res A 67(4):1329–1337
68. Bao QB, Bagga CS, Higham PA (1997) Swelling pressure of hydrogel: a perceived benefit for a spinal prosthetic nucleus. In: Proceedings 10th annual meeting of the International Intradiscal Therapy Society, Naples, FL
69. Boelen EJ et al (2005) Intrinsically radiopaque hydrogels for nucleus pulposus replacement. Biomaterials 26(33):6674–6683
70. Lau S, Lam K (2007) Lumbar stabilisation techniques. Curr Orthop 21:25–39
71. Laurencin CT et al (1999) Tissue engineering: orthopedic applications. Annu Rev Biomed Eng 1:19–46
72. Nerem RM, Sambanis A (1995) Tissue engineering: from biology to biological substitutes. Tissue Eng 1(1):3–13
73. Halloran DO et al (2008) An injectable cross-linked scaffold for nucleus pulposus regeneration. Biomaterials 29(4):438–447
74. Richardson SM et al (2008) Human mesenchymal stem cell differentiation to NP-like cells in chitosan-glycerophosphate hydrogels. Biomaterials 29(1):85–93
75. Chan S, Lam S, Leung V, Chan D, Luk K, Cheung K (2010) Minimizing cryopreservation-induced loss of disc cell activity for storage of whole intervertebral discs. Eur Cells Mater 19:273–283
76. Ganey T et al (2003) Disc chondrocyte transplantation in a canine model: a treatment for degenerated or damaged intervertebral disc. Spine (Phila Pa 1976) 28(23):2609–2620
77. Gorensek M et al (2004) Nucleus pulposus repair with cultured autologous elastic cartilage derived chondrocytes. Cell Mol Biol Lett 9(2):363–373
78. Bae H, Kanim M, Zhao L (2008) Human fetal chondrocyte transplants for damaged intervertebral disc. Spine J 8(5):928–938
79. Kaneyama S et al (2008) Fas ligand expression on human nucleus pulposus cells decreases with disc degeneration processes. J Orthop Sci 13(2):130–135
80. Takada T et al (2002) Fas ligand exists on intervertebral disc cells: a potential molecular mechanism for immune privilege of the disc. Spine (Phila Pa 1976) 27(14):1526–1530
81. Wolfe HJ, Putschar WG, Vickery AL (1965) Role of the notochord in human intervetebral disk. I. Fetus and infant. Clin Orthop Relat Res 39:205–212
82. Cappello R et al (2006) Notochordal cell produce and assemble extracellular matrix in a distinct manner, which may be responsible for the maintenance of healthy nucleus pulposus. Spine (Phila Pa 1976) 31(8):873–882, discussion 883
83. Erwin WM et al (2006) Nucleus pulposus notochord cells secrete connective tissue growth factor and up-regulate proteoglycan expression by intervertebral disc chondrocytes. Arthritis Rheum 54(12):3859–3867

84. Erwin WM et al (2009) The regenerative capacity of the notochordal cell: tissue constructs generated in vitro under hypoxic conditions. J Neurosurg Spine 10(6):513–521
85. Aguiar DJ, Johnson SL, Oegema TR (1999) Notochordal cells interact with nucleus pulposus cells: regulation of proteoglycan synthesis. Exp Cell Res 246(1):129–137
86. Korecki CL et al (2010) Notochordal cell conditioned medium stimulates mesenchymal stem cell differentiation toward a young nucleus pulposus phenotype. Stem Cell Res Ther 1(2):18
87. Mercuri J, Gill S, Simionescu A, Simionescu D (2010) Xenogenic cues for human mesenchymal stem cell differentiation towads a nucleus pulposus cell-like phenotype. In: Proceedings 25th Annual Meeting of the North American Spine Society. Spine J 10(9): S114–S115
88. Pittenger MF et al (1999) Multilineage potential of adult human mesenchymal stem cells. Science 284(5411):143–147
89. Aggarwal S, Pittenger MF (2005) Human mesenchymal stem cells modulate allogeneic immune cell responses. Blood 105(4):1815–1822
90. McIntosh KR et al (2009) Immunogenicity of allogeneic adipose-derived stem cells in a rat spinal fusion model. Tissue Eng Part A 15(9):2677–2686
91. Steck E et al (2005) Induction of intervertebral disc-like cells from adult mesenchymal stem cells. Stem Cells 23(3):403–411
92. Risbud M, Izzo M, Adams C et al. (2003) Mesenchymal stem cells respond to their microenvironment and in vitro to assume nucleus pulposus-like phenotype. Paper presented at 30th Annual Meeting of the International Society for the Study of the Lumbar Spine. Vancouver, Canada
93. Richardson SM et al (2006) Intervertebral disc cell-mediated mesenchymal stem cell differentiation. Stem Cells 24(3):707–716
94. Sobajima S et al (2008) Feasibility of a stem cell therapy for intervertebral disc degeneration. Spine J 8(6):888–896
95. Sakai D et al (2006) Regenerative effects of transplanting mesenchymal stem cells embedded in atelocollagen to the degenerated intervertebral disc. Biomaterials 27(3):335–345
96. Li X et al (2005) Modulation of chondrocytic properties of fat-derived mesenchymal cells in co-cultures with nucleus pulposus. Connect Tissue Res 46(2):75–82
97. Lu ZF et al (2007) Differentiation of adipose stem cells by nucleus pulposus cells: configuration effect. Biochem Biophys Res Commun 359(4):991–996
98. Kluba T et al (2005) Human anulus fibrosis and nucleus pulposus cells of the intervertebral disc: effect of degeneration and culture system on cell phenotype. Spine (Phila Pa 1976) 30 (24):2743–2748
99. Stokes DG et al (2001) Regulation of type-II collagen gene expression during human chondrocyte de-differentiation and recovery of chondrocyte-specific phenotype in culture involves Sry-type high-mobility-group box (SOX) transcription factors. Biochem J 360 (Pt 2):461–470
100. Tsai TT et al (2007) Fibroblast growth factor-2 maintains the differentiation potential of nucleus pulposus cells in vitro: implications for cell-based transplantation therapy. Spine (Phila Pa 1976) 32(5):495–502
101. Yang SH et al (2005) An in-vitro study on regeneration of human nucleus pulposus by using gelatin/chondroitin-6-sulfate/hyaluronan tri-copolymer scaffold. Artif Organs 29(10): 806–814
102. Yang SH et al (2005) Gelatin/chondroitin-6-sulfate copolymer scaffold for culturing human nucleus pulposus cells in vitro with production of extracellular matrix. J Biomed Mater Res B Appl Biomater 74(1):488–494
103. Li CQ et al (2009) Construction of collagen II/hyaluronate/chondroitin-6-sulfate tri-copolymer scaffold for nucleus pulposus tissue engineering and preliminary analysis of its physicochemical properties and biocompatibility. J Mater Sci Mater Med 21:741–751
104. Calderon L et al (2010) Type II collagen-hyaluronan hydrogel–a step towards a scaffold for intervertebral disc tissue engineering. Eur Cell Mater 20:134–148

105. Sakai D et al (2006) Atelocollagen for culture of human nucleus pulposus cells forming nucleus pulposus-like tissue in vitro: influence on the proliferation and proteoglycan production of HNPSV-1 cells. Biomaterials 27(3):346–353
106. Yu J (2002) Elastic tissues of the intervertebral disc. Biochem Soc Trans 30(Pt 6):848–852
107. Chuang TH et al (2009) Polyphenol-stabilized tubular elastin scaffolds for tissue engineered vascular grafts. Tissue Eng Part A 15(10):2837–2851
108. Tedder ME et al (2009) Stabilized collagen scaffolds for heart valve tissue engineering. Tissue Eng Part A 15(6):1257–1268
109. Addington C, Mercuri J, Gill S, Simionescu D (2011) Stabilized elastin-glycosaminoglycan shape-memory sponge scaffold for nucleus pulposus tissue engineering. In: Transactions 2011 Orthopaedic Research Society Annual Meeting. Long Beach, CA
110. Daamen WF et al (2003) Preparation and evaluation of molecularly-defined collagen-elastin-glycosaminoglycan scaffolds for tissue engineering. Biomaterials 24(22):4001–4009
111. Badylak SF (2007) The extracellular matrix as a biologic scaffold material. Biomaterials 28 (25):3587–3593
112. Wu Z et al (2009) The use of phospholipase A(2) to prepare acellular porcine corneal stroma as a tissue engineering scaffold. Biomaterials 30(21):3513–3522
113. Mercuri J, Gill S, Simionescu D (2011) Novel tissue derived biomimetic scaffold for regenerating the human nucleus pulposus. J Biomed Mater Res A 96(2):422–435
114. Cloyd JM et al (2007) Material properties in unconfined compression of human nucleus pulposus, injectable hyaluronic acid-based hydrogels and tissue engineering scaffolds. Eur Spine J 16(11):1892–1898
115. Seguin CA et al (2004) Tissue engineered nucleus pulposus tissue formed on a porous calcium polyphosphate substrate. Spine (Phila Pa 1976) 29(12):1299–1306
116. Le Visage C et al (2006) Small intestinal submucosa as a potential bioscaffold for intervertebral disc regeneration. Spine 31(21):2423–2430, discussion 2431
117. Baer AE et al (2001) Collagen gene expression and mechanical properties of intervertebral disc cell-alginate cultures. J Orthop Res 19(1):2–10
118. Chou AI, Nicoll SB (2009) Characterization of photocrosslinked alginate hydrogels for nucleus pulposus cell encapsulation. J Biomed Mater Res A 91(1):187–194
119. Chou AI, Akintoye SO, Nicoll SB (2009) Photo-crosslinked alginate hydrogels support enhanced matrix accumulation by nucleus pulposus cells in vivo. Osteoarthr Cartil 17(10): 1377–1384
120. Leone G et al (2008) Amidic alginate hydrogel for nucleus pulposus replacement. J Biomed Mater Res A 84(2):391–401
121. Gaetani P et al (2008) Adipose-derived stem cell therapy for intervertebral disc regeneration: an in vitro reconstructed tissue in alginate capsules. Tissue Eng Part A 14(8):1415–1423
122. Roughley P et al (2006) The potential of chitosan-based gels containing intervertebral disc cells for nucleus pulposus supplementation. Biomaterials 27(3):388–396
123. Reza AT, Nicoll SB (2009) Characterization of novel photocrosslinked carboxymethylcellulose hydrogels for encapsulation of nucleus pulposus cells. Acta Biomater 6(1):179–186
124. Yang SH et al (2008) Three-dimensional culture of human nucleus pulposus cells in fibrin clot: comparisons on cellular proliferation and matrix synthesis with cells in alginate. Artif Organs 32(1):70–73
125. Hamilton DJ et al (2006) Formation of a nucleus pulposus-cartilage endplate construct in vitro. Biomaterials 27(3):397–405
126. Brown RQ, Mount A, Burg KJ (2005) Evaluation of polymer scaffolds to be used in a composite injectable system for intervertebral disc tissue engineering. J Biomed Mater Res A 74(1):32–39
127. Abbushi A et al (2008) Regeneration of intervertebral disc tissue by resorbable cell-free polyglycolic acid-based implants in a rabbit model of disc degeneration. Spine 33(14): 1527–1532

128. Revell PA et al (2007) Tissue engineered intervertebral disc repair in the pig using injectable polymers. J Mater Sci Mater Med 18(2):303–308
129. Nomura T et al (2001) Nucleus pulposus allograft retards intervertebral disc degeneration. Clin Orthop Relat Res 389:94–101
130. Alini M et al (2008) Are animal models useful for studying human disc disorders/degeneration? Eur Spine J 17(1):2–19
131. O'Connell GD, Vresilovic EJ, Elliott DM (2007) Comparison of animals used in disc research to human lumbar disc geometry. Spine (Phila Pa 1976) 32(3):328–333
132. Gillett NA et al (1988) Age-related changes in the beagle spine. Acta Orthop Scand 59(5): 503–507
133. Nuckley DJ et al (2008) Intervertebral disc degeneration in a naturally occurring primate model: radiographic and biomechanical evidence. J Orthop Res 26(9):1283–1288
134. Ziv I et al (1992) Physicochemical properties of the aging and diabetic sand rat intervertebral disc. J Orthop Res 10(2):205–210
135. Masuda K et al (2005) A novel rabbit model of mild, reproducible disc degeneration by an anulus needle puncture: correlation between the degree of disc injury and radiological and histological appearances of disc degeneration. Spine 30(1):5–14
136. Sobajima S et al (2005) A slowly progressive and reproducible animal model of intervertebral disc degeneration characterized by MRI, X-ray, and histology. Spine 30(1):15–24
137. Singh K, Masuda K, An H (2008) Animal models for human disc degeneration. In: Yue J, Bertagnoli R, McAfee P, An H (eds) Motion preservation surgery of the spine: advanced techniques and controversies. Elsevier, Philadelphia, pp 639–648
138. Ulrich JA et al (2007) ISSLS prize winner: repeated disc injury causes persistent inflammation. Spine 32(25):2812–2819
139. Iatridis JC et al (1999) Compression-induced changes in intervertebral disc properties in a rat tail model. Spine (Phila Pa 1976) 24(10):996–1002
140. Kroeber MW et al (2002) New in vivo animal model to create intervertebral disc degeneration and to investigate the effects of therapeutic strategies to stimulate disc regeneration. Spine (Phila Pa 1976) 27(23):2684–2690
141. Kim KS et al (2005) Disc degeneration in the rabbit: a biochemical and radiological comparison between four disc injury models. Spine (Phila Pa 1976) 30(1):33–37
142. Hoogendoorn RJ et al (2007) Experimental intervertebral disc degeneration induced by chondroitinase ABC in the goat. Spine (Phila Pa 1976) 32(17):1816–1825
143. Zhou H et al (2007) A new in vivo animal model to create intervertebral disc degeneration characterized by MRI, radiography, CT/discogram, biochemistry, and histology. Spine (Phila Pa 1976) 32(8):864–872

Adv Polym Sci (2012) 247: 233–276
DOI: 10.1007/12_2011_155

Published online: 28 October 2011

Functionalized Biocompatible Nanoparticles for Site-Specific Imaging and Therapeutics

Ranu K. Dutta, Prashant K. Sharma, Hisatoshi Kobayashi, and Avinash C. Pandey

Abstract The applicability of nanoparticles is determined by their unique size-dependent properties, such as their optical and magnetic properties, which make them very attractive candidates for numerous biomedical applications such as drug delivery nanosystems, diagnostic biosensors, and imaging nanoprobes for magnetic resonance imaging contrast agents. Surface chemistry defines the functional properties and biological reactivity of these nanocrystals. Targeted delivery of therapeutics has the potential to localize therapeutic agents to a specific tissue as a mechanism to enhance treatment efficacy and mitigate side effects. Moieties that combine imaging and therapeutic modalities in a single macromolecular construct may confer advantages in the development and applications of nanomedicine. Here, an insight into the development of various kinds of functionalized biocompatible nanoparticles for site-specific imaging and therapeutics is discussed in detail.

Keywords Biocompatible · Cancer diagnostics · Functionalization · Imaging · Nanoparticles · Surface modification · Targeted drug delivery

Contents

R.K. Dutta (✉), P.K. Sharma, and A.C. Pandey (✉)
Nanotechnology Application Centre, University of Allahabad, Allahabad 211002, India
e-mail: ranu.dutta16@gmail.com; prof.avinashcpandey@gmail.com

H. Kobayashi
Biomaterials Research Group, National Institute of Material Science, Tsukuba, Japan

1 Introduction

In the past, use of nanoparticles has been made without knowledge of their existence, and we now find that the particles used were in the nanoregime. Nanoparticles were used by artisans as far back as the ninth century in Mesopotamia for generating a glittering effect on the surface of pots. These nanoparticles were created by the artisans by adding copper and silver salts and oxides, together with vinegar, ochre and clay, onto the surface of previously glazed pottery.

The last few decades have seen the emergence of nanomaterials for several applications in almost all fields of life, ranging from solid state lighting to biomedical applications [1, 2]. Their properties lie between those of the bulk material and those of atoms as they are only made up of some atoms arranged in an ordered fashion. The small size of these nanoparticles gives them significantly different properties, which in turn result in the widespread applications of these nanomaterials. Various nanoscale materials, such as nanorods, nanowires [3], nanotubes, and nanofibers [4], have been explored in many biomedical applications [5] because of their novel properties, such as the high surface-to-volume ratio, surface tailorability, and multifunctionality. Transmission electron microscopic (TEM) images of plasmonic gold nanostructures such as nanospheres, nanorods, and nanoshells are shown in Fig. 1.

For biomedical applications, nanomaterials have been widely used in the field of tissue engineering, for diagnosis and treatment of certain diseases such as cancer, and for other biomedical applications include targeted drug delivery and imaging, hyperthermia, magneto-transfections, gene therapy, stem cell tracking, molecular and cellular tracking, magnetic separation technologies (e.g., rapid DNA sequencing), and detection of liver and lymph node metastases. The most recent applications of superparamagnetic iron oxide nanoparticles (SPIONs) are in early detection of cancer, atherosclerosis, and diabetes.

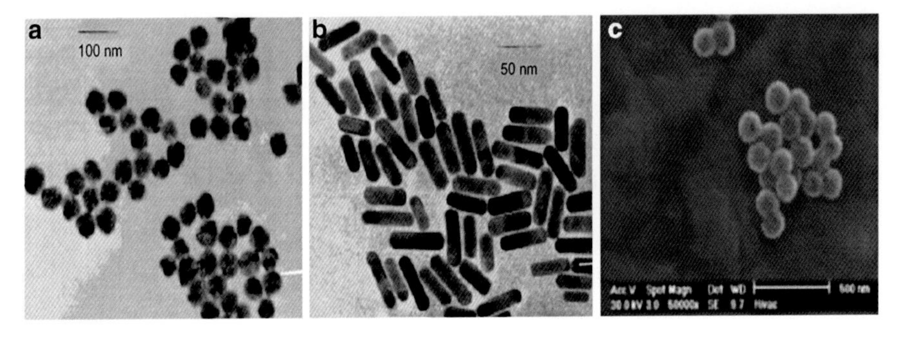

Fig. 1 TEM images of plasmonic gold nanostructures: (**a**) nanospheres, (**b**) nanorods, and (**c**) nanoshells. (Figure adapted from [6])

> The objective of nanotechnology is to detect every cancer cell instead of full grown cancers (Ranu K. Dutta).

Nanoparticles play a very important role in cancer research. Due to their extremely small size, nanoparticles are easily and more readily taken up by the human body. Hence, they can interact well with biological membranes so that cells, tissues, and organs are eligible for entrance of nanoparticles [7]. Nanoparticles are used in the field of medicine for several reasons: (1) improved delivery of poorly water-soluble drugs, which enhances their faster dissolution in the blood stream; (2) transcytosis of drugs across tight epithelial and endothelial barriers; (3) delivery of large macromolecule drugs to intracellular sites of action; (4) co-delivery of two or more drugs or therapeutic modality for combination therapy; (5) visualization of sites of drug delivery by combining therapeutic agents with imaging modalities; and (6) Conjugating nanoparticles with site specific ligands can be useful in site specific imaging, image guided therapy, for specific MRI contrast agents, ablation, hyperthermia, etc.

2 Properties of Nanoparticles

The size of nanoparticles lies between that of bulk material and the atomic structures, which gives them unique properties and makes them eligible candidates for several applications. At the nanoscale, the percentage of atoms present at the surface of a material becomes significantly enhanced, resulting in drastically changed properties of these nanomaterials. For bulk materials larger than 1 μm (1 micron), the percentage of atoms at the surface is insignificant in relation to the number of atoms present in their nanoform. At the small size range (<10 nm) the concept of quantum confinement originates. The wave function associated with these materials will be confined and degenerate, resulting in strong quantum confinement. Several varieties of nanoparticles are available, such as polymeric nanoparticles, dendrimers [8–10], metal nanoparticles, quantum dots (QDs), liposomes, micelles, and other types of nano-assemblies [11–13]. Nanoparticles have a large surface area-to-volume ratio that helps in diffusion process. Use of nanoparticles can also lead to special properties such as increased heat and chemical resistance.

For biomedical applications, the size, charge, and surface chemistry of the nanoparticles are important since they strongly affect both their circulation time and also their bioavailability within the body. In addition, magnetic properties and internalization of particles depend strongly on the size of these magnetic particles. For example, following systemic administration, larger particles with diameters greater than 200 nm are usually sequestered by the spleen as a result of mechanical filtration and are eventually removed by the cells of the phagocyte system, resulting in decreased blood circulation time. On the other hand, smaller particles with diameters of less than 8 nm are rapidly removed through extravasation and renal clearance. Particles of 10–100 nm are optimal for subcutaneous injection and demonstrate the most prolonged blood circulation times. The particles in this size range are small enough both to evade the reticulo-endothelial system (RES) of the body and to penetrate the very small capillaries within the body tissues and, therefore, may offer the most effective distribution in certain tissues. A nanoparticle has emerged as a promising candidate for the efficient delivery of drugs used in the treatment of cancer by avoiding the RES, utilizing the enhanced permeability and retention effect and tumor-specific targeting.

Nanoparticles such as those of the heavy metals, like cadmium selenide, cadmium sulfide, lead sulfide, and cadmium telluride are potentially toxic [14, 15]. The possible mechanisms by which nanoparticles cause toxicity inside cells are schematically shown in Fig. 2. They need to be coated or capped with low toxicity or nontoxic organic molecules or polymers (e.g., PEG) or with inorganic layers (e.g., ZnS and silica) for most of the biomedical applications. In fact, many biomedical imaging and detection applications of QDs encapsulated by complex molecules do not exhibit noticeable toxic effects [16]. One report shows that the tumor cells labeled with QDs survived in circulation and extravasated into tissues

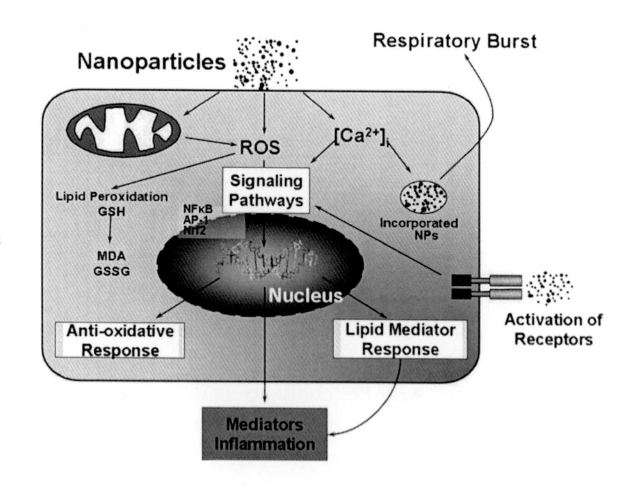

Fig. 2 Possible mechanisms by which nanoparticles cause toxicity inside cells. *GSH* glutathione, *GSSG* glutathione disulfide, *MDA* malondialdehyde, *NFκB* nuclear factor kappa B, *Nrf2* nuclear factor-erythroid 2-related factor 2, *ROS* reactive oxygen species

just as effectively as unlabeled cells; there was no obvious difference in their ability to form tumors in mice after 40 days and QDs had no adverse effects on the physiology of the host animal or labeled cells [17].

Quantum dots are nanoparticulate clusters of semiconductor material (smaller than the Bohr exciton radius) that show quantum confinement effects. The quantum confinement effect means that the optical properties of these nanoparticles are controlled by their size, rather than their composition, which makes them useful optical imaging agents. The size of the band gap of these materials dictates the energy of the photon emitted and, according to Plancks equation (where energy is inversely proportional to the wavelength), also the wavelength of emitted light. QDs have been the focus of great interest recently because of their biological imaging capabilities [18], via their bright fluorescence, photostability, and their narrow and size-tunable emission spectrum [19].

3 Surface Modification

Nanoparticle surface modification is of tremendous importance to prevent nanoparticle aggregation prior to injection, decrease the toxicity, and increase the solubility and the biocompatibility in a living system [20]. Imaging studies in mice clearly show that QD surface coatings alter the disposition and pharmacokinetic properties of the nanoparticles. The key factors in surface modifications include the use of proper solvents and chemicals or biomolecules used for the attachment of the drug, targeting ligands, proteins, peptides, nucleic acids etc. for their site-specific biomedical applications. The functionalized or capped nanoparticles should be preferably dispersible in aqueous media.

Surface modification is necessary in nanoparticles for various reasons: (1) to make them biocompatible and non-immunogenic for biomedical applications, (2) to make them dispersible in aqueous media for most biomedical applications, (3) to stabilize the nanoparticles in water for long period, (4) to prevent agglomeration of nanoparticles by use of some capping agents and surfactant molecules, (5) to render specificity towards their target cell or tissue, and (6) to render sterically accessible functional groups for bioconjugation etc.

There are some important points to be remembered during the choice of materials for surface modifications: (1) Most in vivo biomedical applications need the particles to be well dispersed and stable in water. (2) Most of the synthesis methods that produce highly monodisperse, homogeneous nanoparticles use organic solvents. (3) Capping agent, surfactant, and the surface moieties to be attached on the surface of nanoparticles should be chosen carefully. (4) Biocompatible capping agents should be used that do not show any adverse effects like platelet aggregation upon administration into the blood or thrombosis, stenosis etc. Some studies show that carbon nanotubes can aggregate blood platelets [21]. (5) Capping agents should overcome the RES uptake by macrophages and other cells. Conventional surface non-modified nanoparticles are usually caught in the circulation by the RES (e.g., in the liver and the spleen), depending on their size and surface characteristics. (6) The chemicals used should be of high purity.

4 Surface Modifying Agents

4.1 Capping and Passivating Agents

Capping agents provide biocompatibility besides controlling the particle size. Passivation is achieved mostly by capping the particles with organic materials, but inorganic materials are also used. Organic ligands such as thiopyridines and thiolates have been reported to minimize surface defects and increase luminescence efficiencies. Several capping agents such as silica [22], starch, biotin, citric acid [23], polyvinyl pyrollidine, polyvinyl alcohol, oleic acid, and cytosine [24] have been regularly used for diverse biomedical applications. In view of the diverse applications of these nanoparticles in biology and medicine, hundreds of capping strategies have been studied. Many of these typically involve some kind of polymer such as poly(ethylene glycol) (PEG) [25–29], poly(lactic-*co*-glycolic acid) (PLGA), or a polysaccharide [30–32]; natural macromolecules such as oligonucleotides [33], peptides, and proteins [34–39]; poly(vinyl alchohol), poly(vinyl pyrolline), silica etc. Sharma et al. [40] have shown the surface modification of nanoparticles by embedding ZnO nanoparticles in SiO_2 matrix and measuring the luminescence. The luminescence properties and structure of these ZnO nanoparticles embedded in SiO_2 matrix are shown in Fig. 3.

Sharma et al. [40] have used tetra-ethyl-ortho-silicate (TEOS) for the formation of SiO_2 matrix. The chemical structure of TEOS can be represented as shown in

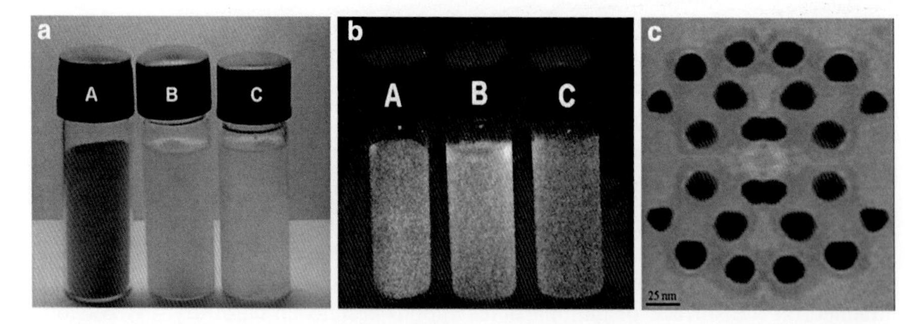

Fig. 3 Prepared ZnO QDs embedded in SiO_2 matrix under (**a**) ordinary and (**b**) UV (365 nm excitation) lamps. (*A*) Solid samples, (*B*) samples dissolved in ethanol, and (*C*) samples dissolved in water. Photographs were taken with a digital camera just after completion of synthesis. (**c**) TEM Image of symmetrically dispersed ZnO QDs embedded in SiO_2 matrix. (Figure adapted from [40])

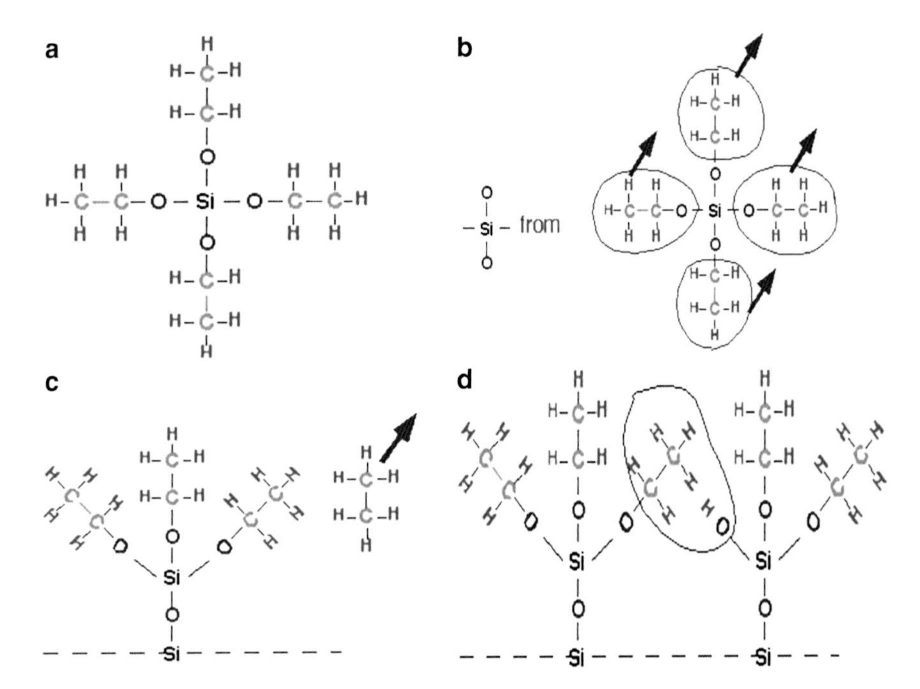

Fig. 4 (**a**) Chemical structure of tetra-ethyl-ortho-silicate (TEOS). (**b**) Removal of two oxygen atoms from TEOS. (**c**) Mechanism depicting how TEOS chemisorbs onto silanol groups (Si-OH) on surface. (**d**) Formation of Si–O–Si bridges by elimination reactions with neighboring molecules. (Figure adapted from [40])

Fig. 4a. TEOS is liquid at room temperature and slowly hydrolyzes into silicon dioxide and ethanol when in contact with ambient moisture. In TEOS, the silicon atom is already oxidized; the conversion of TEOS to SiO_2 is essentially a rearrangement rather than an oxidation reaction. The overall reaction for the SiO_2 matrix requires the removal of two oxygen atoms from TEOS as shown in Fig. 4b.

In this case, formation of SiO_2 matrix was probably the result of TEOS surface reaction. TEOS chemisorbs onto silanol groups (Si-OH) at the surface, as well as at strained surface bonds. The mechanism is depicted in Fig. 4c. TEOS will not adsorb onto the resulting alkyl-covered surface during intermediate reaction steps, so SiO_2 matrix formation was probably limited by removal of these intermediate surface alkyl groups. As shown in Fig. 4d, these groups can undergo elimination reactions with neighboring molecules to form Si–O–Si bridges. This process quickly occurs in the ambient reaction conditions: TEOS can be its own oxygen source, and SiO_2 can be deposited in the form of a matrix from TEOS. However, additional oxygen atoms provided by ZnO increases the matrix formation rate, presumably this is the cause behind the formation of symmetrically dispersed ZnO QDs embedded in the SiO_2 matrix.

4.2 *Role of Surfactants*

The most challenging issue is the synthesis of well-dispersed and stable colloidal synthesis of nanoparticles for biomedical applications. The fundamental problems in use of the nanoparticles is the lack of stability of their dispersions and the generation of spacious aggregates in the dry state, which leads to loss of their special nanoscale properties and also making them unsuitable for biomedical applications. In aqueous media, the nanosized particles are primarily separated by the ionic repulsion forces produced due to adsorption on their surface [41, 42]. Nanoparticles capped by surfactants stay well dispersed in solution for a longer time. The surfactants act as microreactors for inorganic reactions and also act as steric stabilizers to inhibit aggregation of nanoparticles. Surfactants modify the nanoparticle's shape, size, and other surface properties differently depending on their molecular structure, i.e., nature of the head group, length of hydrophobic tail, and type of counterions.

Nanocrystals such as CdS, CdSe, CdTe, or CdSe/ZnS, ZnS, ZnO etc. have been synthesized in organic solvents at high temperatures in the presence of surfactants to yield monodisperse and stable particles [43]. Surfactant molecules coat the surface of nanoparticles, as shown in Fig. 5. The polar surfactant head group is attached to the inorganic particle surface, while the hydrophobic chains protrude into the organic solvent, mediating colloidal stability. Then, to make the particle water soluble or dispersible, the surfactant layer is replaced or a coating is made, with an additional layer introducing either electric charge or hydrophilic polymers for mediating solubility in water. Coulomb repulsion between nanocrystals with surface charge of the same polarity prevents aggregation in water. Hence, synthesis

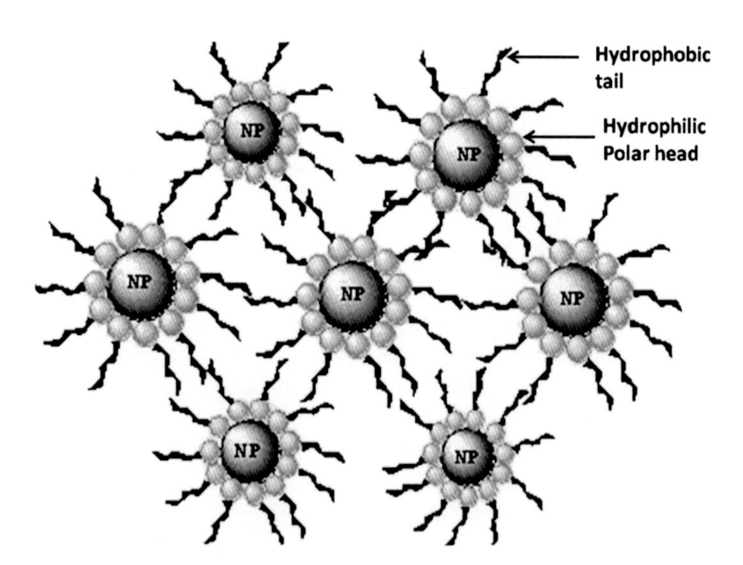

Fig. 5 Surfactant molecules on the surface of nanoparticles

of monodisperse and colloidal suspension in aqueous media can be accomplished. However the exact mechanism that governs formation and stabilization is still not very well understood.

5 Synthesis of Colloidal Nanoparticles

Colloidal nanoparticles are a class of nanomaterials synthesized by wet chemical methods. The reaction chamber is a reactor containing a mixture of liquids that control the nucleation and the growth. The precursors are introduced according to the desired atomic species for the growth of the nanocrystals. The precursors decompose, forming new reactive species or ions that are needed for the nucleation and growth of the nanocrystals. The key factor in the colloidal synthesis of nanocrystals is the surfactant used. A surfactant is dynamically adsorbed to the surface of the growing QDs under the reaction conditions. The surfactant must be mobile enough to provide access for the addition of the reaction precursor atomic or ionic species. It should prevent aggregation of the nanocrystals. A surfactant that binds very strongly to the surface of the QD would not allow the nanocrystal to grow. A weakly co-ordinating molecule would yield large particles or aggregates. Some suitable surfactant molecules include alkyl thiols, phosphines, phosphine oxides, phosphates, phosphonates, amides, amines, carboxylic acids etc. The surfactant molecule should be stable if the reaction is carried out at higher temperatures. By controlling the mixture of surfactant molecules that are present during the generation and nucleation of QDs, control of their size and shape becomes possible. Colloidal nanocrystals are dispersed in solution; hence, they can be functionalized easily with molecules such as proteins and oligonucleotides.

6 Different Types of Nanoparticles

6.1 Functionalized Nanoparticles

6.1.1 Lipid-Functionalized Nanoparticles

Nanoparticles, such as iron oxide particles and QDs, are mostly synthesized in nonpolar organic solvents and capped with a surfactant. If they are to be solubilized in aqueous buffers, their hydrophobic surface components must be replaced by amphiphilic ones. An alternative strategy was developed by Dubertret et al. for TOPO-coated QDs [44]. The hydrophobic particles were dissolved in chloroform together with PEGylated phospholipids. After evaporating the solvent and hydrating the mixed film of QDs and lipids, QD-containing micelles were formed. The same method can be used for encapsulating hydrophobic iron oxide particles in

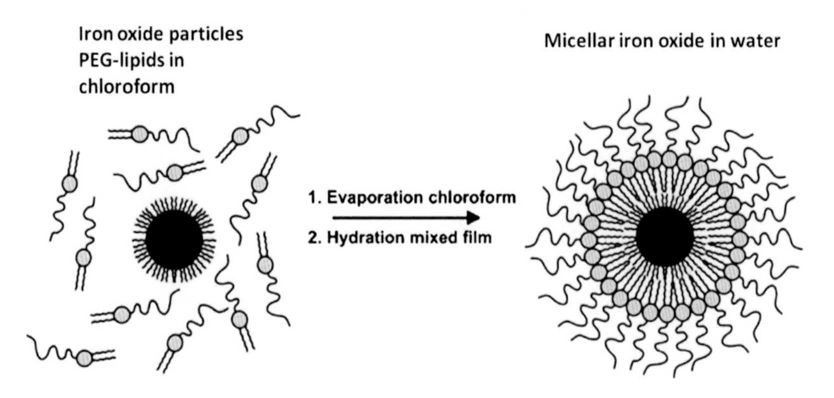

Fig. 6 Encapsulation of hydrophobic nanoparticles in micelles

micelles. With this elegantly simple procedure, one obtains a water-soluble particle of small size, which can easily be functionalized by just mixing the appropriate lipids. A representation of the encapsulating procedure for hydrophobic nanoparticles in micelles is shown in Fig. 6.

Lipids are mixed with the nanoparticles in an apolar solvent. The mixed film obtained is hydrated. Thereafter, the nanoparticle-containing micelles and empty micelles are separated by centrifugation.

6.1.2 Peptide-Conjugated Nanoparticles

Several peptide-conjugated nanoparticles have been synthesized by researchers for targeted delivery into the cell and cellular organelles. Inspired by viruses, Tkachenko and coworkers conjugated peptides to bovine serum albumin (BSA) via an ester linker, and then conjugated the BSA to gold nanoparticles [45]. The four peptides they used were from viral cell entry or targeting proteins, and they were able to achieve targeted entry of the gold nanoparticles into the nucleus of HepG2 cells. Furthermore, it should be noted that the cells were still viable after entry of the gold nanoparticles. Nitin et al. developed a similar approach for solubilizing iron oxide [46]. They conjugated TAT peptides and a fluorescent label to the distal end of the PEG chains of the phospholipids to coat the iron oxide particles. The TAT peptide has been shown to deliver nanoparticles into cells, making it attractive for intracellular labeling. The uptake of this conjugate was demonstrated in vitro with both MRI and fluorescence microscopy. Another study showed the intracellular delivery of peptide-conjugated QDs. The authors selected an insect neuropeptide, allatostatin (AST1; APSGAQRLYG FGL-NH_2), conjugated it to CdSe−ZnS QDs, and investigated the intracellular delivery of the conjugate in living cells such as human epidermoid ovarian carcinoma cells (A431) and mouse embryonic fibroblast cells (3T3) [47], as shown in Fig. 7. Figure 7a shows a fluorescence image of A431 cells incubated with a 5 μM

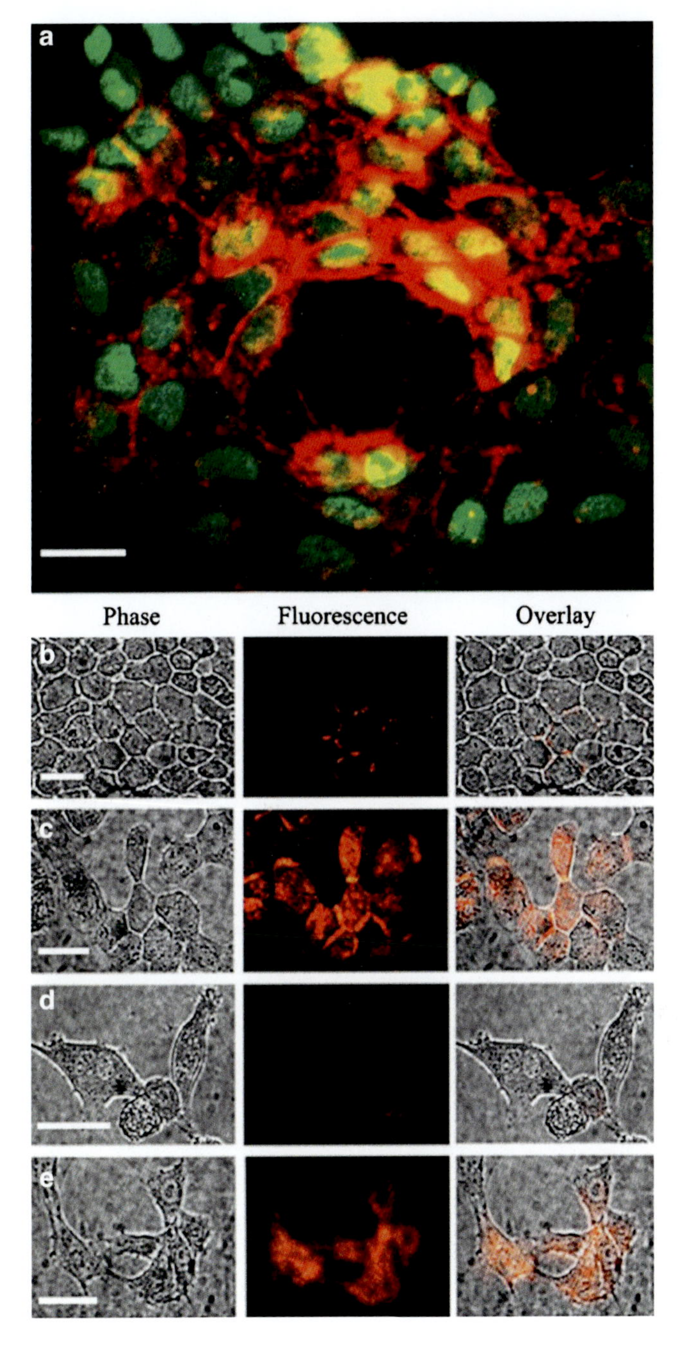

Fig. 7 (**a**) Fluorescence image of A431 cells incubated with a 5 μM Syto16 dye solution for 10 min followed by a 1 nM QD605-AST1 solution for 30 min. Cell nucleus is preferentially stained *green* by Syto16 due to its cell permeability and DNA intercalation. The *yellow-orange*

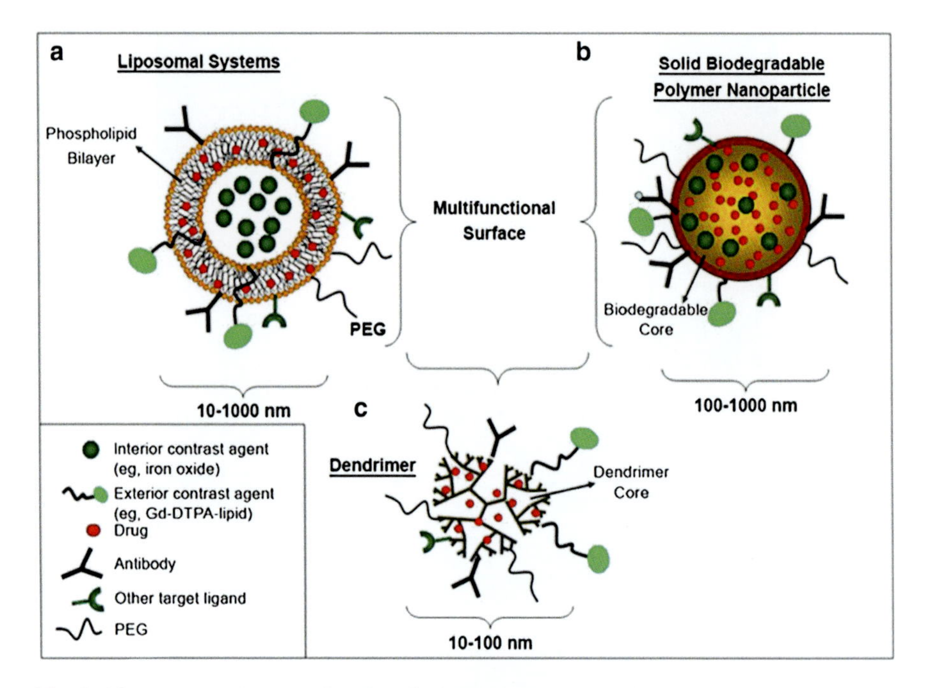

Fig. 8 Nanosystems that may function as simultaneous drug delivery and imaging agents for targeting T cells: (**a**) liposomal systems, (**b**) solid biodegradable nanoparticulates, and (**c**) macromolecular dendrimer complexes. *PEG* polyethylene glycol, *Gd-DTPA* gadolininum-diethylene triamine penta acetic acid. (Adapted from [48])

Syto16 dye solution for 10 min followed by a 1 nM of QD605-AST1 solution for 30 min. In Fig. 7a we see that the cell nucleus is preferentially stained green by Syto16 due to its cell permeability and DNA intercalation. The yellow-orange color indicates colocalized CdSe–ZnS QD-AST1 and Syto16). Phase, fluorescence, and overlay images of A431 and 3T3 cells were labeled using solutions of 0.5 nM QD-streptavidin) and 0.5 nM CdSe–ZnS QD-AST1 (Fig. 7b–e) Some different strategies for nanosystems are illustrated in Fig. 8. Figure 8a shows liposomal systems, Fig. 8b solid biodegradable nanoparticles, and Fig. 8c macromolecular dendrimer complexes.

6.1.3 Aptamer-Conjugated Nanoparticles

Recently, nanoparticle–aptamer bioconjugates, as shown in Fig. 9, have been developed and demonstrated for targeted delivery to cancer cells. The imaging

---

Fig. 7 (continued) color indicates colocalized QD-AST1 and Syto16. (**b–e**) Phase, fluorescence, and overlay images of A431 (**b**, **c**) and 3T3 (**d**, **e**) cells labeled using solutions of 0.5 nM QD-streptavidin (**b**, **d**) and 0.5 nM QD-AST1 (**c**, **e**). *Scale bars*: 25 μm. (Adapted from [47])

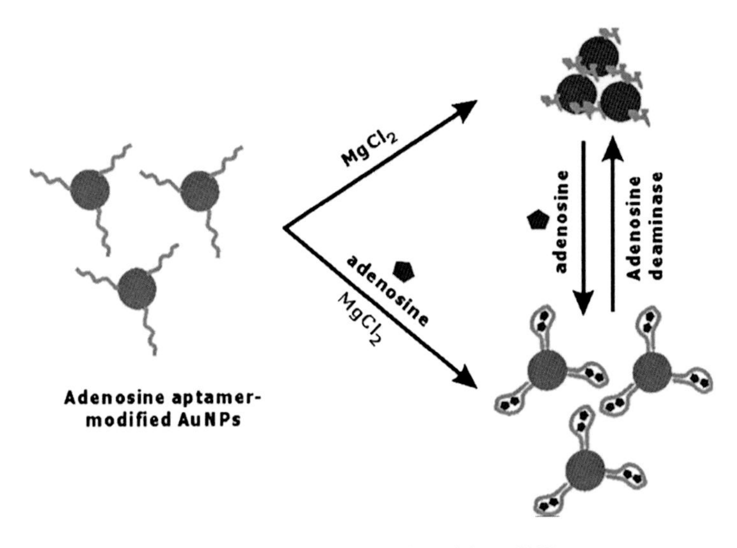

Fig. 9 Aptamers conjugated gold nanoparticles. (Adapted from [49])

and/or monitoring modalities can be based on traditional fluorophores, or QDs and SPIONs. Aptamers are DNA or RNA oligonucleotides that fold by intramolecular interaction into unique three-dimensional conformations capable of binding to target antigens with high affinity and specificity. Aptamers are quickly emerging as a new powerful class of ligands that rival antibodies in their potential for diagnostic and therapeutic application.

Medley et al. have shown the use of aptamer-conjugated nanoparticles for cancer cell detection [50]. Herr et al. [51] have shown the rapid collection and detection of leukemia cells using a novel two-nanoparticle assay with aptamers as the molecular recognition element.

6.1.4 Polymer-Modified Nanoparticles

Biocompatible or biodegradable polymers have been extensively explored in recent years because of their potential biomedical applications, e.g., poly(lactic acid) (PLA), poly(glycolic acid) (PGA), poly(lactic-*co*-glycolic acid) (PLGA), PEG, alginate, etc. A multifunctional biodegradable PLGA nanoparticle attached to moieties such as T-cell antibodies and contrast agents for MRI is shown in Fig. 10.

PEG modification of nanoparticles increases circulation time by evading macrophage-mediated uptake and removal from the systemic circulation. Non-PEGylated nanoparticles are quickly eliminated from the bloodstream because of the adsorption of blood proteins (opsonins) onto their surface, which triggers the recognition of the mononuclear phagocyte system (MPS) by the macrophages. Plasma half-life ($t_{1/2}$) was less than 12 min for amphiphilic poly(acrylic) short chain (750 Da) methoxy-PEG or long chain (3,400 Da) carboxy-PEG QDs. But, plasma $t_{1/2}$ was

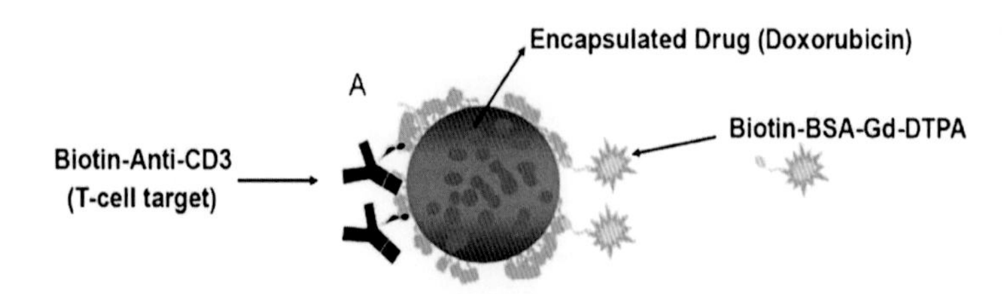

Fig. 10 Multifunctional solid biodegradable PLGA nanoparticles attached to several moieties such as T-cell antibodies, magnetic resonance contrast agent (biotin-BSA-Gd-DTPA) and encapsulated immunosuppressive drug (doxorubicin). (Adapted from [48])

1 h for long-chain (5,000 Da) methoxy-PEG QDs. These coatings determined the pattern of in vivo tissue localization, with retention of some QDs occurring up to 4 months.

PEGylation of nanospheres can be carried out either via a simple adsorption of PEG chains onto the nanoparticles or by a covalent linkage of PEG chains with poly(alkyl cyanoacrylate) (PACA) polymers. The adsorption approach is not preferable since there is no covalent linkage and is not really suitable enough for in vivo applications. It has been demonstrated that these kinds of assemblies (PACA nanoparticles on which poloxamer 388 or poloxamine 908 was adsorbed) are not stable during in vivo administration, resulting in a loss of coating, and have no significant influence on the biodistribution pattern [52]. Biodegradable nanoparticles show an increased residence time in blood vessels that could overcome the post-treatment accumulation of the free drug. In addition, sterilization of the polymeric nanoparticles by membrane filtration is feasible, which offers a great advantage.

Some advantages of PEG surface modification are: (1) water solubility, (2) high mobility in solution, (3) lack of toxicity and immunogenicity, (4) ready clearance from the body, and (5) altered distribution in the body.

PEG polymer has to be suitably functionalized at one or both terminals. PEGs that are activated at each terminus with the same reactive moiety are known as "homobifunctional." If the functional groups present are different, then the PEG derivative is referred as "heterobifunctional" or "heterofunctional." Chemically activated PEG polymer derivatives are used to attach the PEG to the desired molecule or ligand.

Two ways of linking the PEG molecules are: (1) PEG derivatives are obtained by reacting the PEG polymer with a group that is reactive with hydroxyl groups, e.g., anhydride, acid chloride, chloroformate or carbonate; and (2) attachment of functional groups such as aldehydes, esters, amides etc. with the PEG polymer.

The heterobifunctional PEGs are very useful in linking two entities in cases where a hydrophilic, flexible, and biocompatible spacer is needed. Preferred end groups for heterobifunctional PEGs are maleimides, vinyl sulfones, pyridyl disulfides, amines, carboxylic acids, and *N*-hydroxysuccinimide (NHS) esters.

6.2 *Functionalized Carbon Nanotubes*

Several drug delivery approaches have been carried out using carbon nanotubes (CNTs) [53–55]. The biocompatibility of CNTs is a significant issue in many of their proposed bio-applications. For the solubilization of CNTs, the attachment of relatively large functional groups to the nanotubes is required because CNTs are insoluble in water. The first report of CNT functionalization was reported by Haddon and coworkers on the amidation of nanotube-bound carboxylic acids with long-chain alkylamines, such as octadecylamine [56]. A variety of oligomeric and polymeric compounds have been used in the functionalization of CNTs for their solubility in common organic solvents and/or water.

Convenient and effective delivery of functionalized CNTs to a targeted site faces several limitations. Nanocapsules have been proposed as drug carriers that could be used to realize the "magic bullet" concept proposed by Nobel Prize winner Paul Ehrlich (1854–1915). Paul Ehlrich [57] at the beginning of the twentieth century, refers to a drug capable of targeting a particular site and releasing its contents when desired. In order to accomplish this, novel polymeric membrane microcapsule CNT devices have been developed for targeted delivery of therapeutics. CNTs were encapsulated in an alginate–poly(L-lysine)–alginate (APA) membrane to form a polymeric membrane targeted drug delivery device. The nanotubes were embedded in the core or attached to the surface of different types of alginate capsules. The device could be functionalized with targeting ligands for specific diseases or sites of the system, along with the therapeutic moiety to improve the clinical efficiency. The polymeric membrane protects the functionalized CNTs carrying the therapeutics from the harsh external environments encountered during drug delivery. This novel approach facilitates delivery of CNTs and their cargo safely and effectively to target sites.

Folic acid-conjugated CNTs have been investigated for site-specific delivery of drugs. Figure 11 shows an example of folate receptor-mediated targeting of nanoparticles on cancer cell lines. Folate has been used as a targeting ligand because of the presence of overexpressed folate receptors on many cancer cells [58, 59]. The ability of single-walled carbon nanotubes (SWNTs) attached to a therapeutic agent (structure **1** in Fig. 10) to destroy cancer cells was studied by using the MTT assay. Folate receptor-mediated targeting studies were carried out on folate receptor-positive [FR(+)] human choriocarcinoma (JAR) and human nasopharyngeal carcinoma (KB) cell lines. The uptake mechanism of folate-mediated delivery of folate-conjugated CNTs is illustrated in Fig. 11. The study on CNTs as a vector for drug delivery into living cells was carried on HeLa cells [60], as shown in Fig. 12. HeLa cells were incubated with AlexaFluor594-labeled SWNTs for 12 h at 37 °C, and living cells were observed under confocal fluorescence microscope for a CNT uptake study as shown in Fig. 12a, which shows dual confocal detection of AlexaFluor594-SWNT (red) internalized into cells, with the membrane stained by AlexaFluor488 (green).

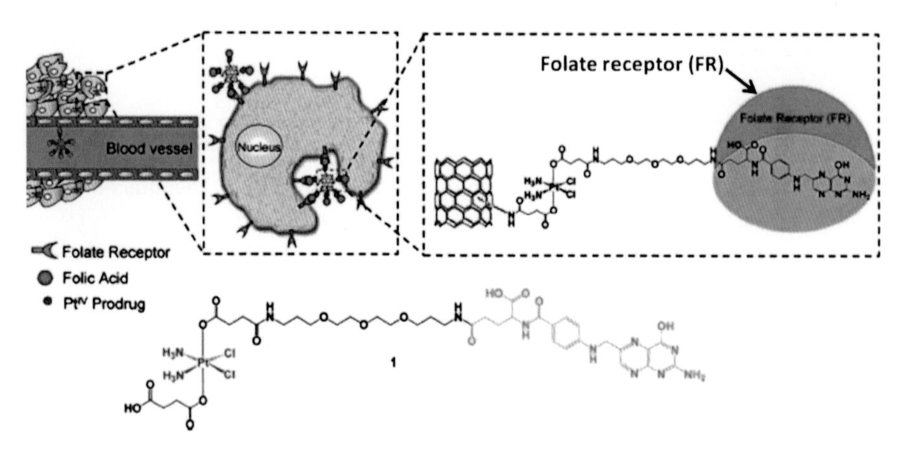

Fig. 11 *Above*: Folate receptor-mediated targeting and SWNT-mediated delivery of structure **1** by endocytosis. *Below*: Structure **1**. (Adapted from [60]).

6.3 Metal Nanoparticles

Faraday, in the mid-nineteenth century, contributed to the study of colloidal metal particles. Turkevich and coworkers in 1951 synthesized and characterized "water-soluble" gold colloids of size 18 nm [62]. Brust and coworkers reported metal particles stabilized by alkanethiols [63, 64]. Murray and coworkers [65] termed these "monolayer-protected clusters" (MPCs), which could be redissolved in common solvents without decomposing or aggregating. MPCs are synthesized using a bottom-up approach from a small number of building blocks. Gold MPCs can range in size from 1 to 10 nm, containing approximately 55–1,000 gold atoms with molecular weights between 10 and 200 kDa. Metallic nanoparticles such as Ag, Au, and Pt have been synthesized by researchers for several centuries. Because of their excellent optical properties, metallic nanoparticles are especially useful for biomedical applications such as optical contrast agents, multimodal sensors combining optical imaging and scattering imaging, and photothermal therapy. Figure 13 shows the UV–vis spectrum of solutions of gold nanoparticles capped with octadecylamine–bis(2-ethylhexyl)-sulfosuccinate (ODA-AOT), and the effect of capping agent concentration on the spectrum.

Water solubility of MPCs is best accomplished by using a thiolated, polar-protecting ligand in a modified Brust reaction. In the Brust reaction, tetrachloroauric acid is reduced from Au^{3+} to Au^{1+} in the presence of the thiol capping ligand, yielding a colorless gold-thiol solution. This is either composed of a gold-thiol polymer or tetramer. Following the initial reduction, the gold is further reduced to Au^0 in the presence of sodium borohydride ($NaBH_4$), yielding a black to dark brown solution. Other potent reducing agents, such as lithium aluminum hydride ($LiAlH_4$) or lithium triethylborohydride, have been used to reduce different metal cores such as palladium and platinum.

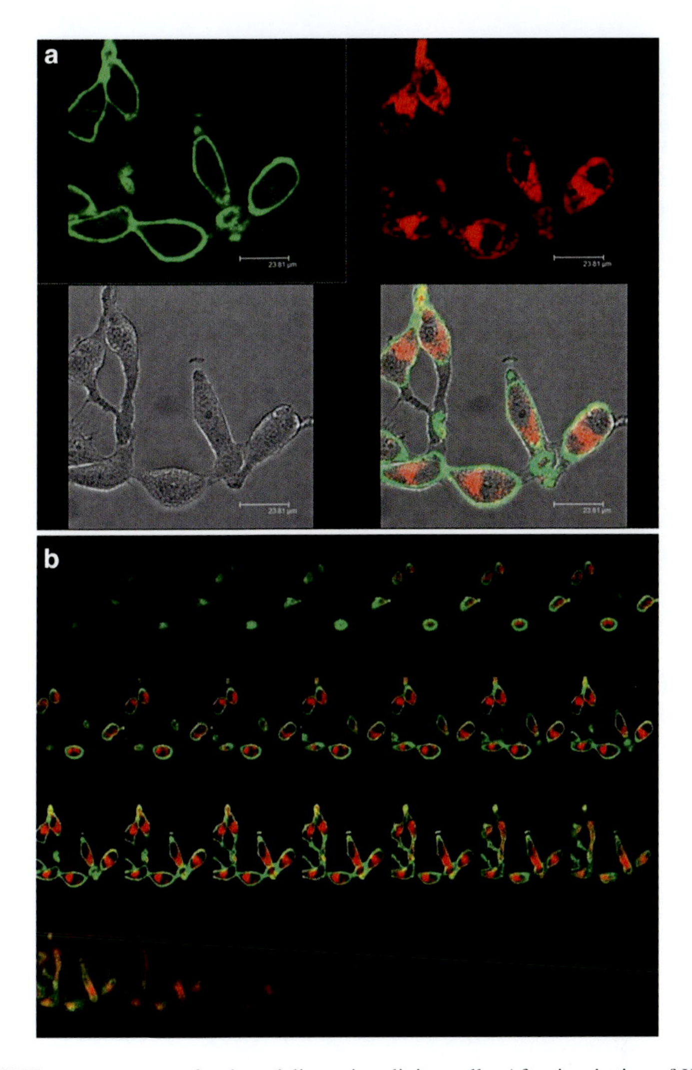

Fig. 12 CNTs act as a vector for drug delivery into living cells. After incubation of HeLa cells with AlexaFluor594-labeled SWNTs for 12 h at 37 °C, living cells were observed under confocal fluorescence microscope for a CNT uptake study. (**a**) Images show dual confocal detection of AlexaFluor594-SWNT (*red*) internalized into cells with the membrane stained by AlexaFluor488 (*green*). (**b**) Series of images of different *z*-focal scanning planes down through cells. (Adapted from [61])

Another method for the synthesis of stable metal nanoparticles involves first mixing the metal hydrosols and an ethanol solution of dodecylamine and then extracting the dodecylamine-stabilized metal nanoparticles into toluene. The ethanol, a water miscible and good solvent for dodecylamine, was used as an intermediate solvent to improve the interfacial contact between citrate-stabilized metal nanoparticles and alkylamine. The extraction of dodecylamine-stabilized metal

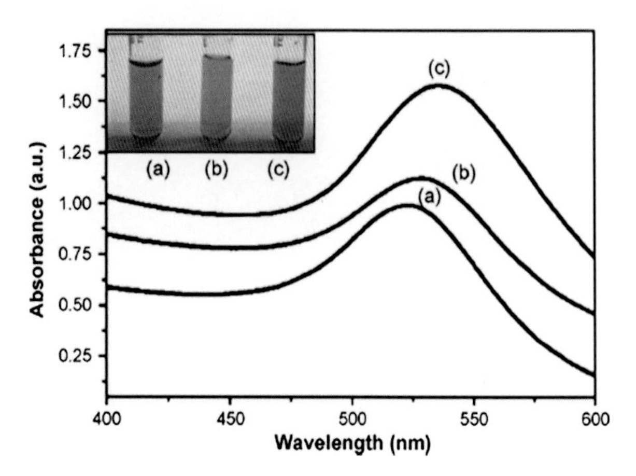

Fig. 13 UV–vis spectrum of ODA-AOT-capped gold nanoparticle solutions with decreasing concentrations of ODA: (*a*) 0.5, (*b*) 0.1, and (*c*) 0.05 M. *Inset* shows the corresponding gold nanoparticle solutions. (Adapted from [66])

nanoparticles to the toluene layer, indicated by a vivid transfer of color from the aqueous phase to toluene, then proceeds quickly and completely to leave a colorless aqueous solution behind. The phase transfer method for preparation of the alkylamine-stabilized nanoparticles of other noble metals, including Ag, Pd, Rh, Ir, and Os, has been investigated by Yang et al. [67].

For the preparation of alkylamine-stabilized noble metal nanoparticles using the phase transfer method, typically 0.8 mL of 40 mM aqueous tri-sodium citrate solution is added to 10 mL of 1 mM aqueous metal salt solution (0.1 mL of concentrated HCl solution was added for the dissolution of $PdCl_2$ in the aqueous environment). Under vigorous stirring, different amounts of 112 mM aqueous $NaBH_4$ solution are introduced dropwise upon the metals to prepare a metal hydrosol in which sodium citrate serves as the stabilizer. The molar ratio of $NaBH_4$ to the valence of the noble metal in their salts is kept above 1.5 to ensure the reduction of the metal to zerovalent state. The hydrosol is then mixed with 10 mL of ethanol containing 200 μL of dodecylamine and the mixture stirred for 2 min. A 5-mL volume of toluene is added and stirring continued for another 3 min. Dodecylamine-stabilized noble metal nanoparticles rapidly extract into the toluene layer, leaving behind a colorless aqueous solution [67]. This general phase transfer protocol for synthesizing alkylamine-stabilized nanoparticles of noble metals has been reported by Yang et al. [67]. TEM images of alkylamine-stabilized Ag and Pd nanoparticles are shown in Figs. 14 and 15, respectively.

Copolymer-stabilized nanoparticles based on Au ($HAuCl_4$ sol), Ag ($AgNO_3$), Pt ($Na_2PtCl_66H_2O$), and Rh (Na_3RhCl_6) have been prepared using a 0.01 wt% solution of the appropriate salt and a 1.0 M aqueous solution of $NaBH_4$ as the reducing agent (molar ratio of $NaBH_4$:dithioester end groups was 25:1). A portion of the reaction mixture was centrifuged for 1 h at 13,000 rpm, and the supernatant removed. The resulting aggregates were re-dispersed in deionized water by agitation. The centrifugation and re-dispersal process was repeated several times to ensure that only covalently bound polymers remained in the colloidal solutions of the stabilized nanoparticles.

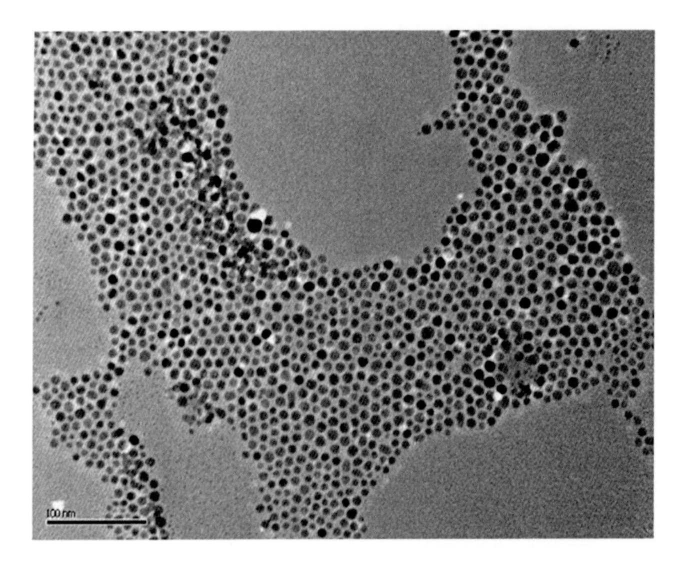

Fig. 14 TEM image of alkylamine-stabilized Ag nanoparticles. (Adapted from [67])

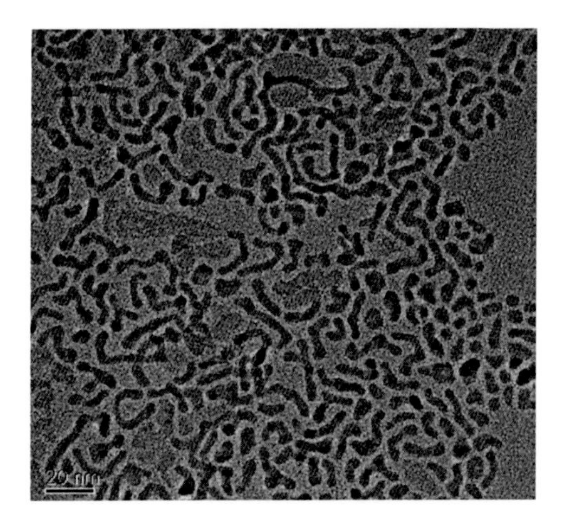

Fig. 15 TEM image of alkylamine-stabilized Pd nanoparticles. (Adapted from [67])

6.3.1 Stabilization of Colloidal Metal Particles in Liquids

Small metal particles are unstable with respect to agglomeration to the bulk. At short interparticle distances, two particles would be attracted to each other by van der Waals forces and, in the absence of repulsive forces to counteract this attraction, an unprotected sol would coagulate. To counteract this, stabilization can be achieved in two ways: electrostatic stabilization and steric stabilization.

Generally, gold nanoparticles have been used for most biomedical applications. Gold nanoparticles with varying core sizes are usually prepared by the reduction of

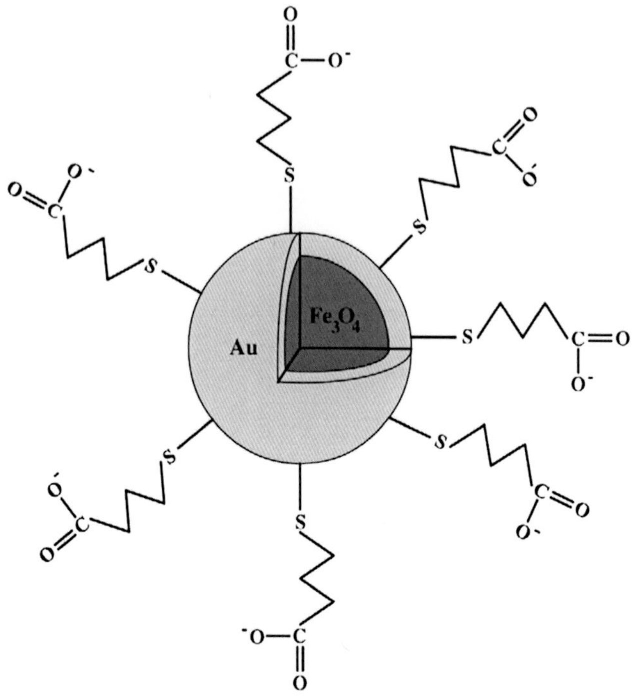

Fig. 16 Gold-coated ferrite nanoparticles can be attached to functional groups through Au–S bonds. (Credit: Charles O'Connor)

gold salts in aqueous, organic phase, or in two phases. However, the high surface energy of gold nanoparticles makes them highly reactive and, as a result, they undergo aggregation. The presence of an appropriate stabilizing agent prevents particle agglomeration by binding to the particle surface to impart high stability and also provides rich linking chemistry if it acts as a functional group for bioconjugation. Nanoparticles based on gold chemistry have attracted significant research and practical attention recently. They are versatile agents with a variety of biomedical applications, including use in highly sensitive diagnostic assays, thermal ablation, and radiotherapy. Surface conjugation of antibodies and other targeting moieties is usually achieved by adsorption of the ligand to the gold surface. Coated ferrite nanoparticles can be attached to functional groups through Au–S bonds, as shown in Fig. 16.

Surface adsorption, however, can denature the proteins or, in some cases, limit the interactions of the ligand with the target on the cell surface due to steric hindrance. Additionally, for systemic applications, long-circulating nanoparticles are desired for passive targeting to tumors and inflammatory sites. Surface functionalization of gold nanoparticles has been carried out by using a hetero-bifunctional PEG spacer for intracellular tracking and delivery [68].

6.4 Semiconductor Nanoparticles

The most common syntheses of semiconductor nanocrystals such as CdS, CdSe, CdTe, or CdSe/ZnS, ZnS, ZnO etc. have been carried out in organic solvents at high temperatures in the presence of surfactants to yield monodisperse and stable particles. This leads to the production of surfactant-coated particles. Here, the polar surfactant head group is attached to the inorganic surface and the hydrophobic chain protrudes into the organic solvent, mediating colloidal stability. The solvents used for the dispersion of these particles (e.g., toluene or chloroform) are not soluble in aqueous media because of their hydrophobic surface layer. For biomedical applications, water-soluble materials are required. Thus, the surfactant layer is replaced or coated by an additional layer that introduces either an electric charge or hydrophilic polymers for mediating solubility in water. Coulomb repulsion between nanocrystals with surface charge of the same polarity prevents aggregation in water.

6.4.1 Non-Oxide Semiconductors

Non-oxide semiconductor nanoparticles include a wide range of nanoparticles such as cadmium chalcogenides (CdE, where E = sulfide, selenide, and telluride). CdE nanocrystals were probably the first material used to demonstrate quantum size effects corresponding to a change in the electronic structure with size, i.e., the increase of the band gap energy with the decrease in size of particles. For example, cadmium selenide (CdSe) QDs dissolved in toluene fluoresce in three noticeably different colors (blue ~481 nm, green ~520 nm, and orange ~612 nm), (as shown in Fig. 17) because the QD bandgap (and thus the wavelength of emitted light) depends strongly on the particle size; the smaller the dot, the shorter the emitted wavelength of light. Hence, it is evident that the "blue" QDs have the smallest particle size, while the "green" dots are slightly larger, and the "orange" dots are the

Fig. 17 Cadmium selenide QDs, dissolved in toluene, fluorescing brightly in the presence of a ultraviolet lamp, in three noticeably different colors (*blue* ~481 nm, *green* ~520 nm, and *orange* ~612 nm). The *blue* QDs have the smallest particle size, the *green* dots are slightly larger, and the *orange* dots are the largest. (Adapted from http://www.amazingrust.com/experiments/current_projects/Misc.html)

largest. These semiconductor nanocrystals are commonly synthesized by thermal decomposition of an organometallic precursor dissolved in an anhydrous solvent containing the source of chalcogenide and a stabilizing material (polymer or capping ligand). Stabilizing molecules bound to the surface of particles control their growth and prevent particle aggregation.

In the case of CdSe, dimethylcadmium $Cd(CH_3)_2$ is used as a cadmium source and bis(trimethylsilyl)sulfide, $(Me_3Si)_2S$, trioctylphosphine selenide (TOPSe), or trioctylphosphine telluride (TOPTe) serve as sources of selenide in trioctylphosphine oxide (TOPO) used as solvent and capping molecule. The mixture is heated at 230–260 °C over a few hours while modulating the temperature in response to changes in the size distribution, as estimated from the absorption spectra of aliquots removed at regular intervals. The particles, capped with TOP/TOPO molecules, were non-aggregated and easily dispersible in organic solvents to form optically clear dispersions. When similar syntheses are performed in the presence of surfactant, strongly anisotropic nanoparticles are obtained, e.g., rod-shaped CdSe nanoparticles can be obtained.

Because $Cd(CH_3)_2$ is extremely toxic, pyrophoric, and explosive at elevated temperatures, other Cd sources have been used. CdO appears to be an interesting precursor. CdO powder dissolves in TOPO and HPA (hypophosphorous acid) or TDPA (tetradecylphosphonic acid) at about 300 °C giving a colorless homogeneous solution. By introducing selenium or tellurium dissolved in TOP, nanocrystals grow to the desired size.

Nanorods of CdSe or CdTe can also be produced by using a greater initial concentration of cadmium as compared to reactions for nanoparticles. This approach has been successfully applied for synthesis of numerous other metal chalcogenides including ZnS, ZnSe, and $Zn_{1-x}Cd_xS$. Similar procedures enable the formation of MnS, PdS, NiS, and Cu_2S nanoparticles, nanorods, and nanodisks [65, 67, 69–73]. An alternative route by which passivation of CdSe QDs is achieved employs one of the nucleobases, cytosine, that makes the CdSe QDs more biocompatible because of its versatility. Cytosine molecules alone do not contribute to fluorescence yield [74, 75] as the changes in the electronic and optical properties of cytosine-capped CdSe QDs are insignificant; rather, cytosine molecules only act as a capping agent for reducing the surface defect density of QDs. The QD was represented by a 32-atom cluster and capping was performed with either four or eight cytosine molecules. The zero of the energy is aligned to the Fermi energy, as shown in Fig. 18a,b. Figure 19a shows the optical absorption and Fig. 19b shows the photoluminescence spectra of CdSe and CdSe QDs capped with different concentrations of cytosine molecules.

6.4.2 Oxide Semiconductors

Oxide semiconductor nanoparticles such as ZnO were surface-modified by Dutta et al. using a different approach [76]. They described the design and fabrication of

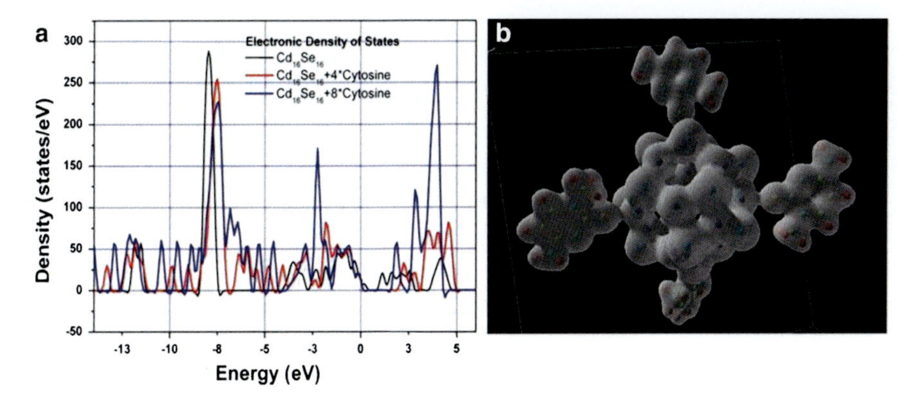

Fig. 18 (**a**) Density of states of CdSe and capped CdSe QDs. (**b**) Charge density of capped CdSe QDs. (Adapted from [75])

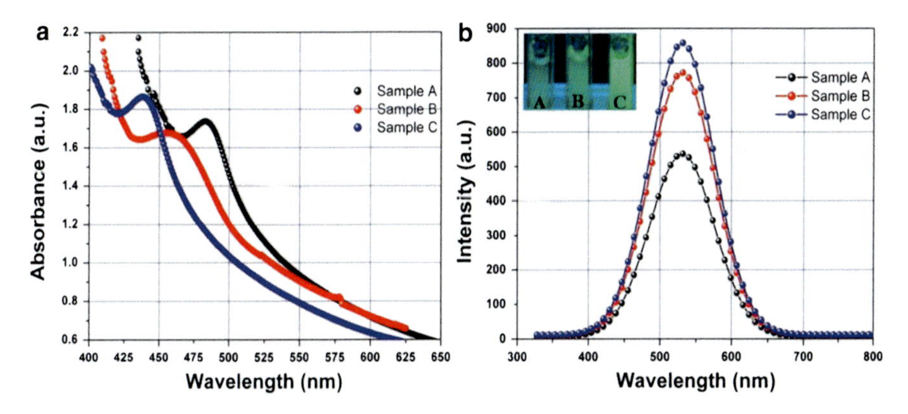

Fig. 19 (**a**) Optical absorption and (**b**) photoluminescence spectra of CdSe and CdSe QDs capped with different concentration of cytosine. Sample *A* is as-prepared QDs and Samples *B* and *C* are capped QDs. (Adapted from [75])

luminomagnets of ZnO:Fe. The luminomagnets were surface-immobilized with the ligand folic acid. The relation between folic acid and cancer has been shown in several published works [77]. For this, the luminomagnets were first surface-modified with (3-aminopropyl)-trimethoxysilane (AEAPS) to form a self-assembled monolayer, and subsequently conjugated with folic acid through amidation between the carboxylic acid end groups on folic acid and the amine groups on the particle surface. A very convenient coupling group is NHS (or its sulfo derivative, sulfo-NHS), which can be easily generated by direct reaction of a carboxylic acid with NHS in the presence of the dehydrating agent, 1-ethyl-3-(3-dimethylaminopropyl) carbodiimide hydrochloride (EDC). The strategy of folic acid conjugation is diagrammatically shown in Fig. 20.

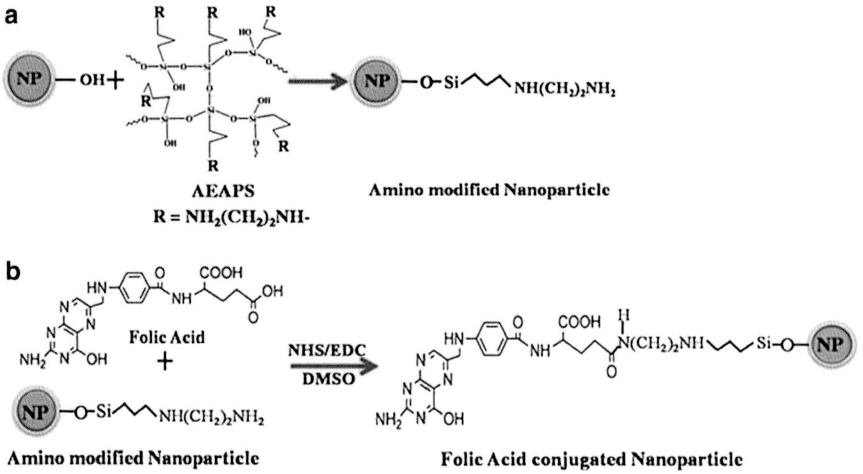

Fig. 20 Reaction mechanism for surface modification of luminomagnets with folic acid. (Adapted from [76])

6.5 Magnetic Nanoparticles

Magnetic nanoparticles (MNPs) functionalized with the drug can serve as potential drug carriers in a new drug delivery strategy based on the application of external magnetic fields. The principle of drug delivery by nanomagnets is based on the use of both constant and high-frequency oscillating magnetic fields. Since these particles are magnetic in nature, they can be targeted to specific areas (e.g., cancer tissues) by a constant external magnetic field. Their size, which is comparable to biological functional units, and their unique magnetic properties allow their utilization as molecular imaging probes. By tuning growth parameters, such as monomer concentration, crystalline phase of the nuclei, choice of solvent and surfactants, growth temperature and time, and surface energy, it is possible to control the size, composition, and magneto-crystalline phase of MNPs. Non-hydrolytically synthesized MNPs are typically coated with hydrophobic ligands. Therefore, it is necessary to exchange these ligands for appropriate ones that will give high colloidal stability in aqueous biofluids, and to avoid the aggregation that can occur under harsh physiological conditions.

Over the past decade, a number of biomedical applications have begun to emerge for magnetic micro- and nanoparticles of differing sizes, shapes, and compositions [78]. Many applications still rely on the use of iron oxide particles (usually Fe_2O_3 or Fe_3O_4), like the original ferrofluids. These particles are available with diameters ranging from ~300 nm to less than 10 nm. They exhibit superparamagnetic behavior, magnetizing strongly under an applied field, but retaining no permanent magnetism once the field is removed. This on/off magnetic switching

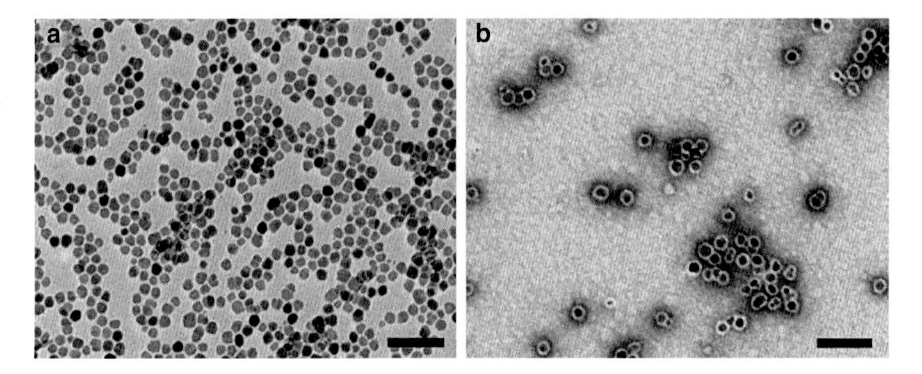

Fig. 21 TEM images of (**a**) as-synthesized iron oxide nanoparticles, (**b**) poly (amino acid)-coated iron oxide nanoparticles in water. *Scale bars*: 80 nm (Adapted from[79])

behavior is a particular advantage in magnetic separation, one of the simplest applications. Magnetic separation is now well established as a viable alternative to centrifugal separation of complex chemical or biological solutions. Iron oxide particles are first encased in a biocompatible coating to form tiny beads. TEM images of some iron oxide nanoparticles are shown in Fig. 21. The beads are then "functionalized," i.e., their surfaces are treated with a biological or chemical agent known to bind to a specific target. On placing the beads in solution, any target cells or molecules will latch onto the functionalized surfaces.

An alternative class of MNPs was produced from pure transition metals, such as Fe, Ni, and Co, which exhibit ferromagnetic behavior. Unlike the SPIO and ultrasmall SPIO (USPIO) particles, these pure metal particles retain their magnetization once an external magnetic field is removed, causing particulate clustering. Ferromagnetic nanoparticles also tend to have a larger magnetic moment than their superparamagnetic counterparts. Ultrasmall Fe particles are, thus, likely to produce a better signal in magnetic sensors or to respond more readily to an applied field gradient than iron oxide particles of the same size. Upon conjugation with the appropriate targeting molecules, MNPs can be utilized for the active detection of cancer. The active targeting can be applicable in diagnosis as well as in therapeutics for cancer [80], artherosclerosis etc.

6.5.1 Magnetite, Maghemite, and Ferrites

Synthesis of PEG-Modified Magnetic Nanoparticles

PEG is hydrophilic and is widely used in biological research because it protects surfaces from interacting with cells or proteins. Thus, coated particles may result in increased blood circulation time. For their preparation, 10-mg magnetite particles were dispersed in 1.0 mL of deoxygenated water by sonication for 30 min. The aqueous dispersion of MNPs was dissolved in the aqueous cores of reverse micelles together with a polymerizable derivative of PEG as a monomer [i.e., maleic anhydride-modified PEG (MA-PEG)] and *N,N′*-methylene bis acrylamide (MBA) as crosslinking agent under nitrogen gas. Additional amounts of water may be added in reverse micellar solution in order to obtain host micellar droplets of desired size. In a typical experimental protocol, 50 mL of 0.05 M AOT solution in hexane, 500 mL of magnetite solution (10 mg/mL), 100 mL of MA-PEG (5 mg/mL), and 10 mL of MBA (0.5 mg/mL) were mixed. The solution was stable and brownish transparent at this stage. Nitrogen gas was bubbled through this solution to remove the dissolved oxygen. After 30 min, 20 mL of 2% ammonium persulfate as an initiator was added. The polymerization of monomers was carried out by a free radical polymerization mechanism at 37 °C for 8 h. After polymerization, the particles were purified from unreacted monomers.

Surface Modification of Magnetic Nanoparticles

Several surface modification methods have been tried for applications in drug delivery and for MRI applications. A multifunctional magnetic nanoparticle probe for deep tissue molecular MRI is shown in Fig. 22. The magnetic nanoparticle (5–10 nm in diameter) has an oligonucleotide hairpin probe on its surface. Cell-penetrating peptides (for deep tissue delivery) and ligands (for targeting specific cell types) are also conjugated on the nanoparticle surface. Binding of two or more nanoparticle probes on an mRNA target inside a cell should generate a measurable

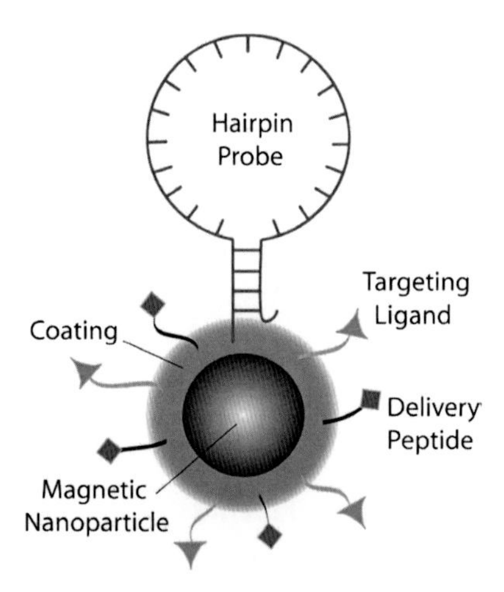

Fig. 22 Multifunctional magnetic nanoparticle probe for deep tissue molecular MRI. The magnetic nanoparticle (5–10 nm in diameter) has an oligonucleotide hairpin probe on its surface. Cell-penetrating peptides (for deep tissue delivery) and ligands (for targeting specific cell types) are also conjugated on the nanoparticle surface. Binding of two or more nanoparticle probes on an mRNA target inside a cell should generate a measurable change in the MRI signal. Cells expressing specific molecular markers of disease, infection, or injury are detected in this way. (Credit: Gang Bao)

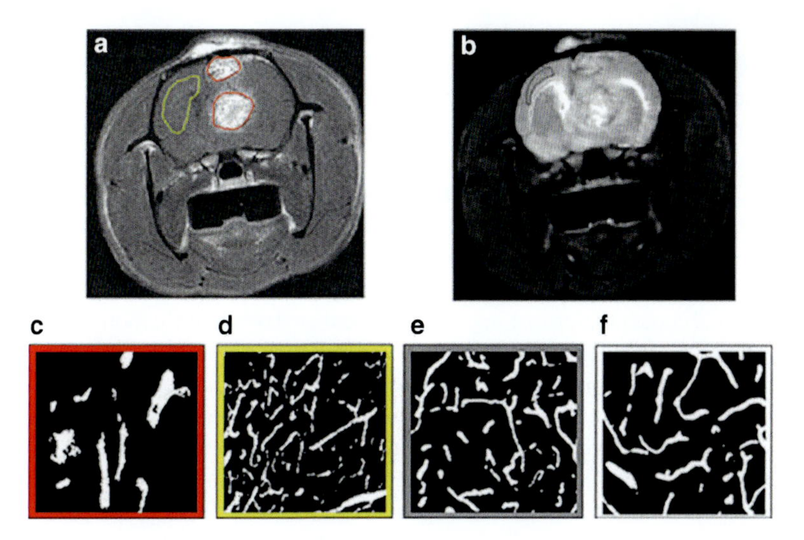

Fig. 23 (**a**) Post-Gd axial MRI of the rat brain illustrating the regions of interest (ROIs) for the tumor (*red*) and contralateral (*yellow*) brain. (**b**) Post-monocrystalline iron oxide nanoparticle (MION) axial MRI of the rat illustrating the ROIs for the gray (*gray*) and white (*white*) matter. (**c–f**) Binarized images of tissue sections of microfilled vessels at 20× magnification of tumor (**c**), contralateral brain (**d**), gray matter (**e**), and white matter (**f**) ROIs. The color of each frame (*red*, *yellow*, *gray* or *white*) corresponds to the color of the ROI from which the tissue sections were sampled for histology. (Adapted from [81])

change in the MRI signal. Cells expressing specific molecular markers of disease, infection, or injury are detected in this way. Previously, monocrystalline iron oxide nanoparticles (MION) had been synthesized for MRI applications. One such application is illustrated in Fig. 23. Josephson et al. [81] have shown the effects of iron oxides on proton relaxivity. The post-MION axial MRI of the rat, illustrating the regions of interest (ROIs) for the gray and white matter, have been demonstrated.

Antibody-Conjugated MNPs

Antibody targeting of drug substances can improve the therapeutic efficacy of the drug, as well as improve the distribution and concentration of the drug at the targeted site of drug action. McCarron et al. studied two novel approaches for the creation of immunonanoparticles with improved therapeutic effect against colorectal tumor cells [82]. They used poly(lactide) polymers and CD95/APO-1 antibody to target nanoparticles. Pan et al. used dendrimer–MNPs for efficient delivery of gene-targeted systems for cancer treatment [83]. Detection of colon cancer is also possible by using Fe_3O_4–rch24 antibody conjugates. FePt–Au nanoparticles conjugated with HmenB1 antibodies, can be used to identify neuroblastoma cells (CHP-134) with a polysialic acid overexpression. These, along with their capability of being manipulated under an external magnetic field, provide controllable means for magnetically tagging of all biomolecules, leading to highly efficient bioseparation and biodelivery, highly sensitive biolabeling, and MRI with enhanced contrast.

Human epidermal growth factor receptor 2 (HER2/neu) antibodies were conjugated to the surface of poly(amino acid)-coated iron oxide nanoparticles (PAIONs) for the detection of breast cancer (Fig. 24) [79]. The method of antibody conjugation was briefly: 0.57 nmol PAION (4 mg) was washed with 4 mL of a phosphate-buffered saline (PBS) solution by vortexing and sonication. The washed particles were collected in a Microcon Centrifugal Filter Device (molecular weight cutoff 100 kDa). After being washed twice, the particles were suspended with 4 mL of 25% glutaraldehyde solution (final concentration of 1 mg/mL) and incubated overnight at room temperature with shaking. Glutaraldehyde-functionalized particles were washed with a PBS solution, as performed in the previous washing step, and this was repeated three times. HER2/neu antibodies were then conjugated to the glutaraldehyde-functionalized nanoparticles as follows: PAION (2 mg, 0.285 nmol) dispersed in 400 μL of PBS was mixed with 100 μL of 200 μg/mL antibody solution (0.054 mM). A BSA-conjugated PAION was used as a negative control which was prepared with the remaining 2 mg of functionalized nanoparticles by suspending them in 500 μL of 3% BSA in PBS. The mixture was kept at 4 °C and maintained overnight to allow stable binding between the protein and PAION. The antibody- and BSA-conjugated particles were then washed with a PBS solution and incubated with excess ethanolamine for 1 h at room temperature to block nonspecific binding sites. Finally, 1 mg/mL antibody- and BSA-conjugated PAIONs in PBS solution were stored at 4 °C until use.

6.6 MR Contrast Agents

6.6.1 Nonspecific Contrast Agents

Nonspecific contrast agents are low molecular weight contrast agents, e.g., Gd–DTPA, and high molecular weight blood pool agents [84], such as high generation dendrimers [85]. These agents can be used for MR angiography and to measure the perfusion and permeability properties of tissues.

6.6.2 Targeted Contrast Agents

Targeted contrast agents are specific and directed to a specific molecular or cellular target with an appropriate targeting ligand molecule. Recombinant high density lipoprotein (HDL)-like nanoparticles, a specific contrast agent for MRI of atherosclerotic plaques, has been reported [86] (Fig. 25).

Sipkins et al. [87] described the detection of tumor angiogenesis with an αvβ3-specific antibody that was conjugated to polymerized paramagnetic liposomes [87]. The red fluorescence represents the liposomes (Fig. 26c) and the green fluorescence represents blood vessels. In Fig. 26, we see that

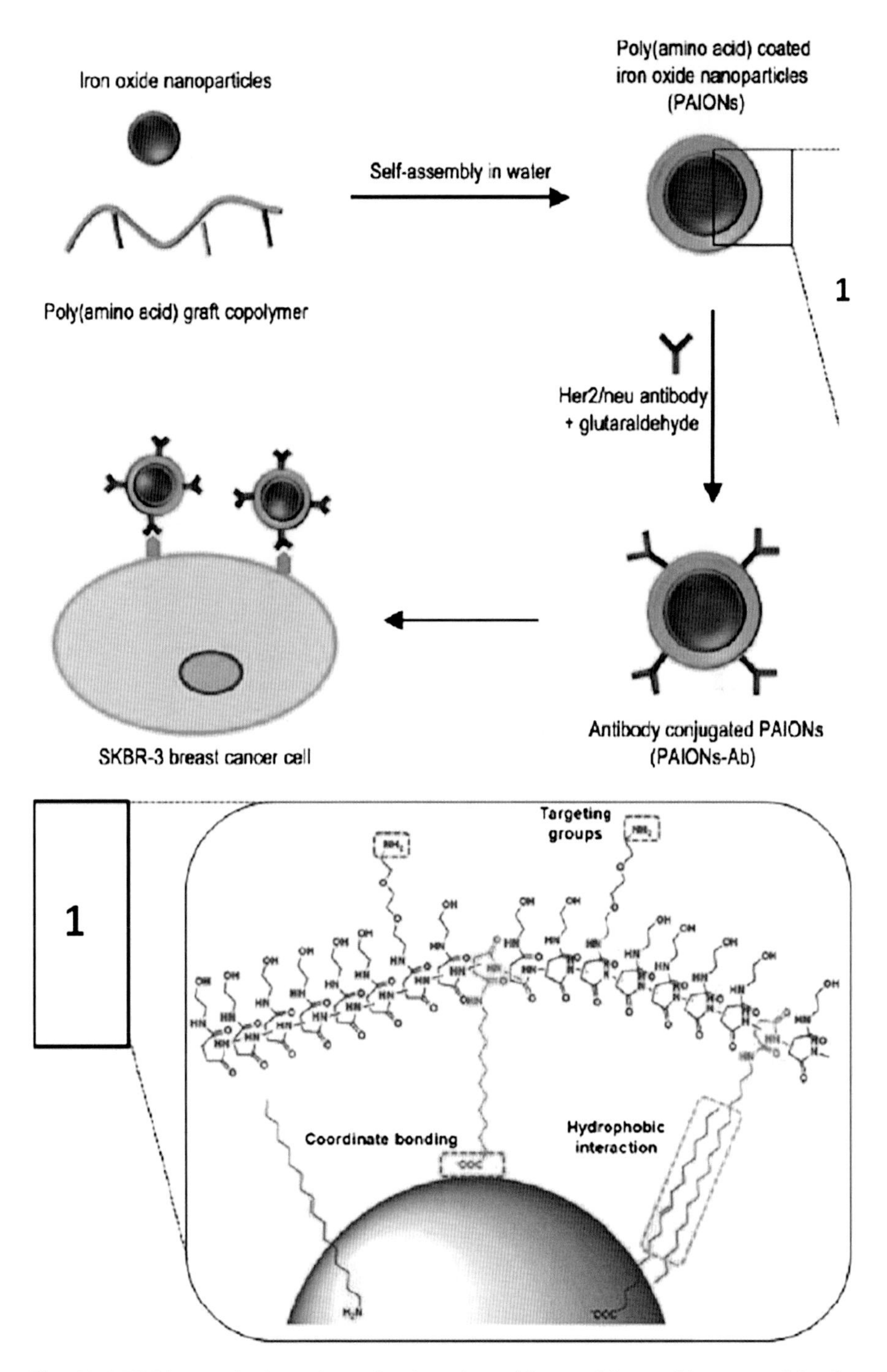

Fig. 24 HER2/neu antibody-conjugated poly(amino acid)-coated iron oxide nanoparticles for breast cancer cell imaging. The part labeled as *1* is shown enlarged. (Adapted from [79])

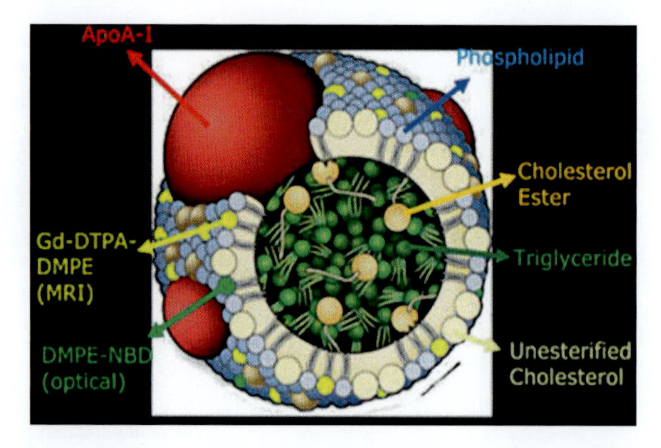

Fig. 25 Different components of the recombinant HDL-like specific contrast agent for MRI of atherosclerotic plaques. (Adapted from [86])

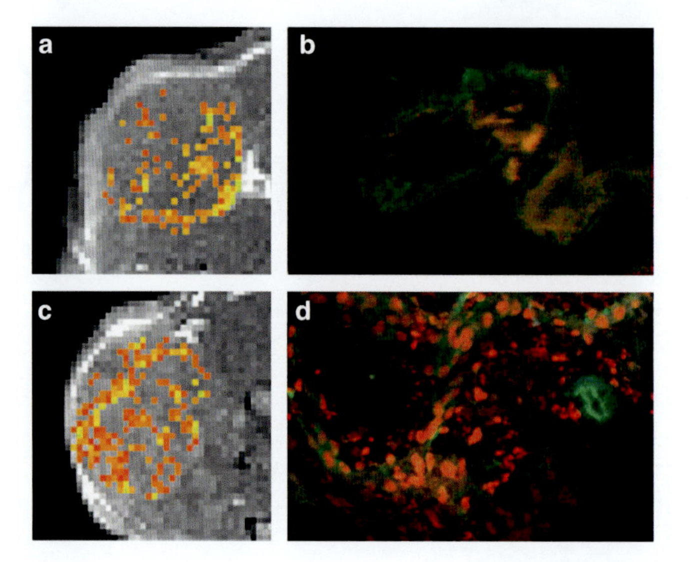

Fig. 26 MR images of tumors of mice after they were injected with (**a**) paramagnetic αvβ3-specific RGD–liposomes and (**b**) nonspecific paramagnetic RAD–liposomes. (**c**, **d**) Fluorescence microscopy of 10 μm sections from dissected tumors revealed a distinct difference between tumors of mice that were injected with RGD–liposomes (**c**) or RAD–liposomes (**d**). Vessel staining was done with an endothelial cell-specific FITC–CD31 antibody. The *red* fluorescence represents the liposomes and the *green* fluorescence represents blood vessels. RGD–liposomes were exclusively found within the vessel lumen or associated with vessel endothelial cells (**c**), whereas RAD–liposomes (**d**) were also found outside blood vessels within the tumor (Adapted from [88])

paramagnetic αvβ3-specific RGD–liposomes were exclusively found within the vessel lumen or associated with vessel endothelial cells Fig. 26c, whereas nonspecific paramagnetic RAD–liposomes Fig. 26d were also found outside blood vessels within the tumor.

6.6.3 Smart Contrast Agents

Smart contrast agents are referred to as activated or responsive agents. One such example is EgadMe, a complex that contains a sugar moiety that prevents water from coordinating with Gd^{3+}. Enzymatic cleavage of this sugar by β-galactosidase improves the accessibility of water to Gd^{3+}, which results in an increase in the relaxivity of the complex [89].

6.6.4 Contrast Agents for Labeling Cells

Contrast agents such as iron oxide particles conjugated to the TAT peptide [90] or Gd–$HPDO_3A$ [gadolinium 1,4,7-tris(carboxymethyl)-10-(2′-hydroxypropyl)-1,4,7,10-tetraazacycl ododecane] [91, 92], fall into this category. HP-DO3A is a neutral (nonionic) gadolinium chelate. In order to meet the diverse requirements sketched above, highly potent, innovative, specific and preferably multimodal contrast agents are required.

6.6.5 MR Molecular Imaging and Drug Targeting of Atherosclerosis with Contrast Agents

Angiogenesis plays an important role in providing the tumor with nutrients. Antiangiogenic therapies are therefore believed to provide a powerful treatment option for cancer. In atherosclerosis, the plaques contain angiogenic microvessels, which are believed to play an important role in plaque development. Imaging plaques and tumors with a αvβ3-specific contrast agent is therefore of importance for early detection, defining the severity of the disease, and following the effect of therapy. Scanning electron micrographs of control fibrin clot and fibrin-targeted paramagnetic nanoparticles bound to clot surface are shown in Fig. 27. Recently,

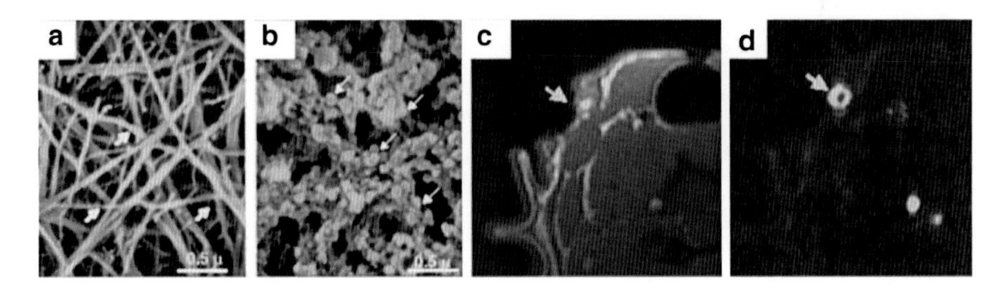

Fig. 27 SEM images of (**a**) control fibrin clot (*arrows* indicate fibrin fibrils), and (**b**) fibrin-targeted paramagnetic nanoparticles bound to clot surface (*arrows* indicate fibrin-specific nanoparticle-bound fibrin epitopes). (**c**, **d**) Thrombus in external jugular vein targeted with fibrin-specific paramagnetic nanoparticles demonstrating dramatic T1-weighted contrast enhancement in gradient-echo image (**c**) with flow deficit (*arrow*) of thrombus in corresponding phase-contrast image (**d**). (Adapted from [93])

Winter and coworkers [93] presented results of a combinatory approach of MR molecular imaging and drug targeting of atherosclerosis with this contrast agent. To that end, they used the αvβ3-specific nanoparticles to target the aortic vessel wall after balloon injury. For therapeutic purposes they included fumagillin in the lipid monolayer of the nanoparticles and observed an anti-angiogenic effect with MRI that was confirmed histologically.

Approximately 15% of cancer patients are diagnosed in stage I or II with conventional diagnostic tools; MR contrast effects may help to improve the rate of cancer diagnosis in its earliest stages. One successful example is the molecular imaging of breast cancer using Fe_3O_4 magnetism-engineered iron oxide (MEIO) nanoparticle probes. Breast cancer cells typically overexpress HER2/neu. When nanoparticles with relaxivity coefficient of 218 mm^{-1} s^{-1} are conjugated with the HER2/neu-specific antibody Herceptin, the SK-BR-3 breast cancer cell lines can be detected (Fig. 28a) [95]. Furthermore, Fe_3O_4–Herceptin probes make the ultrasensitive in vitro detection of cancer cells possible since these probes interact with all HER2/neu-positive cancer cells, including Bx-PC-3 cells which have only a minimal level of HER2/neu (Fig. 28b) [96].

A composite scaffold drug delivery system (CS-DDS) for osteoarticular tuberculosis therapy has been prepared by loading bi-component drugs into a

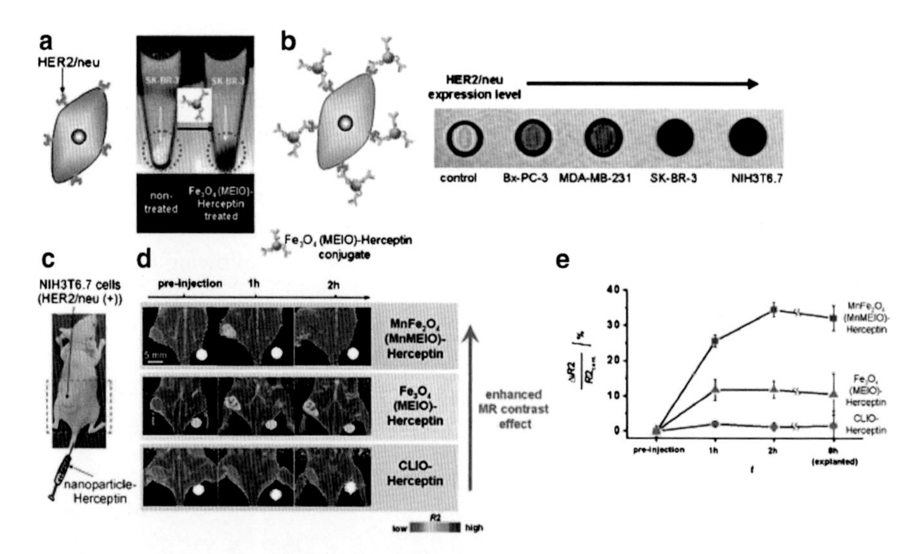

Fig. 28 (**a**) In vitro MR detection of HER2/neu-positive breast cancer (SK-BR-3) by Fe_3O_4 (MEIO)-Herceptin nanoparticle probes. (**b**) MR contrast enhancement effects of various cancer cells with different HER2/neu expression levels, (**c–e**) Highly-sensitive in vivo cancer detection by utilization of $MnFe_2O$ (MnMEIO)-Herceptin nanoparticle probes. (**c**) Intravenous tail-vein injection of the MEIO-Herceptin probes into a mouse with a small (ca. 50 mg) HER2/neu-positive cancer in its proximal femur region. For comparison, MEIO-Herceptin probes and CLIO-Herceptin probes were also tested. (**d**) Color-mapped MR images of the mouse at different times following injection. (**e**) Time-dependent relaxivity ($R2$) changes at the tumor site after injection of the probes (Adapted from [94])

mesoporous silica nanoparticle-coated porous β-tricalcium phosphate scaffold, which was followed by an additional bioactive glass coating [97].

6.6.6 MR Contrast Agents for the Detection of Apoptotic Sites

Annexin V-functionalized crosslinked iron oxide (CLIO) was designed as a contrast agent for MRI, which was additionally labeled with Cy5.5 to allow colocalization with optical imaging techniques [98]. Alternatively, conjugation of multiple Gd–DTPA molecules or SPIO particles to the C2 domain of synaptotagmin I was shown to allow the detection of apoptotic cells in vitro [99]. Zhao et al. [100] were the first to apply a C2 domain-functionalized SPIO and showed very promising results for future in vivo applications of MR contrast agents for the detection of apoptotic sites.

6.7 Lanthanide-Based Nanoparticles

The first staining of biological cells with lanthanides dates back to 1969 when bacterial smears (*Escherichia coli* cell walls) were treated with aqueous ethanolic solutions of europium thenoyltrifluoroacetonate, henceforth appearing as bright red spots under mercury lamp illumination. In the mid-1970s Finnish researchers in Turku proposed EuIII, SmIII, TbIII, and DyIII polyaminocarboxylates and β-diketonates as luminescent sensors for time-resolved luminescent (TRL) immunoassays [101–103]. Dysprosium(III) is another lanthanide ion that has been used in MRI, being classed as a negative contrast agent. Up-converted $NaYbF_4$ microparticles doped with different LnIII, ErIII, TmIII, or HoIII ions, or a combination of them, emit visible orange, yellow, green, cyan, blue, or pink light under near infrared excitation. The emission color can be tuned by modifying either the dopant concentration or the dopant species.

Some lanthanides such as gadolinium(III) remain the dominant starting material for contrast agent design but other lanthanide ions (and other oxidation states, i.e., +2) are also being increasingly investigated as alternatives to gadolinium(III) within laboratory conditions. GdF_3 (or GdF_3/LaF_3) nanoparticles were investigated as T1 contrast agents [104]. The surface of the nanoparticles can be either positively charged by conjugation with 2-aminoethyl phosphate groups or negatively charged by coating with citrate groups. Surface functionalization of the nanoparticles is very important because surface properties play an important role in controlling solubility and retention in specific tissues, and bioactive materials can be conjugated on the surface. The most studied among them are the particles of Gd_2O_3. The Gd_2O_3 samples were synthesized via the polyol route by Ahren et al. [105], whereby 6 mmol $GdCl_3$ and 7.5 mmol NaOH were dissolved in 30 mL of DEG. The two solutions were mixed, magnetically stirred, and heated to about 140 °C. The temperature was held constant for 1 h and then increased to and retained at 180 °C for the next 4 h to obtain nanoparticles of Gd_2O_3.

7 Targeted Drug Delivery Using Nanoparticles

Targeted delivery of drug using nanoparticles is important for the following reasons: (1) the unique pathophysiologic characteristics of tumor vessels enable macromolecules, including nanoparticles, to selectively accumulate in tumor tissues; (2) drug resistance has emerged as a major obstacle limiting the therapeutic efficacy of chemotherapeutic agents, and among several mechanisms of drug resistance, P-glycoprotein is the best known and most extensively investigated; (3) intracellular and organelle-specific strategies are possible; (4) organ- or tissue-specific drug delivery is possible; and (5) targeting.

7.1 Passive Targeting by Nanoparticles

7.1.1 Enhanced Permeability and Retention Effect

Nanoparticles that have the capability of escaping reticuloendothelial system capture can circulate for longer times in the bloodstream and, hence, have a greater chance of reaching the targeted tumor tissues. The enhanced vasculature of a growing tumor enables macromolecules, including nanoparticles, to selectively accumulate in tumor tissues [106], since the existing vessels near the tumor mass need a greater supply of oxygen and nutrients [107]. The resulting imbalance of angiogenic regulators such as growth factors and matrix metalloproteinases makes tumor vessels highly disorganized and dilated, with numerous pores showing enlarged gap junctions between endothelial cells and with compromised lymphatic drainage. These features are called the enhanced permeability and retention effect, which constitutes an important mechanism by which macromolecules, including nanoparticles, with a molecular weight above 50 kDa, can selectively accumulate in the tumor interstitium.

7.2 Active Targeting by Nanoparticles

When the nanoparticle drug delivery vector is made more specific by the addition of some target ligand or antibody, it is called active targeting [108]. One approach suggested to overcome these limitations is the inclusion of a targeting ligand or antibody in polymer–drug conjugates. Initially, direct conjugation of an antibody to a drug was investigated. However, this approach was not successful during clinical trials for the treatment of cancer. One of the reasons for this is that the number of drug molecules that can be loaded on the antibody while preserving its immune recognition is limited.

7.2.1 Internalization of Targeted Conjugates

Whether targeted conjugates can be internalized after binding to target cells is an important criterion in the selection of proper targeting ligands. Internalization usually occurs via receptor-mediated endocytosis. Using the example of the folate receptor (Fig. 29), when a folate-targeted conjugate binds with a folate receptor on the cell surface, the invaginating plasma membrane envelopes the complex of the receptor and ligand to form an endosome. Newly formed endosomes are transferred to target organelles. As the pH in the interior of the endosome becomes acidic and lysozymes are activated the drug is released from the conjugate and enters the cytoplasm, provided the drug has the proper physico-chemical properties to cross the endosomal membrane. Released drugs are then trafficked by their target organelle, depending on the drug. Meanwhile, the folate receptor released from the conjugate returns to the cell membrane to start a second round of transport by binding with new folate-targeted conjugates. Some examples of different types of nanocarriers for drug delivery systems are shown in Fig. 30.

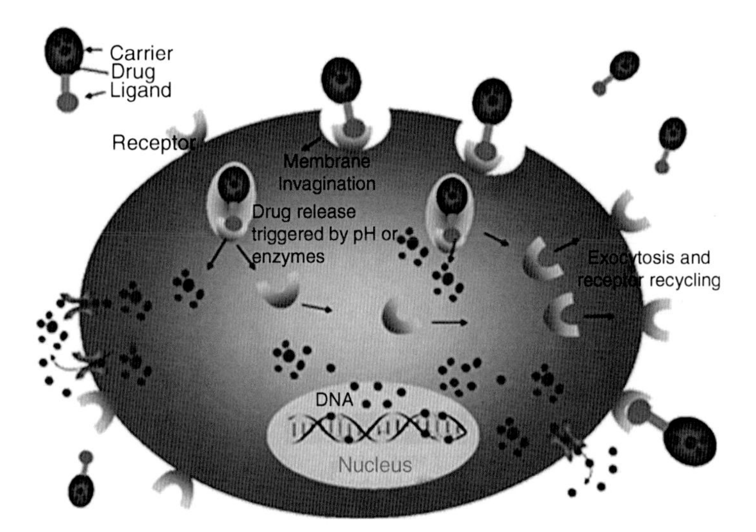

Fig. 29 Internalization of nanoparticles via receptor-mediated endocytosis. Tumor-specific ligands or antibodies on the nanoparticles bind to cell-surface receptors, which trigger internalization of the nanoparticles into the cell through endosome. As the pH in the interior of the endosome becomes acidic, the drug is released from the nanoparticles and goes into the cytoplasm. Drug-loaded nanoparticles bypass the P-glycoprotein efflux pump, not being recognized when the drug enters cells, leading to high intracellular concentration. *MDR* multidrug resistance. (Adapted from [107])

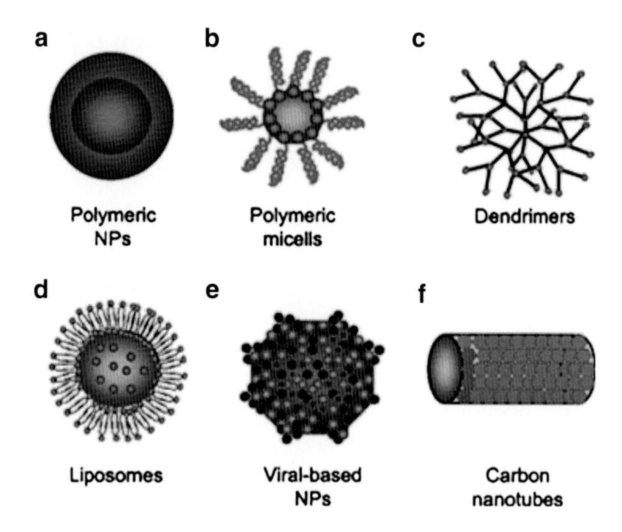

Fig. 30 Types of nanocarriers for drug delivery. (**a**) Polymeric nanoparticles: polymeric nanoparticles in which drugs are conjugated to or encapsulated in polymers. (**b**) Polymeric micelles: amphiphilic block copolymers that form nanosized core–shell structures in aqueous solution. The hydrophobic core region serves as a reservoir for hydrophobic drugs, whereas hydrophilic shell region stabilizes the hydrophobic core and renders the polymer water-soluble. (**c**) Dendrimers: synthetic polymeric macromolecule of nanometer dimensions, which are composed of multiple highly branched monomers that emerge radially from the central core. (**d**) Liposomes: self-assembling structures composed of lipid bilayers in which an aqueous volume is entirely enclosed by a membranous lipid bilayer. (**e**) Viral-based nanoparticles: in general, the structure are the protein cages, which are multivalent, self-assembled structures. (**f**) Carbon nanotubes: carbon cylinders composed of benzene rings. (Adapted from [107])

AlphaRx is one of the leaders in the development of nanoparticulate drug delivery systems to enhance the bioavailability of the drugs towards targeted diseased cells, promoting the required response while minimizing side effects. A good illustration is given in Fig. 31. Nanoscale drug platforms include a class of particles made of FDA-approved polymers or lipids which, because of their size and chemical composition, permit systemic and local treatment. Blood vessels that supply tumors are more porous than normal vessels, making nanoscale drug delivery systems a particularly attractive prospect.

External magnets are being used to guide a novel, intra-arterially administered chemotherapy delivery vehicle directly to the tumor site (Fig. 32). Scott C. Goodwin, chief of vascular and interventional radiology, UCLA Medical Center, reported results of an ongoing phase I/II study of this new regional therapy technique at the annual scientific meeting of the Society of Cardiovascular and Interventional Radiology (http://www.cancernetwork.com/news/display/article/10165/85758). The product, MTC-DOX (magnetic targeted carriers–doxorubicin), currently being tested in primary liver cancer patients, is under development by San Diego-based FeRx Incorporated. It consists of doxorubicin adsorbed to the company's proprietary MTCs.

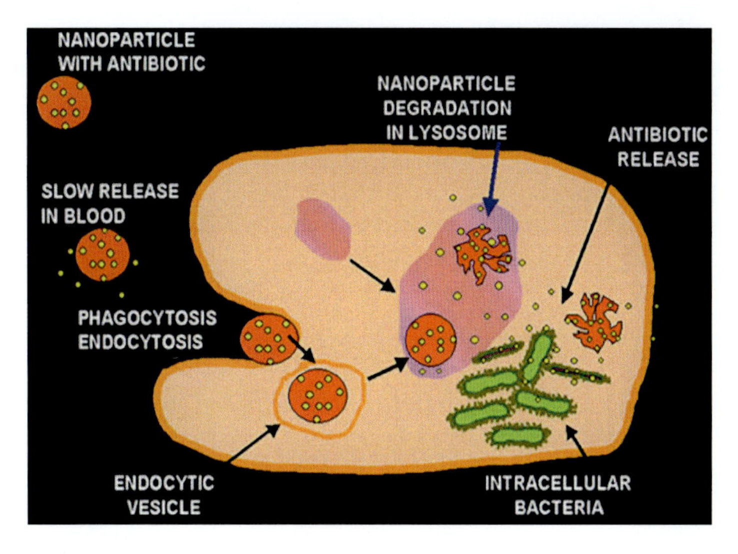

Fig. 31 Therapeutic nanoparticles for drug delivery in cancer

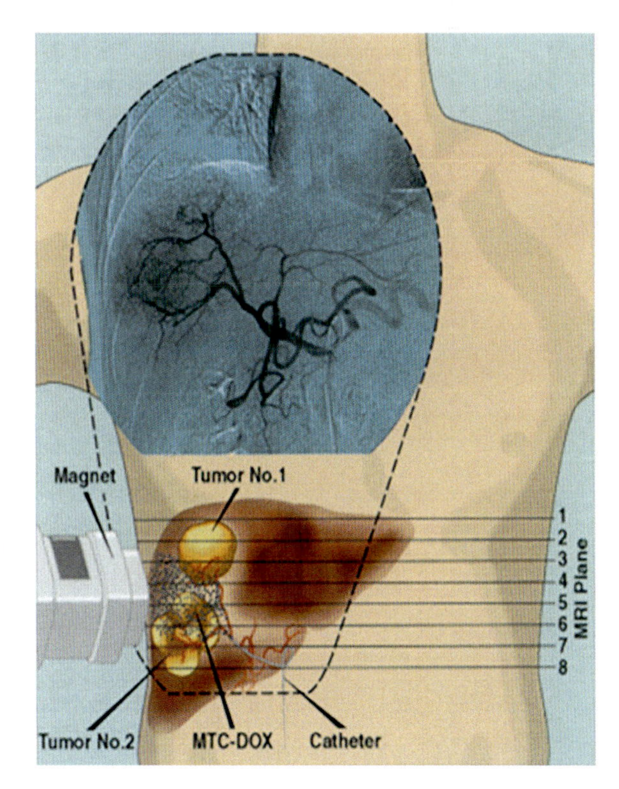

Fig. 32 An investigational product known as MTC-DOX (doxorubicin adsorbed to magnetic targeted carriers) is infused to the hepatic artery feeding the targeted tumors. An external magnet creates a magnetic field that pulls the MTC-DOX out of circulation into the liver tissue. An angiogram verifies patency of the targeted artery and shows that no embolism occurred. The diagram shows the location of eight MRI scans taken through the liver postinfusion

In this study, delivery was targeted to a single lesion in a specific hepatic segment, using a small, externally positioned magnet (5 kilogauss) to create a localized magnetic field within the body over the tumor site. The physical force created by the magnetic field induces transport (extravasation) of the MTCs through the vascular wall. The external magnet remains in place for about 15 min after dosing. Upon removal of the magnet, the MTCs do not recirculate but are retained in the tissue, where the drug then desorbs from the MTCs, leading to sustained release of the particles at the desired site.

7.3 Drug-Infused Nanoparticles Stop Cancer from Spreading

By using tumor-targeting nanoparticles filled with chemotherapy drugs, scientists kept kidney and pancreas cancers from spreading through the bodies of mice (Fig. 33). Researchers led by University of California, San Diego pathologist David Cheresh designed nanoparticles that selectively attached to a protein found on the surface of blood vessels that supply tumors with nutrients and oxygen [109]. The particles were loaded with doxorubicin, an effective but highly toxic anticancer drug with side effects ranging from white cell destruction to fatal heart disease. By targeting blood vessel cells, the researchers needed just one-fifteenth of the amount used in a traditional, system-flooding dose.

Duke University bioengineers have developed a simple and inexpensive method for loading cancer drug payloads into nanoscale delivery vehicles, and have demonstrated in animal models that this new nanoformulation can eliminate tumors after a single treatment [109]. After delivering the drug to the tumor, the delivery

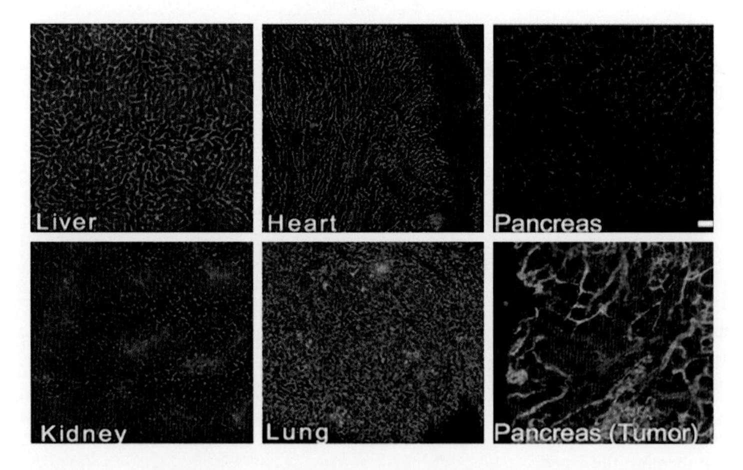

Fig. 33 Prevention of spread of cancer to other body parts (Adapted from [109])

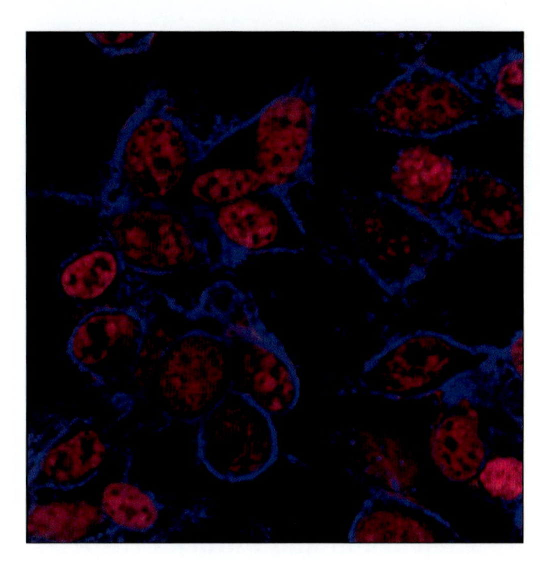

Fig. 34 Mouse tumor cells (*blue*) showing penetration of anticancer drug after 24 h. Cancer cells have taken up a chimeric polypeptide–drug combination, shown in *magenta*. (Adapted from [110])

vehicle breaks down into harmless byproducts, markedly decreasing the toxicity for the recipient.

Nanoscale delivery systems have become increasingly attractive to researchers because of their ability to efficiently get into tumors. Since blood vessels supplying tumors are more porous, or leaky, than normal vessels, the nanoformulation can more easily enter and accumulate within tumor cells. One such example is shown in Fig. 34; mouse tumor cells show penetration of anticancer drug after 24 h. Cancer cells have taken up a chimeric polypeptide–drug combination This means that higher doses of the drug can be delivered, increasing its cancer-killing abilities while decreasing the side effects associated with systematic chemotherapy.

7.4 Mechanism of Nanoparticle Internalization

Peptide coatings can help nanoparticles slip into cells. Anas, Ishikawa, Biju and coworkers with Japan's National Institute of Advanced Industrial Science and Technology had a better idea after thoroughly studying the insect neuropeptide allatostatin and learning that it helps CdSe-ZnS QDs pass through cell membranes by recruiting clathrin, a protein that facilitates endocytosis by forming vesicles around foreign substances [47] (Fig. 35). The researchers suspected that allatostatin might be gaining access to cells via galanin receptors. But when they blocked those receptors, it did not have much influence on the influx of QDs. By inhibiting PI3K, a kinase that is crucial to the formation of clathrin vesicles, they noted that fewer than half of the QDs made their way in, hence indicating that most of the nanoparticle entry is clathrin-mediated.

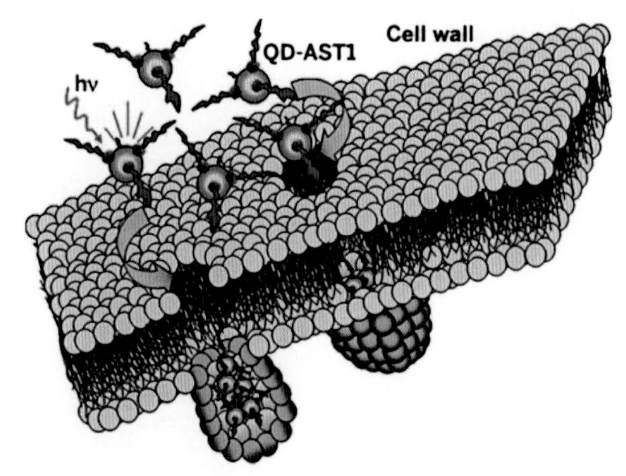

Fig. 35 Clathrin vesicles help allatostatin-coated CdSe-ZnS nanoparticles penetrate cell membranes. (Adapted from [47])

8 Conclusion and Future Prospects

Some nanotechnology for drug delivery and imaging applications is already being marketed. Future efforts in cancer therapy are envisaged to be driven by multifunctionality and modularity, i.e., creating functional modalities that can be assembled into nanoplatforms and can be modified to meet the particular demands of a given clinical situation. These strategies couple targeting and imaging/monitoring modalities, and can selectively deliver therapeutics intracellularly. Targeting modalities can be based on the recognition properties of cell-surface receptor ligands, monoclonal antibodies, peptides, or aptamers. Eventually, multiplex nanoparticles may be capable of detecting malignant cells (active targeting moiety), visualizing their location in the body (real-time in vivo imaging), killing the cancer cells with minimal side effects by sparing normal cells (active targeting and controlled drug release or photothermal ablation), and monitoring treatment effects in real time.

Acknowledgements Authors are thankful to the Department of Science and Technology (DST) and Council of Scientific and Industrial Research (CSIR), India, for supporting the "Nanotechnology Application Centre" under "Nano-Mission" and "NMITLI" schemes.

References

1. McNeeley KM, Karathanasis E, Annapragada AV, Bellamkonda RV (2009) Biomaterials 30:2329–2339
2. Felton EJ, Reich DH (2007) Biological applications of multifunctional magnetic nanowires. In: Labhasetwa V, Leslie-Pelecky DL (eds) Biomedical applications of nanotechnology. Wiley, Hoboken, pp 1–22 doi: 10.1002/9780470152928 ch1

3. Jayakumar R, Prabaharan M, Nair SV, Tamura H (2010) Biotechnol Adv 28(1):142–150, Jan-Feb
4. Pandey AC, Sharma PK, Dutta RK (2007) Recent advances in biomedical applications of multifunctional composites. In: Tiwari A (ed) Recent developments in bio-nanocomposites for biomedical applications.Nova Science, Hauppauge, pp 409–432
5. Sinha R, Kim GJ, Nie S, Shin DM (2006) Mol Cancer Ther 5:1909
6. Chen PC, Mwakwari SC, Oyelere AK (2008) Nanotechnol Sci Appl 1:45–66
7. Liu M, Fréchet JM (1999) Pharm Sci Technol Today 2(10):393, 1
8. James R, Baker J (2009) Dendrimer-based nanoparticles for cancer therapy. Hematology 2009(1):708
9. Swanson SD, Kukowska-Latallo JF, Patri AK, Chen C, Ge S, Cao Z, Kotlyar A, East AT, Baker JR (2008) Int J Nanomedicine 3(2):201–210
10. Nasongkla N, Bey E, Ren J, Ai H, Khemtong C, Guthi JS (2006) Nano Lett 6:2427–2430
11. Kukowska-Latallo JF, Candido KA, Cao Z (2005) Cancer Res 65:5317–5324
12. Kam NW, O'Connell M, Wisdom JA, Dai H (2005) Proc Natl Acad Sci USA 102:11600
13. Correa-Duarte MA, Giersig M, Liz-Marzán LM (1998) Chem Phys Lett 286(5–6):497
14. Green M, Howman E (2005) Chem Commun (1):121–123
15. Miyawaki A, Sawano A, Kogure T (2003) Nat Cell Biol 5 (suppl):S1–S7
16. Voura EB, Jaiswal JK, Mattoussi H, Simon SM (2004) Nat Med 10(9):993
17. Bruchez M Jr, Moronne M, Gin P et al (1998) Science 281:2013
18. Hemmila I (1991) Applications of fluorescence in immunoassays, 1st edn. Wiley Interscience, New York
19. LaConte L, Nitin N, Bao G (2005) Mater Today 8:32
20. Ballou B, Lagerholm BC, Ernst LA, Bruchez MP, Waggoner AS (2004) Noninvasive imaging of quantum dots in mice. Bioconjug Chem 15:79–86
21. Radomski A, Jurasz P, Alonso-Escolano D, Drews M, Morandi M, Malinski T, Radomski MW (2005) Br J Pharmacol 146(6):882
22. Koole R, van Schooneveld MM, Hilhorst J, Castermans K, Cormode DP, Strijkers GJ, De Mello Doneg C, Vanmaekelbergh D, Griffioen AW, Nicolay K, Fayad ZA, Meijerink A, Mulder WJ (2008) Bioconjug Chem 19(12):2471
23. Sharma PK, Ranu K, Dutta MK, Singh PK, Pandey AC, Singh VN (2011) IEEE Trans Nanotech 10(1):163
24. Dutta RK, Sharma PK, Pandey AC (2010) Appl Phys Lett 97:253702
25. Katari JE, Colvin VL, Alivisatos AP (1994) J Phys Chem 98:4109
26. Zhang Y, Kohler N, Zhang M (2002) Biomaterials 23:1553
27. Fortin M-A, Petoral RMP Jr, Soderlind F, Klasson A, Engstrom M, Veres T, Kall P-O, Uvdal K (2007) Nanotechnology 18:395501
28. Bridot JL, Faure AC, Laurent S, Riviere C, Billotey C, Hiba B, Janier M, Josserand V, Coll JL, VanderElst L, Muller R, Roux S, Perriat P, Tillement OJ (2007) J Am Chem Soc 129:5076
29. Kohler N, Fryxell GE, Zhang MJ (2004) J Am Chem Soc 126:7206
30. Park JY, Choi ES, Baek MJ, Lee GH, Woo S, Chang Y (2009) Eur J Inorg Chem:2477
31. Lemarchand C, Gref R, Couvreur P (2004) Eur J Pharm Biopharm 58:327
32. Lacava LM, Lacava ZGM, Da Silva MF, Silva O, Chaves SB, Azevedo RB, Pelegrini F, Gansau C, Buske N, Sabolovic D, Morais PC (2001) Biophys J 80:2483
33. McDonald MA, Watkin KL (2006) Acad Radiol 13:421
34. Xie J, Huang J, Li X, Sun S, Chen X (2009) Curr Med Chem 16:1278
35. Li ZB, Cai W, Chen XJ (2007) J Nanosci Nanotechnol 7:2567
36. Gupta AK, Gupta M (2005) Biomaterials 26:3995
37. Medintz IL, Uyeda HT, Goldman ER, Mattoussi H (2005) Nat Mater 4:435
38. Bahr JL, Tour JMJ (2002) J Mater Chem 12:1952
39. Rosi NL, Mirkin CA (2005) Chem Rev 105:1547
40. Sharma PK, Dutta RK, Kumar M, Singh PK, Pandey AC (2009) J Lumin 129:605

41. Chin SF, Iyer KS, Raston CL (2009) Cryst Growth Des 9:2685
42. Yong K-T, Rui Hu, Roy I, Ding H, Vathy LA, Bergey EJ, Mizuma M, Maitra A, Prasad PN (2009) ACS Appl Mater Interfaces 1(3):710
43. Chen SH, Fan ZY, Carroll DL (2002) J Phys Chem B 106:10777
44. Dubertret B, Skourides P, Norris DJ, Noireaux V, Brivanlou AH, Libchaber A (2002) Science 298:1759
45. Tkachenko AG, Xie H, Coleman D, Glomm W, Ryan J et al (2003) J Am Chem Soc 125(16):4700
46. Nitin N, Laconte LEW, Zurkiya O, Hu X, Bao G (2004) J Biol Inorg Chem 9:706–712
47. Anas A, Okuda T, Kawashima N, Nakayama K, Itoh T, Ishikawa M, Biju V (2009) ACS Nano 3(8):2419
48. Fahmy TM, Fong PM, Park J, Constable T, Saltzman WM (2007) AAPS J 9(2):E171–E180
49. Zhao W, Chiuman W, Lam JCF, McManus SA, Chen W, Cui Y, Pelton R, Brook MA, Li Y (2008) DNA aptamer folding on gold nanoparticles: from colloid chemistry to biosensors. J Am Chem Soc 130(11):3610–3618
50. Medley CD, Bamrungsap S, Tan W, Smith JE (2011) Anal Chem 83(3):727
51. Herr JK, Smith JE, Medley CD, Shangguan D, Tan W (2006) Anal Chem 78:2918
52. Nicolas J, Couvreur P (2009) WIREs Nanomed Nanobiotech 1:111. doi:10.1002/wnan.15, Jan/Feb 2009
53. Foldvari M, Bagonluri M (2008) Nanomedicine 4(3):183
54. Bianco A, Kostarelos K, Prato M (2005) Curr Opin Chem Biol 9(6):674
55. Hilder TA, Hill JM (2008) Current Appl Phys 8:258
56. Chen J, Hamon MA, Hu H, Chen Y, Rao AM, Eklund PC, Haddon RC (1998) Solution properties of single-walled carbon nanotubes. Science 282:95–98
57. Couvreur P, Vauthier C (2006) Pharm Res 23:1417
58. Sudimack J, Lee RJ (2000) Adv Drug Deliv Rev 41:147
59. Gabizon A, Horowitz T, Goren D, Tzemach D, Mandelbaum-Shavit F, Qazen MM, Zalipsky S (1999) Bioconjug Chem 10:289
60. Sun YP, Fu K, Lin Y, Huang W (2002) Acc Chem Res 35:1096
61. Taft BJ, Lazareck AD, Withey GD, Yin A, Xu JM, Kelley SO (2004) Site-specific assembly of DNA and appended cargo on arrayed carbon nanotubes. J Am Chem Soc 126:12750
62. Turkevich J, Stevenson PC, Hillier J (1951) Discuss Faraday Soc 11:55
63. Brust M, Walker M, Bethell D, Schiffrin DJ, Whyman R (1994) J Chem Soc Chem Commun:801–802
64. Brust M, Fink J, Bethell D, Schiffrin DJ, Keily CJ (1995) J Chem Soc Chem Commun:1655–1656
65. Murray CB, Norris DJ, Bawendi MG (1993) J Am Chem Soc 115:8706
66. Wangoo N, Bhasin KK, Boro R, Suri CR (2008) Analytica Chim Acta 610:142–148
67. Yang J, Lee JY, Too HP (2007) Anal Chim Acta 588(1):34, 1
68. Zkar SO, Finke RG (2003) Langmuir 19:6247
69. Shenoy D, Fu W, Li J, Crasto C, Jones G, DiMarzio C, Sridhar S, Amiji M (2006) Int J Nanomed 1(1):51
70. Berry CR (1967) Phys Rev 161:848
71. Healy MD, Laibinis PE, Stupik PD, Barron AR (1989) J Chem Soc Chem Commun:359
72. Manna L, Scher EC, Alivisatos AP (2000) J Am Chem Soc 122:12700
73. Peng ZA, Peng X (2002) J Am Chem Soc 12:3343
74. Callis PR (1979) Chem Phys Lett 61:563
75. Sharma PK, Dutta RK, Pandey AC, Liu CH, Pandey R (2010) Mater Lett 64(10):1183
76. Dutta RK, Sharma PK, Pandey AC (2010) J Nanopart Res 12:4
77. Salazar MD, Ratnam M (2007) Cancer Metastasis Rev 26:141. doi:10.1007/s10555-007-9048-0
78. Zong X, Feng Y, Knoll W, Man H (2003) J Am Chem Soc 125:13559
79. Yang HM et al (2010) Biomacromolecules 11:2866

80. Pankhurst QA, Thanh NKT, Jones SK, Dobson J (2009) J Phys D Appl Phys 4(2):224001
81. Josephson L, Lewis J, Jacobs P, Hahn PF, Stark DD (1988) The effects of iron oxides on proton relaxivity. Magn Reson Imaging 6(6):647–653
82. McCarron PA, Marouf WM, Quinn DJ et al (2008) Bioconjug Chem 19:1561
83. Pan B, Cui D, Sheng Y et al (2007) Cancer Res 67:8156–8163
84. Kroft LJ, de Roos A (1999) J Magn Reson Imaging 10:395
85. de Lussanet QG, Langereis S, Beets-Tan RG, van Genderen MH, Griffioen AW, van Engelshoven JM, Backes WH (2005) Radiology 235:65
86. Frias JC, Williams KJ, Fisher EA, Fayad ZA (2004) J Am Chem Soc 126:16316
87. Sipkins DA, Cheresh DA, Kazemi MR, Nevin LM, Bednarski MD, Li KC (1998) Nat Med 4:623
88. Mulder WJM, Strijkers GJ, van Tilborg GAF, Griffioen AW, Nicolay K (2006) NMR Biomed 19:142–164
89. Louie AY, Huber MM, Ahrens ET, Rothbacher U, Moats R, Jacobs RE, Fraser SE, Meade TJ (2000) Nat Biotechnol 18:321
90. Lewin M, Carlesso N, Tung CH, Tang XW, Cory D, Scadden DT, Weissleder R (2000) Nat Biotechnol 18:410–414
91. Crich SG, Biancone L, Cantaluppi V, Duo D, Esposito G, Russo S, Camussi G, Aime S (2004) Magn Reson Med 51:938
92. Yang L, Peng XH, Wang YA, Wang X, Cao Z, Ni C, Karna P, Zhang X, Wood WC, Gao X, Nie S, Mao H (2009) Clin Cancer Res 15(14):4722
93. Flacke S, Fischer S, Scott MJ, Fuhrhop RJ, Allen JS, McLean M, Winter P, Sicard GA, Gaffney PJ, Wickline SA, Lanza GM (2001) Novel MRI contrast agent for molecular imaging of fibrin: implications for detecting vulnerable plaques. Circulation 104:1280–1285
94. Lee J-H, Huh Y-M, Jun Y-W, Seo J-W, Jang J-T, Song H-T, Kim S, Cho E-J, Yoon H-G, Suh J-S, Cheon J (2007) Nat Med 13:95–99
95. Jun Y, Huh Y-M, Choi J-S, Lee J-H, Song H-T, Kim S-J, Yoon S, Kim K-S, Shin J-S, Suh J-S, Cheon J (2005) J Am Chem Soc 127:5732
96. Huh Y-M, Jun YW, Song H-T, Kim SJ, Choi J-S, Lee J-H, Yoon S, Kim K-S, Shin J-S, Suh J-S, Cheon J (2005) J Am Chem Soc 127:12387
97. Zhu M et al (2011) Biomaterials 32(7):1986
98. Schellenberger EA, Sosnovik D, Weissleder R, Josephson L (2004) Bioconjug Chem 15:1062
99. Jung HI, Kettunen MI, Davletov B, Brindle KM (2004) Bioconjug Chem 15:983–987
100. Zhao M, Beauregard DA, Loizou L, Davletov B, Brindle KM (2001) Nat Med 7:1241–1244
101. Soini E, Hemmila I (1979) Clin Chem 25:353
102. Hemmila I, Ståhlberg T, Mottram P (1995) Bioanalytical applications of labelling technologies, 2nd edn. Wallac Oy, Turku, Finland
103. Bünzli JCG (2010) Chem Rev 110(5):2729–2755doi: 10.1021/cr900362e
104. Evelyn Ning Man Cheung (2010) Chem Mater 22(16):4728–4739
105. Ahren M et al (2010) Langmuir 26(8):5753
106. Maeda H (2001) Adv Enzyme Regul 41:189
107. Cho K, Wang X, Nie S, Chen ZG, Shin DM (2008) Clin Cancer Res 14:1310
108. Carmeliet P, Jain RK (2000) Nature 407:249
109. Duke News Service (2009) Duke develops nano-scale drug delivery for chemotherapy. The Herald Sun, 1 Nov 2009. Available at http://www.heraldsun.com/pages/full_story/push?article-Duke+develops+nano-scale+drug+delivery+for+chemotherapy%20&id=4240839. Last accessed 2 Sept 2011
110. MacKay JA, Chen M, McDaniel JR, Liu W, Simnick AJ, Chilkoti A (2009) Self-assembling chimeric polypeptide–doxorubicin conjugate nanoparticles that abolish tumours after a single injection. Nat Mater 8:993–999

Index

R

S